Methods for the Mycological Examination of Food

NATO ASI Series

Advanced Science Institutes Series

A series presenting the results of activities sponsored by the NATO Science Committee, which aims at the dissemination of advanced scientific and technological knowledge, with a view to strengthening links between scientific communities.

The series is published by an international board of publishers in conjunction with the NATO Scientific Affairs Division

A	**Life Sciences**	Plenum Publishing Corporation
B	**Physics**	New York and London
C	**Mathematical and Physical Sciences**	D. Reidel Publishing Company Dordrecht, Boston, and Lancaster
D	**Behavioral and Social Sciences**	Martinus Nijhoff Publishers
E	**Engineering and Materials Sciences**	The Hague, Boston, Dordrecht, and Lancaster
F	**Computer and Systems Sciences**	Springer-Verlag
G	**Ecological Sciences**	Berlin, Heidelberg, New York, London,
H	**Cell Biology**	Paris, and Tokyo

Recent Volumes in this Series

Volume 115—Mechanisms of Secondary Brain Damage
edited by A. Baethmann, K. G. Go, and A. Unterberg

Volume 116—Enzymes of Lipid Metabolism II
edited by Louis Freysz, Henri Dreyfus, Raphaël Massarelli,
and Shimon Gatt

Volume 117—Iron, Siderophores, and Plant Diseases
edited by T. R. Swinburne

Volume 118—Somites in Developing Embryos
edited by Ruth Bellairs, Donald A. Ede, and James W. Lash

Volume 119—Auditory Frequency Selectivity
edited by Brian C. J. Moore and Roy D. Patterson

Volume 120—New Experimental Modalities in the Control of Neoplasia
edited by Prakash Chandra

Volume 121—Cyst Nematodes
edited by F. Lamberti and C. E. Taylor

Volume 122—Methods for the Mycological Examination of Food
edited by A. D. King, Jr., J. I. Pitt, L. R. Beuchat,
and Janet E. L. Corry

Series A: Life Sciences

Methods for the Mycological Examination of Food

Edited by

A. D. King, Jr.

United States Department of Agriculture
Albany, California

J. I. Pitt

Commonwealth Scientific and Industrial Research Organization
North Ryde, New South Wales, Australia

L. R. Beuchat

University of Georgia
Experiment, Georgia

and

Janet E. L. Corry

Ministry of Agriculture, Fisheries, and Food
London, England

Plenum Press
New York and London
Published in cooperation with NATO Scientific Affairs Division

Proceedings of a NATO Advanced Research Workshop on
Standardization of Methods for the Mycological Examination of Foods,
held July 11–13, 1984,
in Boston, Massachusetts

Library of Congress Cataloging in Publication Data

NATO Advanced Research Workshop on Standardization of Methods for the
Mycological Examination of Foods (1984: Boston, Mass.)
  Methods for the mycological examination of food.

  (NATO ASI series. Series A, Life sciences; v. 122)
  "Proceedings of a NATO Advanced Research Workshop on Standardization of
Methods for the Mycological Examination of Foods, held July 11–13, 1984, in
Boston, Massachusetts"—T.p. verso.
  "Published in cooperation with NATO Scientific Affairs Division."
  Includes bibliographies and index.
  1. Food—Microbiology—Congresses. 2. Mycology—Technique—Congresses.
3. Food—Analysis—Congresses. I. King, A. D. II. North Atlantic Treaty Organiza-
tion. Scientific Affairs Division. III. Series.
QR155.N38   1984   576′.163   86-25428
ISBN 0-306-42479-7

© 1986 Plenum Press, New York
A Division of Plenum Publishing Corporation
233 Spring Street, New York, N.Y. 10013

Printed in the United States of America

The desirability, indeed the necessity, for standardization of methods
for the examination of foods for contaminant and spoilage mycoflora has been
apparent for some time. The concept of a specialist workshop to address
this problem was borne during conversations at the Gordon Research
Conference on "Microbiological Safety of Foods" in Plymouth, New Hampshire,
in July 1982. Discussions at that time resulted in an Organizing Committee
of four, who became the Editors, and a unique format: all attendees would
be expected to contribute and, in most cases, more than once; and papers in
nearly all sessions would be presented as a set of data on a single topic,
not as a complete research paper. Each session would be followed by general
discussion, and then a panel would formulate recommendations for approval by
a final plenary session. The idea for this format was derived from the
famous "Kananaskis I" workshop on Hyphomycete taxonomy and terminology
organized by Bryce Kendrick of the University of Waterloo, Ontario in 1969.
Attendance would necessarily be limited to a small group of specialists in
food mycology.

The scope of the workshop developed from answers to questionnaires
circulated to prospective participants. To generate new data which would
allow valid comparisons to be drawn, intending participants were given a
variety of topics as assignments and asked to bring information obtained to
the workshop. Because of time constraints and other factors, the Organizing
Committee decided that some subjects which might be considered to be
appropriate for a workshop of this type lay outside its scope. The most
obvious of these was the question "Should we count fungi in foods?" We felt
this was a philosophical question, for which we had neither the time needed
nor the information available to provide the answer. We are aware that
people are enumerating fungi in foods and that they will continue to do so.
So the assumption was made that enumeration is a worthwhile system, and the
major aim set was to standardize techniques currently in use. Given the
information gained in this way, more fundamental questions can be tackled at
some future date.

It was agreed to concentrate attention on dilution plating, a
universally accepted procedure, and direct plating, in which particulate
foods or commodities such as grains are incubated directly on media. It was
also agreed that the estimation of fungi in foods by other methods, such as
chitin or ATP assays, and consideration of the taxonomy of food spoilage
fungi should be presented in review sessions without attempting to reach
consensus. The microscopic counting of mold fragments and analysis for
mycotoxins in foods were judged to be outside the scope of the workshop.

The First International Workshop on "Methods for Mycological Examination
of Foods" was held in Boston on July 11-13, 1984, just two years after it
was conceived. Twenty-six people attended from Australia, Denmark, England,
Hungary, the Netherlands, Turkey and the United States. Some eighty formal

presentations in ten organized sessions, and seemingly endless formal and informal discussion sessions were held during a very full three-day program.

The aims of the Workshop were in large measure accomplished. Agreement was reached on the standardization of a wide range of methods and media suitable for particular aspects of isolating and enumerating fungi in foods. In this Proceedings are published the edited manuscripts, the general discussion following various sessions and the recommendations approved at the final plenary session. Thus a reasonably complete picture is provided of the proposed standard methods and the data on which they are based.

This workshop was sponsored by grants obtained from the United States National Science Foundation and the Australian Department of Science and Technology under the United States - Australia Cooperative Science Program and from the North Atlantic Treaty Organization as an Advanced Research Workshop. Additional support was received from Difco Laboratories, Detroit, Michigan, Oxoid Ltd., Basingstoke, England and the Pillsbury Company, Minneapolis, Minnesota. We express our gratitude to these sponsors for making the workshop possible.

Special thanks are expressed to Ailsa D. Hocking, Commonwealth Scientific and Industrial Research Organization, North Ryde, New South Wales and Donald E. Conner, University of Georgia, Experiment, Georgia, for their assistance in arranging the workshop and recording its proceedings.

A. D. King, Jr.
J. I. Pitt
L. R. Beuchat
J. E. L. Corry

CONTENTS

CHAPTER 1

SAMPLE PREPARATION

Sample preparation is the first step in the mycological analysis of foods. It is obvious that sample preparation needs to be standardized if comparable results are to be obtained in different laboratories.

Responses to questionnaires returned by participants prior to the workshop indicated that a variety of sample preparation techniques were in use. For instance, about equal numbers of laboratories used stomaching or blending, spread or pour plating, and incubating plates upright rather than inverted. A wide variety of diluents were used for dilution plating and at least nine different techniques were reported for surface disinfection in preparation for direct plating of food pieces or grains. Responses from most people indicated that they used plastic Petri dishes and incubation temperatures of about 25°C. Their replies thus resulted in the assignment of a series of experiments that were reported at the workshop and are described in this chapter.

Equally important but not considered here is sampling, i.e., the selection of samples for analysis. The organizers believe that this subject is adequately handled by the ICMSF publication "Microorganisms in Foods. 2. Sampling for Microbiological Analysis: Principles and Specific Applications" (University of Toronto Press 1974). In consequence, sampling was not addressed at this workshop.

► Effect of Sample Size on Variance of Mold and Yeast Counts from Hard Cheese and Skim Milk Powder

Because of intrinsic variability in the distribution of microbes in foods, results of enumeration procedures cannot be representative unless the sample size examined is as large as may conveniently be handled (Kilsby & Pugh 1981). The relevance of this observation to enumeration of molds was demonstrated by Jarvis et al. (1983) for cereal products. This investigation extends our observations to dairy products (cheese and skim milk powder).

Materials and methods

Sample. Sufficient quantities of cheese and skim milk powder were used to provide replicate 10- and 50-g subsamples of cheese and 2- and 10-g subsamples of milk powder.

Media. DRBC and DG18 media were supplied by Oxoid and prepared according to instructions. The glycerol added to DG18 medium was BDH Analar grade.

Diluent. The diluent used was 0.1% peptone (Oxoid L37) in distilled water.

Sample size. For hard cheese, 10-g and 50-g samples were examined in duplicate. For skim milk powder, 2-g and 10-g samples were examined in duplicate.

Examination of samples. Samples were blended in a Stomacher using 10 volumes of diluent. Appropriate dilutions (0.1 ml) were spread-plated in triplicate. Plates were incubated upright at 25°C for 5 days and mold and yeast colonies were counted.

Statistical analysis. After logarithmic transformation, the mean colony count ($\bar{x}$) and variance (s^2) were derived by normal methods. Estimates of log average counts (α) were derived using the equation, $\log \alpha = \bar{x} + \ln s^2/2$ where ($\bar{x}$) = estimate of the true mean log count (μ) and s^2 = estimate of the log population variance (s^2) as described by Kilsby & Pugh (1981).

Results and discussion

The data are summarized in Tables 1 and 2. Previous experience (Kilsby & Pugh 1981; Jarvis et al. 1983) suggested that as sample size was increased the mean log propagule (or colony) count would also increase and the variance would decrease. This phenomenon was observed also in three of the four comparisons made in the present investigation but not for the fourth investigation (mold propagule count on milk powder). The reason for this discrepancy cannot be identified specifically but could well have arisen through chance effects.

Comparison of the mean log propagule or colony counts with the derived log average counts ($\log \alpha$) indicates a closer comparability for the derived yeast counts than for mold counts. A similar situation was observed in previous studies. The data in Table 3, derived from the data of Jarvis et al. (1983), show similar variability in derived log average counts. This may indicate that the primary distribution of fungal propagules follows an even more disparate system than do bacterial or yeast colony counts.

Table 1. Effect of sample size on mean log and log average propagule count of molds

Sample	Sample size (g)	No. of tests (n)	Mean $\log_{10}$ count ($\bar{x}$) and (range)	Variance (s^2)	$\log_{10}$ ave. count ($\log \alpha$)
Cheese	10	11	5.81 (5.00–6.27)	0.413	5.98
	50	12	6.05 (5.27–6.63)	0.126	6.20
Milk powder	2	11	3.52 (3.14–3.71)	0.028	3.55
	10	11	3.20 (2.69–3.65)	0.158	3.38

Table 2. Effect of sample size on mean log and log average propagule count of yeasts

Sample	Sample size (g)	No. of tests (n)	Mean $\log_{10}$ counts ($\bar{x}$) and (range)	Variance (s^2)	$\log_{10}$ ave. count ($\log \alpha$)
Cheese	10	8	5.96 (5.47–6.53)	0.113	6.09
	50	12	6.06 (5.77–6.64)	0.075	6.15
Milk powder	2	11	3.16 (2.0–4.12)	0.740	4.01
	10	7	3.55 (2.0–4.04)	0.520	4.15

Table 3. Effect of varying sample size on propagule counts in a dry cereal product   (Data of Jarvis et al. 1983)

Sample size (g)	No. of replicates (n)	Mean log cfu/g ($\bar{x}$)	Variance (s^2)	Log average count ($\log \alpha$)
2	6	3.74	1.96	6.00
50	12	4.16	0.14	4.33
250	12	4.53	0.14	4.70

Conclusions

Increasing sample sizes generally gives a higher degree of reproducibility and a lower variance than can be achieved with small samples. However, the constant "derived log average count" concept of Kilsby & Pugh (1981) which can be applied in control analyses for bacteria and yeasts appears not to be applicable to mold propagules.

References

JARVIS, B., SEILER, D. A. L. OULD, ANGELA, J. L. & WILLIAMS, A. P. 1983 Observations on the enumeration of moulds in food and feeding stuffs. Journal of Applied Bacteriology 55, 325–336.
KILSBY, D. C. & PUGH, M. E. 1981 The relevance of the distribution of micro-organisms within batches of food to the control of microbiological hazards from foods. Journal of Applied Bacteriology 51, 345–354.

B. JARVIS
N. SHAPTON

Evaluation of Blending, Stomaching and Shaking for the Mycological Analysis of Dried Foods

Homogenization by blending is a common method of preparing food samples for microbiological analysis, especially for those laboratories following reference methods such as the U. S. Food and Drug Administration Bacteriological Analytical Manual (1980). The limitations of blending include production of aerosols, purchase and maintenance of blender jars, and decreased recovery of microorganisms due to rising temperatures during blending (Emswiler et al. 1977).

The Colworth Stomacher was developed as an alternative sample homogenizer. Tuttlebee (1975) described the mode of action of the instrument and the benefits derived from "stomaching" samples. Comparative studies on the recovery of bacteria from stomached and blended samples have been published (Tuttlebee 1975; Emswiler et al. 1977). Andrews et al. (1978) evaluated the use of the Stomacher for microbiologial analysis of thirty categories of food. They concluded that the efficiency of the Stomacher is food-specific.

Many of the previous comparative studies evaluated the effects of these homogenization methods on aerobic plate counts (APC) and recovery of specific bacteria. In this study, the effects of three sample preparation methods on enumeration of fungi from dried foods was determined.

Materials and methods

Experimental design. The experiment was planned using the randomized complete block design (RCBD, Dowdy & Wearden 1983) shown in Table 1. This design permitted simultaneous yet independent evaluations of the effects of sample preparation methods and diluents on enumeration of fungi from dried foods. The results of the test on diluents will be reported later in this chapter.

Cultures. The molds used in this study were Aspergillus flavus, A. ochraceus, another Aspergillus sp. and two Penicillium spp. Five Petri dishes of PDA (Difco) were surface-inoculated with each mold. The cultures were incubated at 25°C until heavy sporulation occurred.

Table 1. Randomized complete block design to evaluate the effects of sample preparation methods and diluents on enumeration of fungi from dried foods

Design component	Effect tested	Description
Treatments	Sample preparation methods	Blending Stomaching Shaking
Blocks	Diluents	0.1% peptone Phosphate-buffered water
Replicate observations	Sampling error	3 subsamples per treatment x block evaluation 3 plates at countable dilution per subsample

Inoculation of commodities. Natural contamination of dried foods by molds is sporadic, occurring as hot spots where environmental conditions favor growth. Variation in mold counts between subsamples of the same lot may limit the evaluation of sample preparation methods as sampling error may mask treatment effects. For optimum performance of the RCBD described earlier, the error component due to sampling from a heterogeneously distributed population was reduced by analysis of artificially contaminated foods.

Dried foods and inocula are listed in Table 2. PDA cultures of heavily sporulated molds were placed in disposable glove bags. A one-inch paint brush was used to gently mix a small portion of the commodity with the culture (predominately conidia) on each the five plates. The stock inoculum was added to 2 kg of the commodity in an eight-quart blender (Cross Flow, Patterson-Kelley) and mixed for 20 min to ensure that the conidia were homogeneously distributed in the food. Lab personnel wore dust masks while operating the blender.

Sample preparation by blending. For each commodity, six 25-g subsamples were weighed into sterile one-quart stainless steel cups. To three of the subsamples, 100 ml of 0.1% peptone diluent were added; 100 ml of Butterfield's diluent (0.3 mM KH_2PO_4, pH 7.2) were added to the other three subsamples. Samples were blended for 2 min on a Waring Commercial Blender at low speed. Further dilutions were prepared in the same diluent.

Sample preparation by stomaching. Samples and diluents were the same as described for blending except they were weighed into sterile 18 cm x 30 cm polyvinylidene chloride bags (Cryovac, Duncan, SC). Samples were mixed in a Stomacher 400 (Tekmar Co., Cincinnati, OH – distributor) (30 sec). Samples were double-bagged to prevent spillage from broken bags. The diluted rice samples were held 1 h before stomaching to soften the rice.

Sample preparation by shaking. Subsamples (25 g), diluent (100 ml) and 1 g of 3-mm sterile glass beads were distributed by hand through sterile funnels into sterile 250-ml Wheaton bottles. The bottles were shaken for 2 min.

Table 2. Effect of sample preparation method on enumeration of molds from foods

		Mean $\log_{10}$ mold count/g[a]		
Commodity	Inoculum	Blending	Stomaching	Shaking
Rice	Aspergillus flavus	4.56a[b]	4.50a	4.35b
Cocoa	A. flavus	5.37b	5.49a	5.51a
Cocoa	Penicillium 1	5.84b	6.32a	6.31a
Cocoa	Penicillium 2	5.66b	6.35a	6.30a
Dextrose	A. ochraceus	4.27b	4.23b	4.47a
Almond meal	Aspergillus sp.	4.02a	3.97a	3.97a

[a]Represents the mean of 18 counts
[b]Mean $\log_{10}$ mold counts within a commodity inoculated with a given mold which are followed by the same letter were not significantly different ($P \leq 0.05$)

Enumeration medium. DRBC modified to contain 15 µg of rose bengal per ml (DRBC-15) instead of 25 µg, was autoclaved at 121°C for 20 min. Chloramphenicol selective supplement (Oxoid, 100 µg/ml) was added to tempered medium and then dispensed into 15 x 100 mm plastic Petri dishes for inoculation by the spread-plate technique.

Plating technique. One preliminary subsample per commodity was homogenized by blending, serially diluted, spread-plated at various dilutions and the optimum dilution that produced 10-100 colonies per plate was determined. This reduced the volume of media and number of plates needed for the actual experiment.

The eighteen subsamples per commodity were prepared and diluted as described earlier. For each subsample, triplicate plates of DRBC-15 were spread-plated with 0.1-ml portions of the optimum dilution. Plates were incubated upright for 5 days.

Statistical evaluation. Statistical analysis of the data was made from $\log_{10}$ counts of individual plates since this provided additional degrees of freedom to the randomized complete block analysis of variance (RCB ANOVA). Replication on triplicate plates was good, and low variation between triplicate subsamples of each homogenization method and diluent combination showed that the commodity was homogeneously contaminated by conidia.

Differences were tested by the RCB ANOVA. Means for commodities exhibiting significant ($P \leq 0.05$) treatment effects were subjected to Duncan's Multiple Range Test to determine which means differed significantly.

Results

Five of the six studies exhibited significant differences between sample preparation as determined by the RCB ANOVA. The differences between the means were analyzed by Duncan's Multiple Range Test and are reported in Table 2. The differences were independent of block (diluent) effects and treatment-block interactions.

Discussion

Only two of the five differences reported in Table 2 may have practical significance. Blending considerably reduced the number of conidia recovered from the two cocoa samples inoculated with Penicillia. Perhaps the hydrophobic conidia of the Penicillia may have been drawn into the foam layer above the blended samples. Incorporation of the foam layer into the rest of the diluted sample was usually difficult. Other disadvantages of the blending technique have been identified elsewhere in this report.

Sample preparation by stomaching and shaking resulted in similar recoveries of conidia. Problems this laboratory has experienced with the Stomacher include breakage of both polyethylene bags and heavier polyvinylidene chloride bags, unavailability of bags and replacement costs of the paddle flexible couplings which will deteriorate with repeated use.

Sample preparation by shaking was the most time-consuming procedure. The samples were probably agitated longer than necessary for this study. The effects on recovery of eliminating the glass beads and reducing agitation time were not determined. Shaking samples also produced foam, but there was no apparent effect on recovery of conidia.

Conclusions

We concur with Andrews et al. (1978) that choice of a sample preparation method is food-specific. Lower recoveries of Penicillia from blended samples indicate that the selection of a method may also be organism-specific. To test this hypothesis, it will be necessary to run more extensive comparative studies of available methods than reported here. Morever, if mold contamination of a sample is suspected, the procedure that most effectively limits release of antigenic or infectious fungal propagules into the air should be selected.

We are grateful for the assistance of Richard K. Smith in the statistical analysis of the data.

References

ANDREWS, W. H., WILSON, C. R., POELMA, P. L., ROMERO, A., RUDE, R. A., DURAN, A. P., McCLURE, F. D. & GENTILE, D. E. 1978 Usefulness of the Stomacher in a microbiological regulatory laboratory. Applied and Environmental Microbiology 35, 89–93.
DOWDY, S. & WEARDEN, S. 1983 Statistics for Research. New York: John Wiley & Sons.
EMSWILER, B. S., PIERSON, C. J. & KOTULA, A. W. 1977 Stomaching vs. blending. Food Technology, Champaign. 31, 40–42.
TUTTLEBEE, J. W. 1975 The Stomacher – its use for homogenization in food microbiology. Journal of Food Technology 10, 113–122.
U. S. FOOD AND DRUG ADMINISTRATION. 1980 Bacteriological Analytical Manual, 5th ed. 1980 supplement. Washington, D. C.: Association of Official Analytical Chemists.

L. M. LENOVICH
J. L. WALTERS
D. M. REED

▶ Comparison of Stomaching Versus Blending in Sample
Preparation for Mold Enumeration

Sample preparation for dilution plating may consist of shaking with a diluent, grinding to a fine powder and then blending with a diluent, direct blending with a diluent, or pummelling with a Stomacher. Studies have shown that higher counts are usually obtained by stomaching or blending then by rinsing or shaking techniques (Jarvis et al. 1983). These same studies have shown that differences were not statistically significant. The present study compared blending with stomaching for a range of foods.

Materials and methods

Samples. Ten food samples were tested, of which seven were of commercial origin (dry gravy mix, chopped walnuts, dry split peas, bleached flour, yellow corn meal, pecan pieces and "organic" corn meal) and three were bleached wheat flour inoculated with three levels (10^2, 10^4 and 10^6/g) of a mixture of conidia from nine different fungi. These were Aspergillus flavus, A. niger, A. ochracus, Penicillium aurantiogriseum, P.

roquefortii, P. viridicatum, Fusarium graminearum, Cladosporium sp. and Alternaria sp. Dry spores of each fungus were mass produced on bread cubes in quart Mason jars (Sansing & Ciegler 1973), and allowed to dry. Spores were harvested by adding 100-200 g of flour to each jar, blending, then thoroughly mixing for use as a combined inoculum. The total number of spores in the mixture was determined using surface plate count and plate MPN techniques (Tan et al. 1983) and appropriate amounts of the mixture were used to inoculate flour samples to obtain approximate counts of 10^2, 10^4 and 10^6 propagules/g.

Sample preparation. Samples were prepared by combining 11 g of sample with 99 ml of sterile phosphate buffer diluent (Speck 1976) and mixing by either blending or stomaching for 3 min. Subsequent serial dilutions were then made. Test aliquots were removed and plated within 1 min of completing the dilution sequence.

Plate counts. Counts were made by surface plating 0.1 ml aliquots in duplicate. Two media were used, PDA (Difco) + 40 µg/ml tetracycline and DRBC (Oxoid). Fungal populations in each sample were determined in triplicate; thus each count represents a mean of six determinations. Plates were incubated upright at 25°C for 5 days and then mold and yeast colonies were counted and reported as combined values. Only plates containing 10-100 colonies were counted. If countable plates could not be obtained using a 0.1-ml aliquot, analyses were repeated using a 1.0-ml aliquot. Actual counts were converted to $\log_{10}$ counts for reporting.

Statistical design analysis. There were two experimental factors in this study: (1) method of preparation of sample (blending vs. stomaching) and (2) culture media (see Chapter 2). A split plot design was used. The method for preparation of inoculum was considered as block. Mean $\log_{10}$ values of counts for each food sample were analyzed by analysis of variance (Steel & Torrie 1980). The α-value was chosen to be 0.05.

Results

Results are given in Table 1. There appears to be good correlation of

Table 1. Mean yeast and mold counts obtained by blending vs stomaching using DRBC and PDA + tetracycline with a surface plate count technique (mean $\log_{10}$ counts ± standard deviation)

Commodity	Blending		Stomaching	
	DRBC	PDA	DRBC	PDA
Dry gravy mix	2.50 ± .20	2.52 ± .37	2.23 ± .13	2.22 ± .10
Chopped walnuts	3.31 ± .05	3.08 ± .51	3.43 ± .13	3.32 ± .11
Dry split peas	1.34 ± .25	1.32 ± .42	1.23 ± .19	1.34 ± .23
Bleached flour	1.84 ± .27	2.46 ± .26	2.17 ± .16	2.42 ± .27
Yellow corn meal	2.48 ± .13	2.63 ± .07	2.60 ± .33	2.45 ± .14
Pecan pieces	1.90 ± .18	2.10 ± .09	1.84 ± .17	1.94 ± .07
Organic corn meal	5.41 ± .07	5.42 ± .07	5.36 ± .11	5.39 ± .09
Inoculated flour (10^6)	5.09 ± .40	5.19 ± .10	5.33 ± .08	5.34 ± .07
Inoculated flour (10^4)	3.16 ± .14	3.13 ± .12	3.27 ± .32	3.34 ± .14
Inoculated flour (10^2)	2.20 ± .21	2.27 ± .13	2.26 ± .44	2.36 ± .05

counts obtained between stomaching and blending. Some differences were obtained between the two methods of sample preparation, but in the commercial food samples there appeared to be no consistent pattern. With the four inoculated samples, consistently higher counts were obtained with stomaching.

Statistical analyses. The difference in yeast and mold counts between blending and stomaching obtained on both media were not statistically significant in any commodity except the flour inoculated at 10^2/g. In this case, counts from stomached samples were significantly higher than counts from blended samples.

Conclusions

In this study, there were no significant differences observed between mold counts obtained from samples prepared by either blending or stomaching. This was true with both commercial food samples and inoculated flour samples except from the low (10^2/g) counts.

Recommendation

Sample preparation by blending or stomaching appears to give comparable results on the types of samples tested in this study. Therefore, either method may be used depending upon individual circumstances, preferences and availability of equipment.

References

JARVIS, B., SEILER, D. A. L., OULD, ANGELA J. L. & WILLIAMS, A. P. 1983 Observations on the enumeration of molds in food and feedstuffs. Journal of Applied Bacteriology 55, 325-336.
SANSING, G. A. & CIEGLER, A. 1973 Mass propagation of conidia from several Aspergillus and Penicillium species. Applied Microbiology 26, 830-831.
SPECK, M. L. (ed). 1976 Compendium of Methods for the Microbiological Examination of Foods. Washington, D. C.: American Public Health Association.
STEEL, R. G. & TORRIE, J. H. 1980 Principles and Procedures of Statistics, a Biometrical Approach. 2nd edition. New York: McGraw-Hill Book Co. p. 235.
TAN, S. Y., MAXCY, R. B. & STROUP, W. W. 1983 Colony forming unit enumeration by a plant-MPN method. Journal of Food Protection 46, 836-841.

J. W. HASTINGS
W. Y. J. TSAI
L. B. BULLERMAN

Published as Paper No. 7610, Journal Series, Agricultural Research Division, Lincoln, NE. Research reported was conducted under project 16-029.

► Evaluation of Blending, Stomaching and Shaking for Mold
  Counts in Flour

In conjunction with the comparison of media for enumeration of molds
presented in Chapter 2, methods of sample preparation were investigated. We
looked at two factors: (1) preparation of suspensions by blending,
stomaching and shaking and (2) the time of settling of suspensions.

Materials and methods

Twelve laboratories (see our paper in Chapter 2) collaborated in two
series of investigations using samples of wheat flour and meal. Initial
suspensions were prepared by adding 10 g of sample to 90 ml of diluent
(peptone, 0.1% and NaCl, 0.8%, in tap water). Four methods of preparation
were used:

  a. Shaking by hand for 3 min
  b. Shaking by a horizontal shaker for 20 min
  c. Blending by MSE Atomix blender for 3 min
  d. Stomaching by Stomacher 400 for 1 min

Decimal dilutions were made immediately after preparation of suspensions, as
well as after 10, 20 and 40 min of settling. Four different enumeration
media were used and results were pooled for mathematical analysis. Log_{10}
transformed data were used for the analysis of variance.

Results and discussion

Table 1 summarizes the counts obtained from samples dispersed by
different methods. Significant differences were not detected. The method
of preparation of suspension exerted no effect on the repeatability and
reproducibility of results. Stomaching was carried out by one laboratory on
flour only. A mean of (log_{10}) 2.44 cfu/g was obtained, a value comparable
to the other methods. In contrast, the time of settling of suspensions
strongly influenced both the numerical results and their precision and
reliability (Table 2).

Table 1. Effect of suspension technique on the enumeration of molds in flour

| | Method of Shaking | | | | | |
| | Hand, 3 min | | Shaker, 20 min | | Blender, 3 min | |
Measure	Series 1	Series 2	Series 1	Series 2	Series 1	Series 2
Mean value[a]	2.60	4.64	2.50	4.61	2.53	4.42
r[b]	0.40	0.32	0.50	0.39	0.47	0.44
R[c]	0.68	0.51	0.78	0.80	0.70	0.72

[a]Data are expressed in log cfu/g. Data represent means of 44 tests (11
 laboratories, 2 replications, 2 media)
[b]r = repeatability (within-laboratory error)
[c]R = reproducibility (between-laboratory error)

Table 2. Influence of settling times on mold counts in flour[a]

Settling time (min)	$\log_{10}$ cfu/g	r[b]	R[c]
0	3.70	0.267	0.427
10	3.43	0.299	0.357
20	3.31	0.307	0.430
40	3.20	0.306	0.499

[a]Data represent means of 144 tests (12 laboratories, 2 samples, 2
 replications, 3 media)
[b]r = repeatability (within-laboratory error)
[c]R = reproducibility (between laboratory error)

It is obvious that the number of cfu/g decreased with increasing
settling time. Significantly smaller counts were obtained even after the
shortest settling period. However, the precision of the method increased
with increasing settling time up to 20 min. At the longer settling time,
the degree of separation within the suspension became high. Consequently,
the accuracy and precision decreased.

Conclusions

    The method of sample homogenization, i.e., shaking by hand or by shaker,
blending or stomaching, did not influence the sensitivity and precision of
the enumeration of molds, at least in easily suspended products such as
flour and meal.
    On the other hand, the time of settling has a significant effect on both
attributes of enumeration. Without a drastic decrease in sensitivity,
reliable results can be achieved using a settling time of 10 min.

                                                                T. DEÁK
                                                       V. TABAJDI-PINTER
                                                               I. FABRI

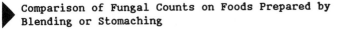

Comparison of Fungal Counts on Foods Prepared by
Blending or Stomaching

    To compare recovery of mold propagules using these two sample
preparation methods, a series of counts was made on a diverse group of
foods. The foods were primarily commercial products purchased in local
stores. Foods analyzed (and number of samples) were: spices (7), fruits
(12), nuts (5), dry grains or vegetables (16), meats (3), dairy (1), and dog
food (1). Fifty grams of product were weighed into either a Waring Blendor
jar or Stomacher bag (Sharpe & Jackson 1972) and 450 g of water were added.
After either 5 or 10 min of soaking, the samples were either stomached or
blended for 2 min and surface-plated immediately on DRBC medium using 0.1%
peptone as a diluent where appropriate. Plates were incubated for 5 days at
25°C. Spices were not always available in 50-g quantities. In these
instances, samples were combined with water at a 1:9 ratio.

Table 1. Effect of stomaching time on fungal counts and comparison of counts with blending

Type of mixing	Time (min)	Mean	Standard deviation	N
Stomaching	2	3.318	0.966	8
	3	3.590	0.864	8
	4	3.539	1.089	8
	5	3.664	1.064	8
	6	3.819	0.998	8
	7	3.652	1.043	8
Blending	2	3.994	0.620	7

Populations (cfu/g) of fungi detected in the 45 foods prepared by the two methods were statistically analyzed. The mean $\log_{10}$ counts were 3.020 for blending and 2.762 for stomaching with variances of 2.282 and 2.252, respectively. Comparison of these means by Student's test indicated no significant difference in the mean values. There appeared to be a higher count on the individual foods prepared by blending compared to stomaching; the mean count was 1.8-fold higher, but not great enough for statistical significance ($p = 0.05$).

In a separate test, blending for 2 min was compared with stomaching for 2, 3, 4, 5, 6 or 7 min. The results show that increasing time of stomaching results in higher counts (Table 1). The increase in counts with longer stomaching time was not significant ($p = 0.05$) but there was a positive linear slope to the calculated line ($r = 0.80$). There was no significant difference between blending time and stomaching for 2 min ($p = 0.05$).

Blending is a more labor intensive procedure than is stomaching, requiring more equipment preparation and cleaning. Samples prepared with the stomacher are easier to pipette. However, bags used in stomaching can be punctured by hard or sharp food pieces thus limiting their usefulness. The ease of use and preparation and lower cost make stomaching an attractive alternative to blending some foods in preparation for enumerating fungi.

Recommendation

The data presented here show no statistical difference between blending and stomaching, indicating that either sample preparation technique can be used.

Reference

SHARPE, A. N. & JACKSON, A. K. 1972 Stomaching; A new concept in bacteriological sample preparation. Applied Microbiology 24, 175-178.

A. D. KING, JR.

Evaluation of Diluents Used in the Mycological Analysis of Dried Foods

According to Gerhardt (1981), solutions for diluting microbial cells should be buffered and osmotically balanced to maintain metabolic activity and viability of cells. Yet, selection of a diluent for the routine microbiological examination of foods is more often influenced by ease of preparation and widespread acceptance, especially if the diluent is included in published reference methods. Furthermore, separate suspensions are usually not prepared for yeast and mold enumeration and the microbial aerobic plate count and standard plate count procedures.

Diluents described in reference methods include 0.1% peptone (Gilliland et al. 1976; Anon. 1981), Butterfield's phosphate buffered diluent (Gilliland et al. 1976; U. S. Food and Drug Administration 1980), and Butterfield's diluent amended with 0.1 mM $MgSO_4 \cdot 7H_2O$ (Anon. 1981; Jensen et al. 1978). This paper will describe the results of a comparison of 0.1% peptone and Butterfield's diluent for recovery and enumeration of molds from dried foods.

Materials and methods

Techniques for inoculation of dried foods, sample preparation, diluents, media and plating have been reported by the authors elsewhere in this chapter. Inocula and commodities analyzed are listed in Table 1.

Log_{10} mold counts were subjected to a randomized complete block analysis of variance (RCB ANOVA) for evaluation of treatment (sample preparation method) effects, block (diluent) effects and treatment-block interactions. Means for commodities exhibiting significant ($P < 0.05$) block effects were subjected to Duncan's Multiple Range Test to determine which means differed significantly.

Table 1. Effect of diluent on enumeration of molds from dried foods

Commodity	Inoculum	Mean log_{10} mold count/g[a]	
		Phosphate	Peptone
Rice	Aspergillus flavus	4.15a[b]	4.44b
Cocoa	A. flavus	5.44a	5.47a
Cocoa	Penicillium 1	6.21a	6.11a
Cocoa	Penicillium 2	6.15a	6.06b
Dextrose	A. ochraceus	4.31a	4.33a
Almond meal	Aspergillus sp.	3.96b	4.01a

[a]Represents the mean of 27 counts
[b]Mean log_{10} mold counts within a commodity inoculated with a given mold which are followed by the same letter are not significantly different ($P < 0.05$)

Results

Three of the six studies exhibited significant block (diluent) effects as determined by the RCB ANOVA. The differences between the means for each commodity were analyzed by Duncan's Multiple Range Test and are reported in Table 1.

Discussion

Although there were three instances of statistical differences in this study, we doubt that they are of practical significance. Also, due to the limited scope of this study, the results may not have general applicability.

References

ANON. 1981 Standard Methods for the Examination of Water and Wastewater, 15th ed. Washington, D.C.: American Public Health Association.

GERHARDT, P. 1981 Diluents and biomass measurements. In Manual of Methods for General Bacteriology ed. Gerhardt, P. Ch. 25, pp. 504-507. Washington, D.C.: American Society for Microbiology.

GILLILAND, S. E., BUSTA, F. F., BRINDA, J. J. & CAMPBELL, J. E. 1976 Aerobic Plate Count. In Compendium of Methods for the Microbiological Examination of Foods ed. Speck, M.L. Ch. 4, pp. 107-131. Washington, D.C.: American Public Health Association.

JENSON, J. P., HUHTANEN, C. N. & Bell, R. H. 1978 Culture media and preparation. In Standard Methods for the Examination of Dairy Products, 14th edition ed. Marth, E.H. Ch. 4, pp. 55-75. Washington, D.C.: American Public Health Association.

U. S. FOOD AND DRUG ADMINISTRATION 1980 Bacteriological Analytical Manual, 5th ed., 1980 supplement. Washington, D.C.: Association of Official Analytical Chemists.

L. M. LENOVICH
J. L. WALTERS
D. M. REED

A Comparison Between Peptone and Ringer Solution with Polysorbitan Diluents on the Enumeration of Molds and Yeasts from Hard Cheese and Skim Milk Powder

In earlier work comparing other media for enumerating molds and yeasts, a Ringer-polysorbitan 20 diluent was used (Jarvis 1973). The polysorbitan, included as a wetting agent, is thought to be beneficial in the analysis of dried materials. Because one of the food samples tested in the present study is skim milk powder, a comparision between 0.1% peptone and a Ringer-polysorbitan 20 diluent was included.

Materials and methods

Diluents. Two diluents [0.1% peptone (Oxoid L37) in distilled water and 1/4 strength Ringer solution incorporating 0.5% v/v polysorbitan 20 (Tween 20, Sigma; Jarvis 1973)] were used to prepare homogenates of cheese and skim milk powder.

Table 1. Comparison of peptone (P) and Ringer-polysorbitan (RP) diluents for enumeration of mold propagules

Sample type	No. of tests	Mean and range of log10 propagule count after dilution in:		Mean and range of differences in log count	S.D. difference	No. of tests where:	
		Peptone	Ringer-polysorbitan			P > RP	P < RP
Cheese	20	6.58 (4.9–8.2)	6.59 (4.9–8.6)	−0.059 (−0.38– +0.56)	0.263	5	11
Milk powder	16	2.64 (2.0–3.9)	2.66 (2.0–3.9)	+0.074 (−0.15– +0.46)	0.150	9	3

Table 2. Comparison of peptone (P) and Ringer-polysorbitan (RP) diluents for enumeration of yeasts

Sample type	No. of tests	Mean and range of log10 propagule count after dilution in:		Mean and range of differences in log count	S.D. difference	No. of tests where:	
		Peptone	Ringer-polysorbitan			P > RP	P < RP
Cheese	12	6.55 (4.72–8.09)	6.56 (4.45–8.19)	−0.268 (−0.66– +0.54)	0.962	6	6
Milk powder	5	3.35 (2.39–3.88)	3.77 (2.98–4.64)	+0.348 (−0.76– +0.10)	0.337	1	4

Media. DRBC and DG18 media were supplied by Oxoid and prepared according to instructions. The glycerol added to DG18 medium was BDH Analar grade.

Examination of samples. In all, 36 samples of cheese and milk powder were examined for molds and 17 were examined for yeasts. Samples (10 g) were stomached in 90 ml of diluent, and 0.1-ml aliquots of the appropriate dilutions were spread over the surface of the medium. Duplicate plates were made for each dilution. The plates were incubated upright at 25°C. Mold and yeast colonies were counted.

Statistical analysis. The hypothesis was tested that the mean difference between colony counts, after logarithmic transformation, did not differ significantly from zero, using Student's 't' test.

Results

Data from the enumeration of mold and yeast propagules are summarized in Table 1 and Table 2, respectively. None of the comparisons indicated a statistically significant advantage for selecting either Ringer-polysorbitan or peptone diluent for determining mold and yeast populations in cheese and milk powders. Subjective differences in colonial morphology were observed but could not be quantified.

Conclusion

For the samples tested, no advantage could be discerned for using Ringer-polysorbitan diluent when compared with peptone diluent.

We are indebted to C. Foster for technical assistance.

Reference

JARVIS, B. 1973 Comparison of an improved rose bengal-chlortetracycline agar with other media for the selective isolation and enumeration of moulds and yeasts in foods. Journal of Applied Bacteriology 36, 723-727.

B. JARVIS
N. SHAPTON

 Comparison of Four Diluents for the Mycological Examination of Cereals

A wide range of diluents can be used for the microbiological examination of foods and feedstuffs. Many of them include peptone in water, with or without salt and buffers. This short study was conducted on wheat grains in order to determine to what extent such diluents are suitable for enumeration of fungi from cereals and other dry products.

Table 1. Mold counts (mean, range and S.D.) from wheat grains with four
          dilution fluids

		Diluent number[a]		
1	2	3	4	4 (60)
2.81	2.43	2.72	3.43	2.91
(2.60–3.10)	(1.95–2.86)	(2.46–2.85)	(2.54–4.59)	(2.71–3.20)
S.D. = 0.21	S.D. = 0.43	S.D. = 0.17	S.D. = 1.21	S.D. = 0.20

[a]For key to diluents, see Materials and methods

Materials and methods

Samples. All determinations were carried out on the same sample of
wheat grains, carefully mixed prior to subsampling. For analysis, two
subsamples of 20 g were continuously shaken in 180 ml of each unbuffered
diluent for 30 min, then serially diluted and pour-plated on malt agar (2%)
with reduced a_w (0.95) achieved by the addition of ca 15% glycerol. The
medium also contained 100 μg/ml of chloramphenicol as an antibacterial
agent. Plates were incubated for 5 days at 25°C.

Diluents. Diluents under comparison had the following compositions:
   1. 0.1% peptone in water
   2. 0.1% peptone and 0.85 g NaCl in water
   3. Diluent 2 plus 0.033 g/l of polysorbitan 80
   4. Diluent 3 plus an additional resuscitation time of 60 min.

Results and discussion

Results (Table 1) are given as mean, range and standard deviation (S.D.)
of log_{10} cfu/g. The same species were detected in approximately the same
proportions, regardless of diluent. They include field fungi (Cladosporium,
Alternaria, Aureobasidium) as well as more xerotolerant species belonging to
the Aspergillus glaucus group and to the Penicillia.

The results show no difference in mold counts due to dilution fluids or
resuscitation time. All diluents are suitable for cereal examination. The
more simple, the better.

                                                        D. RICHARD-MOLARD

▶ Influence of the Addition of a Wetting Agent (Polysorbitan
  80) in Diluent on Fungal Counts from Cereals

Fungal spores of species which contaminate cereals and their products
are frequently hydrophobic. Conidia of Penicillium often show water
repellant properties, probably due to the physical conformation of the
external layers of the spore wall (Fisher & Richmond 1970). At least some

field fungi like <u>Alternaria</u> <u>tenuis</u> and <u>Rhizopus</u> <u>stolonifer</u> are also reported to be difficult to wet (Fisher et al. 1972). Thus, the addition of a wetting agent to diluents used for mycological examination of foods may produce better dispersions in diluents. A wide range of diluents can be used (Jarvis 1978) but the effect of an addition of polysorbitan 80 as a wetting agent is not well known.

In the present preliminary study, the influence of increasing levels of polysorbitan 80 in diluent has been tested on three cereal products, i.e., wheat, wheat flour and maize, and a suspension of <u>Penicillium</u> <u>implicatum</u> conidia.

Materials and methods

<u>Diluents</u>. The reference diluent was 0.1% peptone and 0.85% NaCl in water. Test diluents had the same composition with the addition of 1, 2, and 4 drops (i.e., about 0.03, 0.06 and 0.12 g/liter) of polysorbitan 80.

<u>Samples</u>. Products were examined in triplicate using each of the four diluents, i.e., twelve subsamples were separately serially diluted 1:10 in diluent, and then pour-plated on 2% malt agar. Samples of grain (wheat and maize, 100 g in 400 ml of diluent) were soaked for 30 min and ground for 90 sec in a Waring Blendor. Wheat flour (20 g in 180 ml of diluent) was continuously shaken for 30 min before diluting.

Each diluent was also inoculated with 2 ml of a suspension of <u>Penicillium</u> <u>implicatum</u> conidia in 0.1% peptone water. All plates were incubated upright at 25°C and counted at 5 days.

Table 1. Influence of added polysorbitan 80 on viable counts on three commodities and <u>Penicillium</u> <u>implicatum</u> conidia. Values are mean, range and standard deviation

	Sample			
Diluent	Wheat	Maize	Wheat flour	<u>P</u>. <u>implicatum</u>
Reference (0.1% peptone + 0.85% NaCl)	4.24 (4.00–4.41) 0.18	6.22 (6.04–6.34) 0.31	2.58 (2.45–2.69) 0.21	3.39 (3.20–3.51) 0.07
Reference + 0.03 g/L polysorbitan 80	4.34 (4.17–4.54) 0.20	6.43 (6.30–6.51) 0.16	2.62 (2.57–2.68) 0.16	3.57 (3.51–3.62) 0.06
Reference + 0.06 g/L polysorbitan 80	4.03 (3.90–4.11) 0.08	6.36 (6.34–6.39) 0.08	2.53 (2.50–2.56) 0.09	3.85 (3.83–3.92) 0.06
Reference + 0.12 g/L polysorbitan 80	3.86 (3.47–4.30) 0.22	6.62 (6.51–6.73) 0.11	2.37 (2.17–2.57) 0.14	3.77 (3.65–3.88) 0.08

Results

Results are given in Table 1 as mean, range and standard deviation (S.D.) of $\log_{10}$ cfu/g.

Discussion

When evaluating results from 0.1% peptone water inoculated with P. implicatum, differences between the reference diluent and diluent supplemented with 2 drops (0.06 g/L) of added polysorbitan 80 are highly significant ($P < 0.01$). Higher counts were obtained when diluent containing polysorbitan was used. This may result from disruption of short conidial chains. For other products, differences were in general not significant ($P < 0.05$) but counts obtained with one drop of polysorbitan 80 (0.03 g/L) were always higher than with the reference. On the other hand, the diluent with 4 drops (0.12 g/L) probably has an inhibitory effect on spore germination, at least for some species. The addition of one drop of polysorbitan 80 per liter of diluent seems to be beneficial and could be recommended.

References

FISHER, D. J., HOLLOWAY, P. J. & RICHMOND, D. V. 1982 Fatty acid and hydrocarbon constituents of the surface and wall lipids of some fungal spores. Journal of General Microbiology 57, 51-60.

FISHER, D. J. & RICHMOND, D. V. 1970 The electrophoretic properties and some surface components of Penicillium conidia. Journal of General Microbiology 64, 205-214.

JARVIS, B. 1978 Methods for detecting fungi in foods and beverages. In Food and Beverage Mycology. ed. Beuchat, L.R. pp. 471-504. Westport, CT: AVI Publ. Inc.

D. RICHARD-MOLARD

▶ Impromptu Contribution: Blending, Diluents and Sedimentation

Briefly summarized, our experience since 1960 with simple methods for inoculum preparation in the mycological examination of foods of a_w values over 0.95 has led to the following conclusions:

1. An adequate release of propagules from virtually all foods can be achieved by vigorous shaking for 2 min at ambient temperature of a 1:10 macerate in 500-ml screw-cap bottles with about twenty ca. 3-mm diameter glass beads per bottle.

2. Sterile distilled water is the most satisfactory diluent for general purposes, because sufficient soluble constituents are extracted from the majority of foods to avoid plasmoptysis, chelate any traces of toxic metal ions that might be present, and protect injured cells. For the monitoring of foods of acceptable microbiological quality, no further dilution of the primary macerate is required. However, where this might be necessary, it is prudent to use peptone saline instead of distilled water. Also, examination of cheese requires the addition of sodium citrate to the diluent and monitoring of fatty foods sometimes requires the addition of polysorbitan.

Sedimentation for about 1 min will be of no consequence with respect to recovery, while it effectively avoids obscuring of plates by food debris. This experimental observation has been corroborated by calculations based on Stokes' law of spontaneous sedimentation and van der Waal's theorem of adherance of organisms to a food matrix.

D. A. A. MOSSEL

 Effect of Incubating Plates Inverted or Upright When Enumerating Yeasts and Molds in Dry Seed-based Foods

Bacteriologists customarily incubate Petri plates in an inverted position, but food mycologists have not reached agreement on which way plates should be incubated. The study reported here was designed to evaluate the influence of inverted versus upright incubation on viable counts.

Materials and methods

A total of 109 bulk-stored, dry food samples were purchased from eight retail grocers in the Atlanta, GA area. Foods examined included barley, beans, coconut, cereal and legume flours, granola, lentils, dehydrated milk, nuts, oats, peas, popcorn, pumpkin seed, rice, sesame seed, sunflower seed and wheat.

Samples (50 g) were combined with 450 ml of 0.1 M potassium phosphate buffer (pH 7.0) containing 0.1% polysorbitan 80 and vigorously shaken for 2 min. Serial dilutions (0.1 ml) of samples were plated in quadruplicate on plates of surface-dried test media.

Duplicate plates of all dilutions on DRBC, OGY, PCA, PDA and RBC media were incubated in inverted and upright positions for 5 days at 25°C before yeast and mold colonies were enumerated.

Statistically significant ($P \leq 0.05$) differences between mean values were determined by subjecting data to analysis of variance and Duncan's multiple range test.

Results and discussion

Mean values of populations of fungi enumerated in 1090 tests (109 samples, 5 media, 2 plate positions) were calculated and significant differences between means for inverted versus upright plates were determined. Significantly higher populations were detected in 11 (2.0%) tests when plates were incubated in inverted position; conversely, significantly higher populations were detected in 17 (3.1%) tests when plates were incubated in an upright position.

Data for composites of mean values of fungal populations incubated in inverted and upright positions are listed in Table 1. Comparisons reveal that populations of fungi detected on test foods were not influenced by the position of the plate during incubation.

Table 1. Comparison of inverted versus upright position of plates during
          incubation for detection of yeasts and molds in a composite of 109
          foods

Plate position	Population (cfu/g)[a]
Inverted	3814
Upright	3712

[a]Mean values of composite yeast and mold populations recovered from 109
foods using five media (DRBC, OGY, PCA, PDA and RBC)

Table 2. Comparison of stores for mycological quality of composites of foods

Store no.	Number of foods analyzed	Population (cfu/g)[a]
1	4	3054b
2	8	435c
3	3	1202bc
4	16	723c
5	34	1199bc
6	14	514c
7	6	321c
8	24	13725a

[a]Mean values of composite yeast and mold populations recovered from 109
foods using five media (DRBC, OGY, PCA, PDA and RBC). Values which are
not followed by the same letter are significantly different ($P \leq 0.05$)

It was of interest to compare composite data for total fungal populations
by their source, i.e., the store from which they were purchased. These data
are presented in Table 2. While populations could be described as low
(stores 2, 4, 6 and 7), medium (stores 1, 3 and 5) and high (store 8), foods
from store number 8 clearly were contaminated with the highest populations
of fungi. The highest population detected in the 109 foods analyzed was
4.8×10^7 in buckwheat flour. It is interesting to note that store number
8 was the only store included in the study which is strictly a health food
outlet. One cannot argue that customers making purchases at this store
would not get what they paid for – naturally contaminated foods.

In summary, the position of the plate during incubation did not
significantly influence the number of colonies developing on test media.

L. R. BEUCHAT

The Influence of Dilution on the Enumeration of Yeasts from Beverages

This study was undertaken to evaluate the influence of dilution on enumeration of yeasts in naturally contaminated beverage products.

Materials and methods

Media. Three media (acidified PDA, DRBC and OGY) were used to enumerate yeasts in beverages, fruit concentrates and fruit purees. OGY differed from the formula listed in Appendix I in that the glucose concentration was increased to 0.2%.

Enumeration. Undiluted samples were enumerated by both pour-plate and spread-plate techniques. Pour plates (0.1 ml undiluted sample) were mixed with approximately 18 ml of tempered (50°C) medium. Aliquots of 0.1 ml were also applied to spread plates. Samples for dilution plating (0.1 ml) were serially diluted 1:10 in 0.1% peptone water. All plates were incubated for 5 days at 25°C.

Results

One hundred and twenty-four samples of noncarbonated high-acid beverages, beverage concentrates and fruit purees were sampled. Yeasts were not detected in 104 of these samples; results from analysis of the 20 samples in which yeasts were detected are presented in Table 1.

While visual examination of the mean values indicate that slightly higher counts were detected when samples were not diluted, the difference in counts was not statistically significant ($P \leq 0.05$ in diluted or undiluted samples). Mean counts on OGY were higher, but the large standard deviations indicated no significant difference between media.

Conclusion

The results of this study showed no statistical effect of dilution on the enumeration of yeasts in noncarbonated high-acid beverages, beverage concentrates and fruit purees. Dilution may of course be required if the

Table 1. A comparison of plating undiluted versus diluted samples for detecting yeasts in naturally contaminated beverages[a]

Medium	Log_{10} cfu/ml of sample[b]	
	Undiluted samples	Dilution plating
DRBC	2.0 ± 2.3	1.8 ± 1.9
PDA	2.1 ± 2.4	2.1 ± 2.1
OGY	2.4 ± 2.7	2.3 ± 2.5

[a]Twenty samples of noncarbonated, high-acid beverages, fruit purees and fruit concentrates; 104 additional samples had no detectable yeasts by either dilution or direct plating
[b]Mean ± standard deviation of yeast populations detected in 20 samples

microbial history of a sample indicates a possible high level of fungal contaminants.

C. B. ANDERSON
L. J. MOBERG

▶ Causes of Dilution Errors When Enumerating Molds in Food

When determining populations of molds in foods using a pour-plate technique, it is usually found that the counts do not correctly follow a strict decimal progression. Often the population at the countable dilution may be 4- or 5-fold that at the next higher dilution (e.g., 90 colonies per plate at the 10^2 and 20 colonies per plate at the 10^3 dilution).

In a recent collaborative excercise involving five laboratories which carried out counts on ten replicate plates with two consecutive decimal dilutions of different food samples, the counts based on the higher dilution of sample were, on average, about twice as high as those based on the lower dilution. These tests and others suggest that errors tend to be lower in media where the colony size is deliberately restricted by compounds such as rose bengal.

Two possible explanations for this dilution error can be put forward. One explanation is that fragmentation of mycelium or breaking of spore chains or clumps takes place during the process of making serial dilutions of the sample. Secondly, competitive inhibition occurs when large numbers of mold colonies are present on the plates. This paper describes tests carried out to examine these two possibilities.

Materials and methods

To determine the effect of method and time of mixing on disruption of spore clumps, spore suspensions of Aspergillus flavus, A. niger, Cladosporium resinae, Penicillium notatum and Trichothecium roseum were prepared from slant cultures. The preparations were examined microscopically for the percentage of single spores present initially and after shaking (1 shake per sec) for 5, 10, 50 and 100 sec, Vortex stirring (2,600 rev/min) for 0, 10, 30 and 60 sec and pummelling with the Stomacher 400 for 0, 20, 60 and 300 sec.

To investigate the competitive inhibition theory, tests were carried out on samples of flour to compare mold counts using a pour-plate technique with both 9- and 14-cm diameter Petri dishes. In each test, a dilution of flour was selected which was known to result in 10-40 mold colonies per plate when 1 ml was added. Using this dilution, increasing amounts were added per plate to give a range of counts. Ten replicate plates of each size were used for each selected amount of diluted flour sample added. The plates were poured with RBC at 44°C and counts were determined after incubation at 27°C for 5 days.

Results

Change in the average percentage of single spores present in suspensions of the five test fungi as a result of mixing using the different methods is

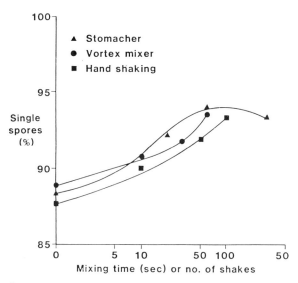

Fig. 1. Change in average percentage of single spores in suspensions of
five test organisms using different mixing methods

shown in Fig. 1. It is apparent that the percentage of single spores
increases with mixing time, the effect being similar for the three methods
of mixing employed. The results with individual mold species were
surprisingly similar, the percentage of single spores only increasing by 4%
with Penicillium and up to 7% with Cladosporium after prolonged mixing.

Figure 2 shows the relationship between the mean count of mold colonies
per plate and the inoculum volume of the selected dilution of flour using 9-
and 14-cm diameter plates (average of three separate tests). Using the
large 14-cm plates, there is a linear relationship between mold count and
the amount of flour sample added. This holds true for counts of up to 180
colonies per plate. With the 9-cm plates, on the other hand, the
relationship deviates significantly from linearity.

Conclusions

There is a progressive breaking of spore clumps as a result of shaking
or mixing suspensions of molds. However, the effect is small and
insufficient indicating that the method of preparing secondary dilutions of
the sample could account for the dilution errors observed. The
non-linearity of counts with increased inoculum volume using 9-cm diameter
plates (Fig. 2) suggests that competitive inhibition is the more likely
explanation for these errors. A medium which reduces colony size (RBC) was
used in these tests. Even larger deviations from linearity might be
expected using less inhibitory media.

24

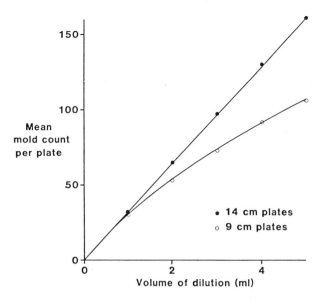

Fig. 2. Relationship between the mean count of mold colonies per plate for
various inoculum volumes of the same dilution of flour using
different sized plates

    To avoid underestimation of mold populations due to competitive
inhibition, there would appear to be three approaches. The first is to use
large 14-cm diameter plates and count up to 180 mold colonies per plate.
The second is to continue to use 9-cm diameter plates but only take into
consideration those plates with 40 or less colonies when calculating
counts. Another approach is to use a medium which drastically restricts
colony size so that the possibility of competitive inhibition on 9-cm
diameter plates is largely eliminated. With the latter procedure there is a
danger that, in restricting colony size, the medium may become inhibitory to
certain species and identification will be difficult. With the second
approach, plates containing 10-40 colonies may not always be obtained using
a conventional decimal dilution series. Moreover, the species present may
not give an overall picture of the mold flora in the food. On balance,
therefore, the use of large 14-cm diameter plates in conjunction with a
slightly inhibitory medium is recommended for pour-plate counts of molds in
foods. I should be mentioned that, in a recent international collaborative
exercise comparing counts of fungi in flour, there was some evidence to
suggest that dilution errors were less pronounced using a spread-plate
technique. This aspects warrants further investigation.

                                                    D. A. L. SEILER

Effect of Presoaking on Recovery of Fungi from Cereals and Cereal Products

Soaking foods in diluent prior to mixing may have at least three effects. First, it may allow resuscitation of sublethally damaged cells from dried, intermediate moisture or low-pH foods. Second, it may allow better release of organisms which are present within tissues. Third, with hard and sharp materials such as cereals, the softening effect of soaking will help to prevent damage to the bags when using the Stomacher for mixing the primary dilution of the sample.

This communication presents the results from a collaborative exercise involving four laboratories undertaken to examine the effects of soaking and mixing times using the Stomacher on recovery of fungi from a variety of cereals and cereal products.

Materials and methods

Ten commodities were examined. For each, nine primary dilutions were prepared, each containing a 40-g subsample in 360 ml of saline (0.85%) peptone (0.1%) diluent. Groups of three dilutions were allowed to soak at ambient temperatures (20°C) for 0, 30 and 60 min, respectively. Subsamples from each of these groups were mixed using the Stomacher 400 for 2, 5 and 10 min. This procedure resulted in nine different combinations of soaking and mixing times. Counts of molds and yeasts were determined using a pour-plate method with OAES (Ohio Agricultural Experimental Station) medium in three laboratories, and OGY in one laboratory, after 5-7 days incubation at 25-27°C. Also included are results of separate tests in which populations of molds in six samples of flour were compared with and without soaking for 50 min in tryptone salt broth prior to homogenizing with the Colworth Stomacher for 2 min.

Results and discussion

Figure 1 shows the mean percentage increase in mold and yeast counts obtained with the various soaking and mixing times compared with no soaking and 2 min mixing. Each block represents data for eight commodities from four laboratories. The counts increase progressively with both soaking and mixing time.

The increases in counts obtained by soaking the various commodities for 60 min prior to mixing with the Stomacher for 10 min are given in Table 1. Also included in Table 1 are the individual results for three samples of wheat and six samples of flour soaked in tryptone salt broth.

It is clear that the increase in count as a result of soaking for 60 min varies considerably from one commodity to the next and also with different samples of the same commodity. Larger increases were obtained with whole cereal grains than with milled cereals or products containing milled cereals. This is not surprising since, with these materials, the soaking and mixing will help to recover fungi which may have penetrated into deep tissues. The effect is more pronounced with some cereal samples than others (e.g., see wheat A as compared with wheats B and C in Table 1). With most milled cereals or products containing milled cereals, a small increase in count was obtained as a result of presoaking. It is not possible to conclude whether this was due to resuscitation or better release of the fungi.

26

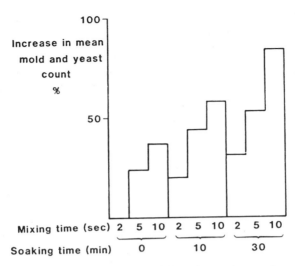

Fig. 1. Mean percentage increase in mold and yeast counts, over that
obtained with no soaking and 2 min mixing, using various soaking
and mixing times for eight commodities tested by four laboratories

Table 1. Effect of soaking prior to mixing on mold and yeast counts in
various cereals, cereal products and cereal based feeds

| Commodity | Mean mold and yeast count (log cfu/g) | | Increase in count (%) |
	No soaking, mixed 10 min	60 min soaking, mixed 10 min	
Wheat	4.75	5.04	95
Barley	5.45	5.76	104
Cow cake	4.39	4.53	39
Hay	7.50	7.56	14
Milled barley	7.50	7.56	14
Pig meal	5.18	5.23	13
Layer crumb	3.61	3.68	17
Sheep feed	5.54	5.50	6
Wheat A	4.90	5.27	137
B	4.78	4.97	58
C	4.48	4.65	50
Flour[a] A	4.04	4.19	45
B	3.72	3.90	52
C	4.34	4.51	50
D	3.72	3.75	8
E	3.53	3.53	0
F	3.67	3.83	43

[a]Soaked in tryptone salt broth for 50 min before mixing for 2 min

27

Recommendations

A standardized method for enumeration of molds and yeasts in cereals should include presoaking in diluent for at least 30 min and preferably 60 min, prior to mixing. A presoaking step should be considered for all foods where the fungi are likely to exist under stressed conditions.

Acknowledgements

The help of the three laboratories from the UK Agricultural Development Advisory Service who took part in these collaborative tests is gratefully acknowledged.

D. A. L. SEILER

Optimization of Conditions for the Surface Disinfection of Sorghum and Sultanas Using Sodium Hypochlorite Solutions

Many foods, especially cereal grains, nuts, fruit and vegetables, have an internal mycoflora which results from air or soilborne contamination at the time of pollination or setting of the fruit. Subsequent fungal invasion may occur depending on growth and harvesting conditions. Any such internal mycoflora may be responsible for fungal spoilage of the product or, more significantly, elaborate mycotoxins in the product.

To assess the internal mycoflora of a foodstuff, it is frequently necessary to surface disinfect the product prior to culturing. Surface disinfection partially eliminates superficial contaminants and fast growing surface saprophytes which may interfere with the determination of the internal mycoflora.

Booth (1971) summarized the various chemicals regularly used for surface disinfection. These included 3% hydrogen peroxide, 2% potassium permanganate, 75% ethanol, 0.001% mercuric chloride and 0.35% sodium or calcium hypochlorite. The recommended contact time with the disinfectant was 1-5 min, followed by rinsing with sterile distilled water. Jarvis (1978) indicated a preference for dipping the sample into 75% ethanol prior to immersion in a 0.1% sodium hypochlorite solution for 2 min, whereas the European Brewing Committee (EBC 1981) recommended 1% sodium hypochlorite (NaOCl) for 1 min to surface disinfect barley and malt. In each case no experimental data were presented to justify the recommended concentrations, contact times or effectiveness on various species of fungi.

This paper describes the effect of free chlorine concentrations and contact times on the surface disinfection of sorghum and sultanas (raisins). As a result of these studies, procedures for surface disinfection are recommended.

Materials and methods

Samples. A sample of rain damaged sorghum badly infected with *Aspergillus niger* and *Aspergillus flavus* was obtained and used for the sorghum studies. Similarly a case of tainted sultanas naturally infected with *A. niger* was obtained and used for some of the sultana studies. An

28

additional lot of sultanas, almost free of fungal contamination was obtained.
Separate subsamples of these sultanas were deliberately contaminated with
conidia of <u>Penicillium</u> <u>chrysogenum</u> and <u>Eurotium</u> <u>repens</u>. An additional sample
was contaminated with mature ascospores of <u>E</u>. <u>repens</u>.

<u>Chlorine solutions</u>. Commercially available NaOCl solutions were
obtained and checked for available chlorine concentration prior to each
experiment by the iodometric-thiosulphate titrimetric method (Vogel 1961).
The hypochlorite solutions were diluted in distilled water to the
concentrations required just prior to the commencement of each experiment.

<u>Surface disinfection</u>. Samples of sorghum (approximately 30 g) and
sultanas (approximately 45 g) were weighed into sterile beakers. To each
sample 100 ml of the appropriate chlorine solution was added and mixed, so
that the sample was thoroughly wetted. After the specified contact time,
the spent chlorine solution was drained, replaced with 100 ml of sterile
distilled water, rinsed and drained again. Grains and fruits were deposited
individually on media with sterile forceps.

<u>Media</u>. Sorghum samples were direct plated onto DRBC medium and incubated
for 5 days at 25°C. Sultana samples were cultured on DG18 medium for 5 days
at 25°C.

<u>Results</u>

The sorghum used in these studies had a dilution plate count on DRBC
medium of 6.5 x 10^6 cfu/g; approximately 38% of the colonies were <u>A</u>. <u>flavus</u>
and 10% were <u>A</u>. <u>niger</u>. When individual sorghum grains were direct plated
onto DRBC, 95% of the grains were contaminated with <u>A</u>. <u>flavus</u> and 89% with
<u>A</u>. <u>niger</u>.

Initially, a 0.37% (w/v) chlorine concentration was used to surface
disinfect the sorghum for various contact times ranging from 0.5-30 min. The
results (Table 1) indicate that a contact time of 2 min is effective when
using 0.37% chlorine.

Various concentrations of chlorine, ranging from 0-1.54%, were then
tested on the sorghum using a 2-min contact time. A rapid decline in the
percentage infection due to <u>A</u>. <u>flavus</u> and <u>A</u>. <u>niger</u> was observed as the
chlorine concentration increased to 0.37% (Table 2).

Table 1. The effect of contact time on the surface disinfection of sorghum
          using a 0.37% (w/v) chlorine solution

Contact time (min) with 0.37% chlorine solution	Percentage of grains infected with	
	<u>A</u>. <u>flavus</u>	<u>A</u>. <u>niger</u>
0	95	89
0.5	60	51
1.0	27	16
2.0	16	8
10.0	11	4
30.0	6	2

Table 2. The effect of chlorine concentration on surface disinfection of sorghum for 2 min

Chlorine concentration (%, w/v)	Percentage of grains infected with	
	A. flavus	A. niger
0	95	89
0.035	54	42
0.070	36	19
0.18	29	24
0.37	20	7
0.73	21	9
1.54	16	9

Further increases in chlorine concentration above 0.37% had only a minimal effect on the percentage infection, probably because not all the Aspergilli were surface contaminants.

The optimal contact time for surface disinfection of sultanas naturally contaminated or infected with A. niger and P. chrysogenum was determined using a 0.35% (w/v) chlorine solution and was found to be 2 min (Table 3).

Sultanas were also treated with various chlorine concentrations and a 0.33% (w/v) solution was required to inactivate most of the surface A. niger spores and mycelium (Table 4).

Since further increases in chlorine concentration failed to significantly reduce the percentage of viable propagules, it is assumed that some sultanas were internally infected with A. niger.

Another batch of sultanas contaminated with P. chrysogenum conidia were treated with a similar range of chlorine concentrations. These spores were more readily inactivated by chlorine than those of A. niger (Table 4); a 0.33% solution disinfected over 90% of the contaminated sultanas. When sultanas contaminated with conidia and ascospores of E. repens were exposed to a similar range of chlorine concentrations (0–0.67% w/v), none of the

Table 3. The effect of contact time on the surface disinfection of sultanas using a 0.35% (w/v) chlorine solution

Contact time (min) with 0.35% chlorine solution	Percentage of grains infected with	
	A. niger	P. chrysogenum
0	100	100
0.5	–	65
1.0	9	33
2.0	18	6
5.0	3	3
10.0	0	1

Table 4. The effect of chlorine concentration on surface disinfection of sultanas

Chlorine conc. (w/v)	Percentages of sultanas infected with		E. repens	
	A. niger	P. chrysogenum	Conidia	Ascospores
0	100	100	100	100
0.035	92	89		
0.070	82	72		
0.11	–	56		
0.17	44	32		
0.22	–	13		
0.33	22	6		
0.67	17	5	100	100
1.0			100	100
1.5			88	98
2.0			35	55
3.0			14	35
4.0			.22	23
5.0			14	16
6.0			18	17
6.7			14	19

sultanas was disinfected. Chlorine concentrations ranging from 1-6.7% were then studied. The ascospores were slightly more resistant than conidia to chlorine (Table 4) but for either type of spore a 3% chlorine solution was required to bring about a substantial reduction in the percentage infection of the sultanas. Further increases in chlorine concentration did not significantly reduce the infection rate.

The resistance of Eurotium spores to chlorine is not restricted to this sterilant as further experiments showed that 3% hydrogen peroxide, 2% potassium permanganate and 75% ethanol failed to disinfect sultanas contaminated with E. repens. The resistance of Eurotium spores to surface disinfection is frequently observed when cereal grains are treated with chlorine and direct-plated onto DG18, a medium on which Eurotium grows rapidly and frequently outgrows other mycoflora.

Discussion

From these studies it is apparent that not all fungi are equally susceptible to inactivation by chlorine. In the absence of surface contamination by Eurotium spp., a chlorine concentration of 0.35% (w/v) with a contact time of 2 min will effectively bring about substantial surface disinfection, permitting isolation of the internal mycoflora.

Disinfection of products heavily contaminated with Eurotium spp. requires either higher concentrations of chlorine or longer contact times at the recommended 0.35% concentration. Treatment of sultanas contaminated with E. repens conidia with 0.35% chlorine reduced the infection rate to an acceptable 29% after a 10-min contact time. On the other hand, the recovery of a high proportion of _Alternaria_ from sorghum after 30 min disinfection with a 0.37% chlorine indicates that chlorine is surface-inactivated and that prolonged exposure is unlikely to dramatically influence the recovery of the internal mycoflora.

When the rates of inactivation of _A. niger_ contaminated sorghum and sultanas were compared, it is apparent that _A. niger_ was more rapidly inactivated on sorghum. It is concluded that the smoother, more inert surface of sorghum facilitates a more rapid rate of surface sterilization than the irregular surfaces of sultanas which consist of organic matter more likely to react with chlorine. Therefore, it is recommended that the exact concentration and contact time for surface disinfection of a product may need to be adjusted according to textural attributes, reactive surface organic matter and species of fungi on the surface of the product.

After surface disinfection, it is the usual practice to drain the disinfectant and rinse the product with sterile water. Rinsing was found not to be essential, as the percentage recovery of the internal mycoflora was not affected, although the percentage infection of the surface contaminants was slightly lower when rinsing was omitted.

Recommendation

It is recommended that most food particles should be effectively surface disinfected by immersion in a 0.35% w/v chlorine solution for 2 min. Rinsing of the treated product with sterile distilled water is not essential, provided the product is adequately drained after disinfection.

References

BOOTH, C. 1971 Introduction to General Methods. In _Methods in Microbiology_, Vol. 4, ed. by Booth, C., London: Academic Press.
EBC 1981 EBC Analytica Microbiologica: Part II. _Journal of Institute of Brewing_ 87, 303–321.
JARVIS, B. 1978 Methods for detecting fungi in foods and beverages. In _Food and Beverage Mycology_, ed. by Beuchat, L. R. Westport, CT: Avi Publ. Co.
VOGEL, A. I. 1961 _A Textbook of Quantitative Inorganic Analysis Including Elementary Instrumental Analysis_, 3rd ed., London: Longman.

S. ANDREWS

▶ Comparison of Methods Used for Surface Disinfection of
Food and Feed Commodities Before Mycological Analysis

Grains and other foods and feedstuffs often contain an actively growing internal fungal flora, a superficial flora consisting of mycelium and spores of the internal invaders, and contaminant spores from the air surrounding the grains. Under favorable conditions the contaminant spores may also give rise to active growth in and on the grain kernels. Different surface disinfection techniques may be used to remove all or part of the superficial flora (Pepper 1961; Booth 1971; Gams et al. 1980). Porter (1944) and Wallace and Sinha (1962) recommended that surface disinfection of grains not be done before mycological analyses, mainly because the surface disinfectants may be

selective and the procedure time consuming. Mislivec & Bruce (1977) showed
that surface disinfection (5% sodium hypochlorite for 1 min) removed a
substantial proportion of conidia from the surface of soybeans: molds grew
from 99.4% of soybeans before surface disinfection but from only 52.8% after
surface disinfection. The aim of this study was to determine if surface
disinfectants would remove all conidia from artificially contaminated kernels
but not inactivate the internal flora in naturally contaminated kernels.

Materials and methods

   Grains. A barley sample of good quality was sterilized in a linear
electron accelerator at Ris (Roskilde, Denmark) with a dose of 1 mrad/h.
Seeds with smooth surfaces (parboiled rice and soybeans) were washed five
times with sterile water. Barley, rice and soybeans were artificially
contaminated with conidia of Penicillium viridicatum CBS 223.71.
Twenty-five samples of barley and green coffee beans were examined by the
direct plating method to compare different surface disinfection methods.
One hundred kernels were used for each test. The first 15 samples (barley
from different farms in Denmark) were plated on PRYES at 20°C (Frisvad 1983)
before and after surface disinfection in 1% NaOCl for 1 min. The surface
disinfected kernels were washed twice in sterile water. Samples 16 and 17
were green coffee; sample 18 was barley kept in an airtight silo for 1 1/2
years; sample 19 was barley of good quality placed in a bucket near a
subsample of sample 18; and samples 20-25 were barley containing ochratoxin
A. Kernels from samples 16-25 were plated on DRBC, AFPA, DG18, PRYES and
DRYES media using two types of surface disinfection: the kernels were
treated with either 1% NaOCl for 1 min and washed twice in sterile water or
treated with 2% formaldehyde for 1 min.

   Media. The media used are listed above. They were all modified to
contain only 100 µg/ml chloramphenicol (added before autoclaving) as an
antibiotic. All media on which samples were deposited were incubated at
25°C except where otherwise noted.

   Surface disinfection methods. The effects of the following surface
disinfection methods were evaluated on kernels artificially contaminated
with conidia of P. viridicatum: (a) All combinations of 1, 2, 5, 10 and 15%
NaOCl (prepared from a 15% solution of active NaOCl in water) with contact
times of 1, 2, 5 and 10 min followed by washing twice in sterile water; (b)
the effect of amount of NaOCl solution (30 ml or 100 ml), temperature (25°C
or 40°C) of the NaOCl solution and the effect of an ultrasonic water bath
treatment (2 min) were examined in all combinations for 2, 5 and 10% NaOCl
solutions; (c) the effect of lowering the pH of the NaOCl solution to 7.0
with dilute HCl and of two washings with sterile water; (d) formalin (HCHO)
treatment (1, 2 or 3%), with or without two washings with sterile water; (e)
flaming 50 kernels with 0.2, 0.4, 0.6 or 0.8 ml of 96% ethanol in a glass
Petri dish; (f) combination of 1 or 2% HCHO and flaming with 0.3 ml of 96%
ethanol and two washings; and (g) surface washing with detergent (0.01%
Triton-X) for 1 min from one to five times.

   Surface structure of kernels. Barley, parboiled rice and soybean
kernels inoculated with P. viridicatum were surface disinfected with 5, 10
or 15% NaOCl for 1, 2 or 5 min and washed twice with sterile water.

   Artificial inoculation. Samples (50 g) of kernels were placed in sterile
polyethylene bags and sprayed with a conidial suspension from a 1-week-old
culture of P. viridicatum using 10 ml 0.85% NaCl diluent to give a final
concentration of 10^7 conidia/ml. Viable counts by dilution plating of the
artificially inoculated barley kernels showed that each kernel was
contaminated with about 2.5 x 10^5 conidia of P. viridicatum. Surface

Table 1. Effect of concentration and contact time of NaOCl on survival of
          Penicillium viridicatum conidia[a]

NaOCl conc. (%)	Contact time (min)	Infection (%)
1, 2	1, 2, 5, 10	100
5	1, 2, 5	100
5	10	98
10	1 & 2	100
10	5	92
10	10	88
15	1	100
15	2	96
15	5	78
15	10	68

[a]30 ml NaOCl solution at 25°C, followed by two washes with sterile water
and incubation for 15 days on PRYES medium.

disinfection experiments were performed immediately after the preparation of
the inoculated samples.

Plates were examined after 5, 7, 10 and 15 days. Results are listed as
percentage of infection after 15 days of incubation in the dark: controls
were 100% infected (50 kernels).

Table 2. Effect of volume and contact temperature of NaOCl solution on
          survival of Penicillium viridicatum conidia on barley kernels

NaOCl conc. (%)	NaOCl solution temperature (°C)	NaOCl solution volume (ml)	Ultrasonic treatment	Infection (%)
2, 5	25	100	−	100
10	25	100	−	94
2	40	30	−	100
5	40	30	−	98
10	40	30	−	78
2	40	100	−	100
5	40	100	−	92
10	40	100	−	72
2	25	30	+	100
5	25	30	+	98
10	25	30	+	96
2, 5	25	100	+	100
10	25	100	+	96
2	40	30	+	98
5	40	30	+	90
10	40	30	+	58
2	40	100	+	100
5	40	100	+	94
10	40	100	+	60

Table 3. Effect of pH adjustment and temperature of NaOCl solutions and of sterile washing on survival of _Penicillium viridicatum_ conidia on barley kernels[a]

NaOCl conc. (%)	NaOCl solution adjusted to pH 7	Washing in sterile water two times	NaOCl solution temperature (°C)	Infection (%)
5	−	−	25	25
5	−	+	25	100
5	−	−	40	90
5	−	+	40	98
2	+	−	25	98
5	+	−	25	86
10	+	−	25	82
2	+	+	25	96
5	+	+	25	94
10	+	+	25	98

[a]Contact time was 2 min in 30 ml of NaOCl solution; incubation was for 15 days at 25°C on PRYES agar.

Results

The results listed in Tables 1, 2 and 3 show that the various treatments with NaOCl did not inactivate all conidia on the surface of the barley kernels. Infection percentages after surface disinfection ranged from 58-100%, except in some extreme cases (e.g., combinations of 10 or 15% NaOCl and 10 min contact time).

Table 4. The influence of grain type on the efficacy as surface disinfectant for conidia of _Penicillium viridicatum_[a]

NaOCl conc. (%)	Contact time (min)	Infection (%) after period (days) of incubation					
		Soybeans			Rice		
		5	10	15	5	10	15
5	1	2	4	10	8	8	8
5	2	0	2	4	0	0	0
5	5	0	2	2	2	2	2
10	1	8	4	14	2	2	2
10	2	4	8	8	4	6	6
10	5	2	8	8	0	0	0
15	1	42	68	74	0	0	0
15	2	18	24	44	0	0	0
15	5	30	42	80	0	0	0

[a]30 ml NaOCl solution at 25°C, followed by two washes with sterile water and incubation for 15 days on PRYES medium

Table 5. The efficacy of formaldehyde as a surface disinfectant for conidia
         of <u>Penicillium viridicatum</u> on barley grains[a]

HCHO conc. (%)	Contact time (min)	Washing in sterile water two times	Infection (%) after incubation (days)		
			5	10	15
	2	−	0	0	0
2	2	−	0	0	0
3	2	−	0	0	0
1	2	+	30	98	100
2	2	+	0	35	60
3	2	+	0	0	0

[a]The kernels were treated with 30 ml formaldehyde at 25°C

The NaOCl treatment was much more effective on kernels with smooth
surfaces (Table 4), but high concentrations of NaOCl (15%) resulted in
damage of the soybean kernels, presumably because skin folding assisted in
penetration of conidia into the internal layers in the soybean.

Formalin was much more effective in removing superficial conidia from
barley kernel surfaces, provided the kernels were not washed after the
treatment (Table 5).

Flaming was also effective in removing superficial conidia from the
surface of barley, provided 0.6 or 0.8 ml of ethanol was used for 50 kernels
(flaming time, 38-40 sec) and provided the kernels were not washed with
sterile water after the treatment (Table 6).

Table 6. Effect of flaming with ethanol on conidia of <u>Penicillium viridi-</u>
         <u>catum</u> on barley kernels as measured by infection precentages on
         PRYES agar

Ethanol (ml)	Flaming time (sec)	Washing in sterile water two times	Infection (%) after incubation (days)		
			5	10	15
0.2	24	−	24	42	94
0.4	29	−	6	6	18
0.6	38	−	0	0	0
0.8	40	−	0	0	0
0.2	25	+	100	100	100
0.4	30	+	100	100	100
0.6	40	+	100	100	100
0.8	40	+	100	100	100

Treatment of kernels with 1 or 2% HCHO followed by ethanol treatment and flaming (0.3 ml, flaming time 25–27 sec, and two washes with sterile water) resulted in infection percentages of 100 and 12%, respectively.

Washing contaminated kernels with detergent (Triton-X) as many as five times was not efficient in removing conidia.

Subsequent experiments showed that flaming also destroyed the internal flora of naturally contaminated barley samples and a barley sample in which P. viridicatum had grown for 3 weeks at 25°C. Furthermore, the flaming treatment was strongly selective. Only Ascomycetes such as Eurotium spp. survived the heat treatment. The flaming treatment may be a good method for isolating Ascomycetes from foods and feedstuffs. Treatment with formaldehyde also removed a significant portion of the internal flora and it seemed to be selective too.

Naturally contaminated foods and feeds. Since NaOCl appeared to be the only treatment not removing parts of the internal flora of barley, we applied a mild treatment (1% NaOCl for 1 min) to 25 samples of barley and green coffee beans. Table 7 shows that NaOCl treatment causes slight decreases in infection percentages of storage fungi and slight increases in infection percentages of field fungi. The NaOCl treatment had the advantages of removing some superficial sporangiospores of the Mucorales and stray conidia which cannot be allocated to any particular kernels on the plate.

Infection of ten samples in which storage molds had grown actively was investigated on DRBC, PRYES, DRYES, AFPA and DG18 media. The barley kernels germinated on DRBC and AFPA, Rhizopus and Mucor overgrowth was observed on DRBC, DRYES and AFPA, and coffee beans discolored the AFPA medium. Table 8 lists the differences in mycoflora recovered on PRYES in comparison with DG18 medium. The surface disinfection treatment with NaOCl caused a reduction in percentage of infection by most molds and altered the recovery ratio of some. For example, Aspergillus versicolor was not detected in sample 22 without NaOCl treatment, probably because Eurotium spp., P. viridicatum and P. aurantiogriseum grew extensively. However, 40% of kernels were positive for the slowly growing A. versicolor after NaOCl treatment in sample 22. One example from Table 8 is especially illustrative: sample 19, a sample of good quality barley, was kept in a bucket near sample 18, which was barley kept

Table 7. The effect of 1% NaOCl treatment on the flora of 15 barley samples[a]

| | Infection (%)[a] | | | |
| | Range | | Mean | |
Mold	Without NaOCl	With NaOCl	Without NaOCl	With NaOCl
P. viridicatum	0 – 11	0 – 6	1.5	0.7
P. aurantiogriseum	1 – 50	0 – 48	15.5	11.7
A. flavus	0 – 17	0 – 20	1.2	1.4
Eurotium spp.	0 – 48	0 – 49	7.1	3.3
Fusarium	0 – 48	0 – 49	26.0	26.5
Dematiaceae	1 – 34	1 – 30	10.3	11.6

[a]Samples were pig feeds; each was comprised of 100 kernels distributed over 10 plates of PRYES and incubated at 20°C for 7 days

in an airtight silo for 1 1/2 years. The specific mycoflora of the latter sample (P. roquefortii, Eurotium spp., A. candidus, Paecilomyces variotii,

Table 8. The effect of 1% NaOCl treatment for 1 min on the mycoflora of samples of green coffee and barley when incubated on PRYES and DG18 at 25°C for 7 days[a]

Substrate	Mold	Infection (%)			
		Range		Mean	
		Without NaOCl	With NaOCl	Without NaOCl	With NaOCl
Green coffee	Eurotium spp.	0 – 34[b]	0	17.0	0
		100	40 – 45	100.0	42.5
	A. niger	100	99 – 100	100.0	99.5
		95 – 100	98 – 100	97.5	99.5
	A. flavus	24 – 80	30	52.0	30.0
		25 – 56	12 – 14	40.5	13.0
	A. tamarii	16 – 19	6 – 12	17.5	9.0
		85 – 90	9 – 35	87.5	22.0
	A. ochraceus	27 – 89	8 – 25	58.0	16.5
		50 – 100	12 – 30	75.0	21.0
	P. citrinum	0 – 2	0	1.0	0
		6 – 23	2 – 8	14.5	5.0
	Wallemia sebi	0	0	0	0
		0 – 10	2 – 3	5.0	2.5
Barley (samples 18 and 19)	Eurotium spp.	0/0[c]	0/6	0	3.0
		83/72	3/3	77.5	3.0
	A. candidus	0/32	3/9	16.0	4.5
		97/70	57/0	83.5	28.5
	Candida spp.	100/61	67/0	80.5	33.5
		0/0	4/0	0	2.0
	P. roquefortii	0/31	36/0	15.5	18.0
		99/28	35/0	63.5	17.5
	Dematiaceae	0/0	0/12	0	6.0
		0/0	0/13	0	6.5
	P. aurantio-griseum	0/36	0/0	18.0	0
		0/0	0/0	0	0
	Fusarium spp.	0/0	0/52	0	26.0
		0/0	0/49	0	24.5

(continued)

Table 8. Continued

Substrate	Mold	Infection (%)			
		Range		Mean	
		Without NaOCl	With NaOCl	Without NaOCl	With NaOCl
Barley (contains ochratoxin A)	A. versicolor	0	0 - 40	0	6.7
		0 - 40	0 - 40	6.7	9.8
	A. niger	0 - 8	0 - 7	1.3	1.2
		0 - 50	0 - 14	8.3	2.3
Barley (contains ochratoxin A)	Eurotium spp.	5 - 74	2 - 46	18.0	16.0
		84 - 100	20 - 100	84.5	56.2
	A. flavus	0 - 95	0 - 80	15.8	13.3
		0 - 14	0 - 41	2.3	6.8
	A. candidus	0	0 - 20	0	7.3
		0 - 76	4 - 20	19.5	9.2
	P. viridicatum	0 - 98	0 - 51	39.2	20.8
		0 - 99	0 - 40	33.0	14.8
	P. aurantio-griseum	17 - 100	9 - 55	66.5	24.2
		20 - 99	0 - 44	54.5	16.2
	Dematiaceae	0 - 44	0 - 21	0.7	5.7
		0	0 - 20	0	3.8
	Fusarium spp.	0 - 1	0 - 11	0.2	2.3
		0	0 - 7	0	1.8

[a]Samples consisted of 100 kernels incubated on 10 plates of PRYES and 10 plates of DG18 medium, and incubated at 25°C for 7 days
[b]The upper row of numbers indicates % infection detected on PRYES and the second row of numbers is the % infection detected on DG18 medium
[c]For the sake of comparison of the two samples of barley, numbers are presented as % infection of samples 18 and 19, respectively

Candida spp.) caused surface contamination of sample 19. Direct plating and dilution plating indicated that the mycoflora of sample 19 was the same as for sample 18. Surface disinfection effectively removed these superficial contaminants. Mycological analysis of the NaOCl-treated kernels from sample 19 showed the typical mycoflora of fresh barley, i.e., dematiaceous fungi (Alternaria alternata and Drechslera spp.), Fusarium spp. (F. poae) and a few A. candidus and Eurotium spp. (see also Table 7). The fungus responsible for production of ochratoxin A in samples 20-25 was present in both disinfected and non-disinfected barley, except in sample 24.

Conclusion

Even though NaOCl treatment did not remove all conidia from heavily contaminated surfaces of kernels, we recommend the inclusion of both NaOCl surface-disinfected and untreated barley kernels in mycological analyses. Surface disinfection results in a reduction in contaminant spores, the appearance of slowly growing fungi masked by ubiquitous fungi such as A. niger, less stray conidia and other spores on the medium surface and a reduction in mucoraceous fungi. Plating of untreated kernels can be of value in detecting superficial fungi which may grow and produce mycotoxins given the right conditions. Mild NaOCl treatment (1% NaOCl for 1 min) is preferable since an increase in concentration or contact time does not result in more effective removal of conidia from heavily contaminated surfaces of kernels and because of the simplicity and nonselectivity of the method.

References

BOOTH, C. 1971 Introduction to general methods. In Methods in Microbiology Vol. 4. ed. Booth, C. pp 1–47. London & New York: Academic Press.

FRISVAD, J. C. 1983 A selective and indicative medium for groups of Penicillium viridicatum producing different mycotoxins in cereals. Journal of Applied Bacteriology 54, 409–416.

GAMS, W. AA., H. A. VAN DER, PLAATS-NITERINK, A. J. VAN DER, SAMSON, R. A. & STALPERS, J. A. 1980 CBS Course of Food Mycology. 2nd ed. Baarn: Centraalbureau voor Schimmelcultures. 109 pp.

MISLIVEC, P. B. & BRUCE, V. R. 1977 Incidence of toxic and other mold species and genera in soybeans. Journal of Food Protection 40, 309–312.

PEPPER, E. A. 1961 The Microflora of Barley, Their Isolation, Characterization, Etiology, and Effects on Barley. East Lansing, MI: Michigan State University. (Thesis).

PORTER, R. H. 1944 Testing the quality of seeds for farm and garden. Iowa Agricultural Experimental Station Research Bulletin 344, 495–586.

WALLACE, H. A. M. & SINHA, R. N. 1962 Fungi associated with hot spots in farm stored grain. Canadian Journal of Plant Science 42, 130–141.

J. C. FRISVAD
A. B. KRISTENSEN
O. FILTENBORG

▶ Dilution Plating Versus Direct Plating of Various Cereal Samples

There are two basic methods of evaluating fungal contamination of foods. One method involves blending or stomaching the product followed by diluting and plating on an appropriate medium to determine the number of fungal propagules/g or ml. In the other method, the food sample is analyzed by direct plating of kernels or subsamples which enables a percentage infection to be determined.

Direct plating will reveal the percentage infection by dominant or surface fungi. Detection of infection levels due to internal mycoflora requires disinfection prior to plating. In either case, the results obtained relate only to the proportion of food pieces infected, and give no indication as to the extent of fungal contamination of the product.

The dilution plating method essentially measures the number of extractable fungal propagules which are formed when a sample of food is milled or homogenized. The viable propagules may be conidia, ascospores, zygospores or hyphal fragments or other types of propagules produced by homogenization of the sample. If homogenization disrupts a sporulating structure, especially if a product is contaminated with a limited number of sporulating colonies, the number of viable fungal propagules measured by dilution plating can be misleading in terms of the fungal quality of a product. In such cases, the information obtained from a direct plating technique may be more appropriate. The question is: which is the more objective method of analysis?

Materials and methods

Sample preparation. Several weather-damaged sorghum samples were obtained for evaluation of the relative merits of dilution plating and direct plating. Samples for dilution plating were first passed through an oilseed mill and crushed to a fine powder. Ten-gram samples of the crushed grain were mixed with 90 ml of 0.1% peptone and subjected to stomaching for 30 sec. Subsequent dilutions were prepared in 0.1% peptone.

Aliquots of 0.1 ml were spread-plated onto DRBC medium, incubated at 25°C for 5 days and then enumerated. The identities of the dominant fungi were ascertained and their proportion in the total fungal population was measured. The mycoflora so determined were then compared to that detected by direct plating of whole grains onto the same medium.

Samples for direct plating were dispensed with sterile forceps onto DRBC plates (10 grains per plate). At least 100 grains were plated for quantitative analysis because the naturally infected sorghum samples were not uniform with regard to distribution of mycoflora.

Samples were surface disinfected by immersion in 0.35-0.37% (w/v) chlorine for 2 min, rinsed in sterile water and drained. For samples that were to be milled subsequently, the grains were dried on sterile filter papers in sterile Petri dishes before being weighed and milled.

Media. DRBC medium was used for standard enumeration. For analysis of the wheat samples for Fusarium graminearum, samples were direct-plated onto Dichloran Chloramphenicol Peptone Agar (DCPA) which has the following formulation: peptone, 15 g; KH_2PO_4, 1 g; $MgSO_4 \cdot 7H_2$, 0.5 g; chloramphenicol, 200 mg; dichloran, 2 mg; agar, 20 g; water to 1 liter.

Results

The first two samples of sorghum examined (Table 1) were heavily infected with fungi (approximately 10^7 propagules/g) of which a large proportion was A. flavus. These samples were implicated in a mycotoxicosis involving broiler chickens and it was therefore assumed that aflatoxins were present and the cause of the mycotoxicosis. Analyses failed to detect any aflatoxins. Direct plating of the grains without chemical treatment indicated a 100% fungal infection level for both samples, predominantly with A. flavus.

When the grains were surface disinfected prior to direct plating however, infection with dematiaceous Hyphomycetes was evident -- 55% and 38% for samples 1 and 2, respectively (Table 1). Alternaria alternata was dominant, together with Drechslera spp. Sample number 1 contained 10 mg/kg alternariol and 3.6 mg/kg alternariol monomethyl ether while sample number 2

Table 1. Dilution plate counts and percentage infections of two sorghum
samples plated on DRBC

Sample number	Dilution plate counts on DRBC (total count)	Percentage grain infection by direct plating on DRBC
1	1.03×10^7 fungal propagules/g, consisting of 5.5×10^6 A. flavus/g, 2.4×10^6 Phoma sorghina/g, 1.2×10^6 Fusarium/g, 1.2×10^6 Cladosporium cladosporiodes/g, A. niger and A. candidus	Direct plating: 100% infection with A. flavus and A. niger; 0% infection with dematiaceous Hyphomycetes Surface disinfection, then direct plating: 100% infection but 55% infection with dematiaceous Hyphomycetes, especially Alternaria
2	1.43×10^7 fungal propagules/g consisting of 4.5×10^6 A. flavus/g, 2.1×10^6 A. terreus/g, 1.4×10^6 A. niger/g, 2.1×10^6 P. citrinum, P. verrucosum and Rhizopus stolonifer	Direct Plating: 100% infection with A. flavus and Rhizopus stolonifer; 0% infection with dematiaceous Hyphomycetes Surface disinfection, then direct plating: 100% infection but 38% infection with dematiaceous Hyphomycetes, especially Alternaria

contained 7.2 mg/kg of both alternariol and alternariol monomethyl ether.
In both samples, Alternaria was not detected on dilution plates, indicating
that either propagules of Alternaria are destroyed by milling and blending,
or that the number of propagules of Alternaria were greatly outnumbered by
the propagules of other fungi.

For such samples direct plating of surface disinfected grains was
essential to recover the mycotoxigenic fungus.

As the internal mycoflora is frequently important in the assessment of
fungal contamination of a product, a study was carried out to determine the
effect of surface disinfection on dilution plate counts. Two additional
sorghum samples were milled and dilutions were plated. Half of each sample
was surface disinfected with 0.35% chlorine for 2 min before milling.

The data in Table 2 demonstrate that surface disinfection reduced the
total fungal propagule count by 85% and 72% for the two samples.
Furthermore, surface disinfection brought about a dramatic change in the
relative proportions of the fungal species present (Table 2). For sample
number 3, the mycoflora was dominated by A. flavus, A. terreus and A. niger;
however, after surface disinfection, a Penicillium sp. predominated and the
three Aspergillus spp. were each reduced by at least 90%. Sample number 4
was considered to be a sorghum sample of high quality as reflected in the
much lower dilution plate count. The mycoflora of this sample, dominated by
A. flavus prior to surface disinfection, was subsequently dominated by a
Penicillium spp. It is concluded that both samples were internally infected
by a Penicillium sp., but in neither case did the Penicillium dominate the
dilution plate count prior to surface disinfection.

Table 2. Dilution plate counts before and after surface disinfection of two sorghum samples

Sample number	Dilution plate count on DRBC (total count)	Dilution plate count on DRBC after surface disinfection (total count)	Percentage reduction
3	1.02×10^7 fungal propagules/g consisting of 6.0×10^6 A. flavus/g, 2.9×10^6 A. terreus/g, 1.3×10^6 A. niger/g and 1.0×10^6 Penicillium spp./g	1.8×10^6 fungal propagules/g consisting of: 5.0×10^5 A. flavus/g 3.0×10^5 A. terreus/g 2.5×10^4 A. niger/g 8.5×10^5 Penicillium spp./g	85% 92% 90% 98% 15%
4	4.7×10^4 fungal propagules/g consisting of 1.5×10^4 A. flavus/g, 4.0×10^3 A. terreus/g, 2.0×10^3 A. niger/g, 7.0×10^3 Penicillium spp./g and 4.0×10^3 Mucor spp./g	1.3×10^4 propagules/g consisting of: A. flavus/g A. terreus/g A. niger/g 6.0×10^3 Penicillium spp./g 3.0×10^3 Mucor spp./g	72% 100% 100% 100% 14% 25%

When the same samples were direct plated, before and after surface disinfection, a significant change in the mycoflora was again observed. Sample 3 was heavily infected (100% infection rate) with both A. flavus and A. niger. After surface disinfection, the grain was still 100% infected, 70% with A. flavus but only 19% with A. niger, indicating that A. niger was mainly a surface contaminant while A. flavus had invaded the grain to a considerable extent. Moreover, 14% of the grains showed dematiaceous Hyphomycetes which had not been observed in previous analyses of this sample. The high quality sorghum (sample 4) was 100% infected, 60% with A. niger and 39% with A. flavus. However, after surface disinfection only 30% of the grains were infected, mainly with Penicillium, Alternaria and Drechslera spp. The initial low count and the low percentage infection of the grain after surface disinfection revealed that most of the fungi were surface contaminants and probably of little significance in terms of fungal deterioration of the product. However, that information would not have been known if a dilution plate count alone or direct plating alone had been carried out.

The question of whether to direct or dilution plate a commodity is of particular relevance for mycotoxigenic fungi. One case has already been described above. A second case concerns a recent investigation of mycotoxins associated with rain-damaged grain. Results (Table 3) showed that direct plating of surface disinfected grains and the determination of percentage infection with the toxigenic fungus provided a good correlation with toxin production. An infection rate of 75% or greater with Fusarium graminearum was consistent with detectable levels of deoxynivalenol. Grain with infection rates of less than 75% contained no detectable deoxynivalenol. There was no such correlation with fungal propagule counts and deoxynivalenol: sample SA-5 had the highest count which consisted predominantly of Cladosporium cladosporioides as a surface contaminant. After surface disinfection and direct plating, 95% of the grains were infected with F. graminearum and the sample contained 2.6 mg/kg deoxynivalenol. Hence the predominant fungus in the dilution plate count need not be the significant fungus.

43

Table 3. Percentage infection with <u>Fusarium graminearum</u> and deoxynivalenol concentrations in weather damaged grains

Sample	Fungal count on DRBC	Percentage infection with F. graminearum	Deoxynivalenol concentration[a] (mg/kg)
Wheat SA-1	1.6×10^4	75	6.7
Wheat SA-2	2.0×10^4	50	0
Wheat SA-3	7.5×10^4	80	0.9
Wheat SA-4	1.0×10^5	75	0.5
Wheat SA-5	1.6×10^5	95	2.6
Wheat SA-7	N.A.[b]	95	2.1
Barley SA-6	1.6×10^5	20	0
Barley SA-8	N.A.	5	0
Triticale SA-9	N.A.	95	1.1
Triticale SA-10	N.A.	100	8.9
Triticale SA-11	N.A.	95	4.5

[a]Deoxynivalenol assays were all carried out by N. Tobin, C.S.I.R.O. Division of Food Research, North Ryde, N.S.W., Australia
[b]Data not available

Discussion

The sorghum samples analyzed in this paper had rather complex mycoflora but have been useful to highlight various approaches to enumeration of fungal contamination of food commodities. For each sample described, neither the dilution plate count nor the percentage infection from direct plating of the grain gave a true description of the nature of the fungal contamination or infection of the product. For all four samples, the predominant fungus was A. flavus and, since the samples had been implicated in a mycotoxicosis outbreak, it was an obvious conclusion to expect to find aflatoxins in the product. The absence of aflatoxins could be interpreted to mean that a nontoxigenic strain of A. flavus was isolated from the product and that other causes of the disease should be sought.

These examples highlight the ever present danger of assuming the dominant organism is the most important organism. If dilution plate counts are to be used, then it is important to emphasize caution in the interpretation of the results, especially once sporulation has occurred on the surface of the product.

Furthermore these studies have adequately demonstrated the value of dilution plating or preferably direct plating after surface disinfection. A combination of dilution plate counts and direct plating after surface disinfection gives a very fair representation of the true mycoflora, surface and internal, of the product. When a product is suspected of containing mycotoxins, it is recommended that determination of the percentage infection with potentially mycotoxigenic fungi is the best parameter to monitor.

It is recommended that, for routine quality control analyses of raw materials or processed products, a dilution plate count be carried out for assessing the fungal status of the product. Variations from the normal baseline level of contamination indicate the need for a more extensive analysis of the product, including examination after surface disinfection.

Recommendation

For routine quality control, dilution plate counting on DRBC medium is recommended. For investigations related to fungal deterioration or production of mycotoxins in foods, it is recommended that, in addition to the dilution plate count, samples of the product be surface disinfected with 0.35% (w/v) chlorine and then direct plated onto DRBC. Determination of the percentage infection of sample with dominant fungi, together with dilution plate counts, gives a more detailed description of the mycoflora usually required for effective fungal investigations.

S. ANDREWS

▶ Comparison of Direct and Dilution Plating for Detecting Penicillium viridicatum in Barley Containing Ochratoxin

Ochratoxin A is a mycotoxin that occurs naturally in Danish barley due to growth of Penicillium viridicatum group II (Frisvad 1983; Lillehoj & Elling 1983). P. purpurescens has also been reported to be present in Danish barley before harvest (Lillehoj & Goransson 1980), and isolates of this fungus produce ochratoxin A in pure culture. This study was designed to determine if fungi capable of producing ochratoxin A could be detected by retrospective analyses of stored barley containing the mycotoxin, to compare dilution plating and direct plating for the detection of these fungi, and to relate mycological counts to chemical analyses of ochratoxin A in barley.

Materials and methods

Seventy samples of barley were collected from farms from different geographic regions in Denmark. Pigs eating barley from the original lots had developed pale kidneys (nephropathy). The samples were stored at 10°C for a period of 1–12 months, followed by 2 weeks at 0.5°C to kill mites before analyses. All samples contained less that 13% moisture except three visibly moldy samples. The samples were examined using dilution plating and direct plating as described by Frisvad (1983) on the selective and indicative medium PRYES. P. viridicatum II was detected by its violet brown reverse color. All other Pencillia were isolated and identified, but none of these fungi produced ochratoxin A in pure culture on YES agar by the method of Filtenborg & Frisvad (1980).

Ochratoxin A in the barley was determined by the method of Nesheim et al. (1973), with development in toluene–ethyl acetate–formic acid (60:30:10, v/v/v). Citrinin was determined qualitatively. All determinations were made in duplicate.

Results and discussion

A summary of the results of the chemical analysis for ochratoxin A, dilution counts and percentage infected with P. viridicatum are presented in Table 1. Three samples that were visibly moldy contained over 18% water and were eliminated from the data in Table 1.

Sixty of the 70 samples of barley causing nepropathic kidneys contained ochratoxin A and three samples contained citrinin in addition. P.

Table 1. Summary of ochratoxin content, counts and percentage infection of
          P. viridicatum in 70 samples of Danish barley causing nephropathy
          in pig kidneys

Measurement	Viable count (x 10^3) of P. viridicatum	Percentage infection (50 kernels)	Ochratoxin (mg/kg)
Mean	206	30	0.59
Range	0-1230	0-100	0-7.38
1st quartile	15	0	0.03
Median	75	18	0.20
3rd quartile	300	60	0.51
No. of observations	67	67	66

viridicatum counts on the 57 visibly sound samples of barley containing
ochratoxin A ranged from 1.2 x 10^3 to 1.2 x 10^6/g; counts on the three
visibly moldy samples ranged from 2.5 x 10^7 to 2.4 x 10^8/g. Counts on
eight samples containing no ochratoxin A ranged from 10^3 to 7.5 x 10^4/g.
In nine samples of barley containing ochratoxin A, P. viridicatum could not
be detected by the direct-plating method. The reason for this result may be
that only 50 kernels were plated for each sample. The remaining 51 samples
containing ochratoxin A yielded from 2-100% kernels infected with P.
viridicatum. Only one sample which contained kernels infected with P.
viridicatum did not contain ochratoxin A.

There was no correlation between content of ochratoxin A in barley
samples and either dilution counts or percentage infection with P.
viridicatum. However, barley samples containing ochratoxin A always had a
positive count of P. viridicatum. Blaser & Schmidt-Lorenz (1981) reached
the same general conclusions regarding aflatoxin content of nuts, almonds
and corn, and dilution counts of Aspergillus flavus on ADM. However, in a
few of their samples containing trace amounts of aflatoxins, A. flavus was
not detected.

Conclusions

No direct correlation was observed between ochratoxin A content of
barley and percentage infection or viable counts of P. viridicatum. This is
due to the problems of retrospective analyses (changes in environmental
factors after ochratoxin A production and uneven distribution of contaminated
kernels). Samples in which P. viridicatum could not be detected by direct
nor dilution plating did not contain ochratoxin A. Nine of 16 samples in
which P. viridicatum was detected by dilution plating, but not by direct
plating, contained ochratoxin A. All samples except one containing P.
viridicatum, detected both by direct and dilution plating, contained
ochratoxin A. Samples of barley containing ochratoxin A always had counts
of P. viridicatum greater than 1000/g. In mycological analyses of cereals,
we recommend both direct and dilution plating, especially in preventive and
research work.

References

BLASER, P. & SCHMIDT-LORENZ, W. 1981 Aspergillus flavus kontamination von nussen, mandeln und mais mit bekannten aflatoxin-gehalten. Lebensmittel Wissenschaft und Technologie 14, 252-259.
FILTENBORG, O. & FRISVAD, J. C. 1980 A simple screening-method for toxigenic moulds in pure cultures. Lebensmittel Wissenschaft und Technologie 13, 128-130.
FRISVAD, J. C. 1983 A selective and indicative medium for groups of Penicillium viridicatum producing different mycotoxins in cereals. Journal of Applied Bacteriology 54, 409-416.
LILLEHOJ, E. B. & ELLING, F. 1963 Environmental factors that facilitate ochratoxin contamination of agricultural commodities. Acta Agriculturae Scandinavica 33, 113-128.
LILLEHOJ, E. B. & GORANSSON, B. 1980 Occurence of ochratoxin- and citrinin-producing fungi on developing Danish barley grain. Acta Pathologica et Microbiologica Scandinavica Section B 88, 133-137.
NESHEIM, A., HARDIN, N. F., FRANCIS, O. J. & LANGHAM, W. S. 1973 Analysis of ochratoxin A and B and their esters in barley, using partition and thin layer chromatography. 1. Development of the method. Journal of the Association of Official Analytical Chemists 56, 817-821.

J. C. FRISVAD
B. T. VIUF

▶ Comparison of Pour Versus Surface Plating Techniques
for Mold Enumeration

The purpose of this study was to compare pour plating with surface plating using flour samples containing low, intermediate and high numbers of mold propagules.

Materials and methods

Samples. Samples consisted of bleached white flour that had been inoculated with 10^2, 10^4 and 10^6 mold spores/g. Spores of nine different molds (Aspergillus flavus, A. niger, A. ochraceus, Penicillium aurantiogriseum, P. roquefortii, P. viridicatum, Fusarium graminearum, Cladosporium sp. and Alternaria sp.) were produced on bread cubes in quart Mason jars according to the method of Sansing & Ciegler (1973). Mature cultures were allowed to dry. Spores were collected from dry, mature cultures by adding 100-200 g of dry flour to each jar and dry blending; the preparations were then combined and counted. Flour samples were inoculated with the mixture to give 10^2, 10^4 and 10^6 spores/g.

Pour and surface counts were made on two media, PDA + 40 μg/ml tetracycline and DRBC. Samples were prepared by two techniques, blending and stomaching. The samples (11 g) were added to phosphate buffer (99 ml) and mixed for 3 min. Aliquots (1 ml for pour plate and 0.1 ml for spread plates) were plated within 1 min after mixing. If 0.1 ml inoculum on surface plates was too low, analysis were repeated using a 1.0-ml aliquot. Counts were made in duplicate and the study was repeated three times on the same samples. Each count represents the mean of six determinations. Plates were incubated upright at 25°C for 5 days before colonies were counted.

Table 1. Mean mold counts ($\log_{10}$ counts $\pm$ S.D.) obtained from inoculated
flour using surface and pour plate techniques on DRBC and PDA +
tetracycline

Medium	Level of inoculum (per/g)	Blending		Stomaching	
		Surface	Pour	Surface	Pour
PDA + tetracycline	10^6	5.19 ± .10	5.11 ± .16	5.34 ± .07	5.10 ± .10
	10^4	3.13 ± .12	3.32 ± .06	3.34 ± .14	3.40 ± .18
	10^2	2.27 ± .12	2.31 ± .10	2.36 ± .05	2.32 ± .07
DRBC	10^6	5.09 ± .40	5.34 ± .10	5.33 ± .08	5.34 ± .15
	10^4	3.16 ± .14	3.30 ± .27	3.27 ± .32	3.49 ± .54
	10^2	2.20 ± .21	2.05 ± .09	2.62 ± .44	2.31 ± .08

Statistical design and analyses. Three experimental parameters were
studied: (1) method of preparation of sample (blending vs. stomaching); (2)
method of inoculation of plates (surface vs. pour); and (3) culture media
(PDA + tetracycline vs. DRBC). The design used and method of preparation of
samples was considered as a block effect. Mean $\log_{10}$ values of counts
were used for analysis of variance (Steel & Torrie 1980). The α-value was
chosen to be 0.05.

Results and discussion

There appeared to be little difference between populations detected on
pour plates and surface plates (Table 1). No consistent trends were
observed and, except for the difference of stomaching versus blending at the
10^2 inoculum level, the differences were not statistically significant.
Both types of plating techniques gave similar results on each medium tested,
and worked equally well when either stomaching or blending was used to
homogenize the samples. A disadvantage of surface plating is that samples
with low counts cannot be enumerated. The main disadvantage observed with
the pour plate was the difficulty in identifying mold colonies that develop
deep within the agar.

Statistical analysis. Mold counts obtained with flour inoculated with
10^2 propagules/g using stomaching as the method of sample preparation were
significantly higher than counts obtained using blending. This difference
in method of sample preparation diminished as the level of mold propagules
increased. The difference between the two types of culture media was
insignificant at all mold levels. The difference between surface- and
pour-plate techniques was also insignificant at all mold populations.

Conclusions

In this study no statistical differences were observed between mold
counts obtained by pour-plate or surface-plating techniques. Both
techniques worked equally well with PDA or DRBC media, and with stomaching
or blending. There were disadvantages associated with both techniques, but
surface plating was preferred because it more readily allowed tentative
identification of mold genera based on colony type. The possible thermal

injury inherent in pour-plate techniques was not a problem for mold counts of uninjured spores such as the mold spores used in inoculating flour in this study. With low numbers (10^2/g) of mold propagules, stomaching gave significantly higher counts than blending.

Recommendation

The surface-plating technique is recommended over pour plating if some idea of colony type and ability to distinguish molds from yeasts is desired. The problem of enumerating samples with low counts can be overcome by increasing the volume of the aliquot plated from 0.1 to 1.0 ml or by adding results from several plates for a single count.

References

SANSING, G. A. & CIEGLER, A. 1973 Mass propagation of conidia from several Aspergillus and Penicillium species. Applied Microbiology 26, 830–831.
STEEL, R. G. & TORRIE, J. H. 1980 Principles and Procedures of Statistics, A Biometrical Approach. 2nd edition. New York: McGraw-Hill Book Co. 235 pp.

J. W. HASTINGS
W. Y. J. TSAI
L. B. BULLERMAN

Published as Paper No. 7609, Journal Series, Agricultural Research Division, Lincoln, NE. Research reported was conducted under Project 16–029.

Comparison of Spread and Pour Plate Methods for Enumerating Yeasts and Molds in Dry Foods Using DRBC Medium

In this study, spread and pour plating methods are compared using DRBC medium. The food samples use in this study were chosen as providing significant numbers and species of molds, recognizing that the yeast population might be very low in some of these dry foods.

Materials and methods

Food samples. Food samples were purchased in one-half pound amounts from retail stores specializing in handling bulk quantities. Samples were stored under conditions similar to those at point of purchase. Raw pecan halves and raw whole almonds were stored at 4°C. The remaining samples of flour, meal and spices were held at room temperature.

Sample preparation. Five replicates of each food sample were analyzed. Samples (22-g) were homogenized with 198 ml of 0.1% peptone water for 30 sec in a Colworth Stomacher 400. The mixed samples were suitably diluted further in 10-fold increments using peptone water and conventional diluting procedures.

Plating. When plating, 0.1- and 0.3-ml quantities of the 10^{-2} dilution were used, while 0.1 ml was used at higher dilutions. Plates of DRBC medium were prepared 18-24 h in advance of use to allow for partial drying of the surface. Fifteen to 18 ml of medium tempered to 46°C was added to each plate. When the agar had hardened, plates were inverted and stored in the dark at room temperature until used. A single glass hockey stick was used to distribute the inoculum (0.1 or 0.3 ml) over the surface, starting at the highest dilution in a series of plates.

Pour plates were prepared by adding 15-18 ml of melted and tempered (46°C) medium to the inoculum (0.1 or 0.3 ml) in each plate. After mixing and solidifying, all plates were incubated at 25°C, undisturbed and upright, for 5 days. Yeast and mold counts were recorded separately.

Table 1. Mold counts using DRBC medium

		Sample		Log_{10} count/g		
Food	Method	Replicates[a]	Dilution	Mean	S.D.	Confidence[b]
Whole wheat	spread	5	10^{-2}	2.92	0.04	NSD
flour	pour	5	10^{-2}	2.94	0.24	
White wheat	spread	5	3×10^{-2}	2.53	0.23	NSD
flour	pour	5	3×10^{-2}	2.51	0.14	
Corn meal	spread	5	10^{-4}	5.52	0.20	S
	pour	5	10^{-4}	5.17	0.24	
Corn flour	spread	5	3×10^{-2}	2.66	0.13	NSD
	pour	5	3×10^{-2}	2.57	0.13	
Pecans	spread	5	3×10^{-2}	1.01	0.94	NSD
	pour	5	3×10^{-2}	1.15	1.06	
Chili powder	spread	4	3×10^{-2}	2.96	0.09	NSD
	pour	3	3×10^{-2}	2.85	0.16	
Cayenne	spread	5	10^{-3}	3.76	0.21	NSD
	pour	5	10^{-3}	3.86	0.12	
Paprika	spread	5	10^{-3}	3.92	0.37	S
	pour	5	10^{-3}	4.37	0.09	
Black pepper	spread	5	10^{-2}	3.94	0.05	NSD
	pour	5	10^{-2}	3.97	0.07	
Almonds	spread	5	10^{-2}	3.67	0.38	NSD
	pour	5	10^{-2}	3.57	0.42	

[a]Replicates less than 5 are a consequence of uncountable plates due to spreading molds; 3% were uncountable
[b]NSD = No significant difference; S = significant difference

Results and discussion

Mold counts. The data obtained from plating ten food types by the spread and pour methods (Table 1) show a broad range of counts that included a wide variety of mold species. Mold counts ranged from a low of 14/g on pecans to 3.3×10^5/g in corn meal. Spreading fungi prevented counting of 3% of the plates. This is significantly lower than the 20% observed on OGY medium (see following paper). The spreaders occurred with equal frequency on both spread and pour plates.

Geometric mean values indicated neither method as being superior. Only two of the ten sample pairs showed a significant difference at the 95% confidence level. One favored the spread method, the other the pour method.

Yeast counts. Yeast counts obtained from eight food samples are summarized in Table 2. Counts ranged from less than 20/g in corn flour, pecans and chili powder to 1.2×10^4/g in paprika. Spreading fungi prevented 5% of the plates from being countable, significantly lower than the 18% observed on OGY (see following report). The frequency of spreading was not related to the method of plating.

Table 2.   Yeast counts using DRBC medium

Food	Method	Replicates[a]	Dilution	Mean	S.D.	Confidence[b]
		Sample		Log_{10} count/g		
Whole wheat flour	spread	5	10^{-2}	2.61	0.34	S
	pour	5	10^{-2}	0.0	X	
White wheat flour	spread	5	3×10^{-2}	1.84	0.25	S
	pour	5	3×10^{-2}	0.0	X	
Corn meal	spread	5	10^{-3}	3.58	2.02	NSD
	pour	4	10^{-3}	3.19	2.19	
Corn flour	spread	5	3×10^{-2}	0.30	0.68	NSD
	pour	5	3×10^{-2}	0.30	0.68	
Pecans	spread	5	3×10^{-2}	0.79	1.09	NSD
	pour	5	3×10^{-2}	0.80	0.98	
Chili powder	spread	4	3×10^{-2}	0.84	0.97	NSD
	pour	3	3×10^{-2}	0.0	X	
Cayenne	spread	5	10^{-3}	3.69	0.12	S
	pour	5	10^{-3}	1.26	1.73	
Paprika	spread	5	10^{-3}	4.06	0.43	NSD
	pour	5	10^{-3}	3.75	0.48	

[a]Replicates less than 5 are a consequence of uncountable plates due to spreaders; 3% were uncountable
[b]NSD = No significant difference; S = significant difference

Table 3.   Totals of yeast and mold populations enumerated on DRBC (from Tables 1 & 2)

Type	Method	Log$_{10}$ count/g $\Sigma \bar{x}$	Log$_{10}$ count/g $\bar{x}$	Arithmetic count/g $\bar{x}$
Yeast	spread	17.71	2.21	160
	pour	9.26	1.15	14
Mold	spread	29.13	3.24	2000
	pour	29.10	3.23	2000

Geometric mean counts were considerably higher by the spread plate method. These data indicate that the spread method was superior for enumerating yeasts, in spite of the fact that geometric mean count differences are significant at the 95% level in only three of the sample pairs. This contradiction arises because of very low numbers or absence of yeast colonies on some of the plates. The high standard deviations resulting from this observation have a marked negative influence on confidence limits. Microscopic examination of representative colonies revealed that only yeasts were present and no evidence of bacterial growth was observed.

Combined yeast and mold data. In Table 3, geometric mean counts from all samples are combined and treated as a single sample. These tabulations again demonstrate the superiority of spread plating for enumerating yeasts in these samples. Molds, on the other hand, are enumerated equally well by either method.

Conclusion

A comparison of spread and pour plating methods using DRBC medium demonstrated that yeast recovery from dry foods is greatly enhanced by use of the spread plate method. On the average, samples showed an 11-fold increase in counts. When enumerating molds, both plating methods perform equally well.

The demonstrated superiority of spread plating for yeasts but not molds can probably be explained by differences in numbers of sublethally injured cells. Dry food products contain a preponderance of mold spores present in their normal dry state; vegetative yeast cells on the other hand, may find the dry state less hospitable.

Spreading mold growth was not a problem with either of the plating methods. Only 3 to 5% of the plates were not countable for this reason. This is a low percentage considering the prevalence of Rhizopus and Mucor spp. in some of the samples.

In comparison with OGY medium (see following report), fungal colonies on DRBC medium were easier to count because of reduced size and distinctive coloration imparted by dichloran and the rose bengal dye.

R. B. FERGUSON

Comparison of Spread and Pour Plate Methods for
Enumerating Yeasts and Molds in Dry Foods Using
OGY Medium

This study compares spread- and pour-plating techniques for enumerating yeasts and molds on OGY medium. Methods have been outlined in the preceding paper.

Results and discussion

Mold counts. The data obtained from testing ten food types by the spread and pour methods (Table 1) show a broad range of counts that included a wide variety of mold species. Mold counts ranged from a low of 60/g on pecans to 3.6×10^5/g in corn meal. Many more spreading molds, consisting

Table 1. Mold counts using OGY medium

| Food | Method | Sample | | Log_{10} count/g | | Confidence[b] |
		Replicates[a]	Dilution	Mean	S.D.	
Whole wheat flour	spread	5	10^{-2}	3.39	0.22	NSD
	pour	5	10^{-2}	3.24	0.18	
White wheat flour	spread	5	3×10^{-2}	2.66	0.06	NSD
	pour	5	3×10^{-2}	2.57	0.14	
Corn meal	spread	2	10^{-4}	5.56	0.06	NSD
	pour	5	10^{-4}	5.55	0.22	
Corn flour	spread	5	3×10^{-2}	2.67	0.14	NSD
	pour	5	3×10^{-2}	2.67	0.10	
Pecans	spread	4	3×10^{-2}	1.79	0.20	NSD
	pour	4	3×10^{-2}	1.91	0.27	
Chili powder	spread	4	3×10^{-2}	3.09	0.08	NSD
	pour	4	3×10^{-2}	3.00	0.07	
Cayenne	spread	5	10^{-4}	1.60	2.19	NSD
	pour	4	10^{-4}	2.08	2.40	
Paprika	spread	5	10^{-4}	3.57	2.00	NSD
	pour	4	10^{-4}	3.27	2.20	
Black pepper	spread	3	10^{-2}	3.87	0.05	NSD
	pour	5	10^{-2}	3.85	0.07	
Almonds	spread	4	10^{-3}	3.89	0.30	NSD
	pour	5	10^{-3}	3.70	0.00	

[a]Replicates less than 5 are a consequence of uncountable plates due to spreading molds; 20% were uncountable
[b]NSD = No significant difference

mainly of <u>Rhizopus</u> spp., were encountered than anticipated. As a result, 20% of the plates were uncountable and at times necessitated counting plates at higher dilutions with few colonies per plate.

Geometric means of mold counts from spread plates were slightly higher than those obtained from pour plates. However, at the 95% confidence level there was no significant difference between methods for any of the samples tested.

<u>Yeast counts</u>. In Table 2, results from enumerating yeasts in eight food samples are tabulated. Counts ranged from less than 20/g in corn flour and pecans to 6.4×10^3/g in corn meal. Spreading molds were a definite problem and frequently obliterated the plates on which yeast counts would be in an acceptable range for optimum accuracy. Eighteen percent of the plates were uncountable because of spreading mold growth and, as noted earlier, this forced one to count colonies on plates at higher dilutions, some with few or no colonies per plate.

Geometric mean counts for each of the food samples tested were considerably higher by the spread plate method. The data clearly show that the spread method was superior for enumerating yeasts in these samples. This was in spite of the fact that mean differences are significant at the

Table 2.   Yeast counts using OGY medium

Food	Method	Sample Replicates[a]	Dilution	Log_{10} count/g Mean	S.D.	Confidence[b]
Whole wheat flour	spread[c]	5	10^{-2}	2.67	0.26	NSD
	pour	5	10^{-2}	1.30	1.20	
White wheat flour	spread[c]	5	3×10^{-2}	2.31	0.16	NSD
	pour	5	3×10^{-2}	0.30	0.64	
Corn meal	spread	2	10^{-4}	4.81	0.05	NSD
	pour	5	10^{-4}	4.46	0.28	
Corn flour	spread	5	3×10^{-2}	1.10	1.03	NSD
	pour	5	3×10^{-2}	0.76	0.88	
Pecans	spread	4	3×10^{-2}	1.22	0.82	NSD
	pour	5	3×10^{-2}	0.38	0.76	
Chili powder	spread	4	3×10^{-2}	1.62	1.10	NSD
	pour	3	3×10^{-2}	0.0	X	
Cayenne	spread[c]	5	10^{-4}	3.51	1.96	NSD
	pour	4	10^{-4}	3.15	2.12	
Paprika	spread	5	10^{-4}	3.49	1.96	NSD

[a]Replicates less than 5 are a consequence of uncountable plates due to spreaders; 18% were uncountable
[b]S = Significant difference; NSD = No significant difference
[c]Count corrected for presence of Gram negative bacteria

Table 3. Totals of yeast and mold populations enumerated on OGY medium
         (from Tables 1 & 2)

Type	Method	Log$_{10}$ count/g $\Sigma\ \overline{x}$	Log$_{10}$ count/g $\overline{x}$	Arithmetic count/g $\overline{x}$
Yeast	spread	20.81	2.60	400
	pour	13.66	1.71	50
Mold	spread	32.09	3.21	1600
	pour	31.84	3.18	1500

95% level in only one of the sample pairs. This contradiction arises
because of very low numbers of yeast colonies and the presence of negative
results on some of the plates. The high standard deviations resulting from
this have a marked negative influence on confidence limits.

Three of the samples tested contained bacteria that were not inhibited
by the chloramphenicol in OGY medium. The bacteria were characterized as
Gram-negative rods, highly aerobic and nonfermentive, and formed colonies on
only the spread plates from whole wheat flour, white wheat flour and cayenne.

Combined yeast and mold data. In Table 3, geometric mean counts from
all samples were combined and treated as a single sample. These tabulations
again demonstrate the superiority of spread plating for the enumeration of
yeasts. Molds, on the other hand, are enumerated equally well by either
method.

Conclusion

A comparison of spread and pour plating methods using OGY medium
demonstrated that yeast recovery from dry foods is greatly enhanced by using
the spread plate method. On the average, samples showed an 8-fold increase
in counts. When enumerating molds, both plating methods performed equally
well.

The demonstrated superiority of spread plating for yeasts but not molds
can probably be explained by differences in numbers of sublethally injured
cells. Dry food products contain a preponderance of mold spores present in
their normal dry state; vegetative yeast cells on the other hand, may find
the dry state less hospitable.

Spreading molds were a problem affecting both the spread and pour
plates. Eighteen to twenty percent of the plates were not countable. This
affected the precision of yeast counts to a greater extent because they are
present in lower numbers in the foods tested. Shorter incubation times
and/or reduced temperatures would minimize the problem.

The antibiotic present in OGY did not prevent growth of Gram negative
bacteria found in three of the samples tested. Bacterial colonies were
observed only on plates prepared by the spread method.

R. B. FERGUSON

Discussion

Howell

It should be pointed out that the sample may be more important than the sample handling technique. Secondly, the purpose of the analysis will dictate to some extent the methodology to be used.

Koburger

Dr. Andrews, how does the 0.37% chlorine relate to sodium hypochlorite or bleach concentration?

Andrews

Actually, we used three different hypochlorite solutions. The first was 10% hypochlorite diluted from the bottle which titrated to less than 5% free chlorine. The 12% sodium hypochlorite titrated out to 6.7% free chlorine. The free chlorine appeared to be about half of the sodium hypochlorite concentration. We observed that sodium hypochlorite generally oxidizes, so therefore it is dangerous just to recommend sodium hypochlorite without titrating. The sodium hypochlorite solutions should be checked routinely.

Pitt

Dr. Andrews and I started off by using normal household bleach which is normally 4% hypochlorite, using it as a 1:10 solution. I think in practice one can go close to recommending this as long as it is a reasonably fresh bleach solution. It seems that this will work out close to the concentration that Dr. Andrews was talking about, near 0.2% active chlorine, which is close to the plateau level that he mentioned. This is the direction that the recommendations should go. I think it will be impossible to ask people to titrate their bleach solutions.

Splittstoesser

Of course pH enters into this also, and pH will determine the activity of the chlorine solution. If you don't take pH into consideration, then the recommendation won't mean very much.

Pitt

The range of pH in foodstuffs with which people are going to be working with won't be terribly wide and variable.

Splittoesser

Well, of course, in a high-acid food it would be more effective than in other foods. All of this work has been published.

Seiler

May I ask why you surface-sterilize raisins?

Pitt

We wanted to see if surface disinfection worked.

Seiler

You are not necessarily looking for the internal microflora.

Andrews

In practice, we sometimes try to isolate the internal microflora. In work involving mycotoxicoses, without surface disinfection we couldn't effectively recover the mycotoxigenic fungi.

Pitt

This question is in regard to settling times. I got the impression from some of the speakers this morning that they were allowing their materials to settle for ten minutes before pipetting, whereas David Seiler said that he allowed his material to settle for one minute. May I get clarification on what should be a good settling time? We did some settling experiments which we did not present here and we found that we got quite rapid settling of fungal spores.

King

In a study in which a sonicator was put directly into the dilution bottle, it was found that the counts increased greatly. This was due primarily to clumping and adherence of bacterial cells. I am sure that a similar phenomenon would occur with yeast and mold cells. This clearly showed that settling time is very important in getting accurate counts.

Seiler

We have looked at bacteria, yeasts and molds. Bacterial counts were reduced very little with setting time. Yeast counts slightly reduced after a settling time of somewhere in the neighborhood of thirty minutes. But mold propagules, especially the larger spores which have a greater attraction for one another, settled out very quickly within five minutes.

Ferguson

I noted that most people blend for two minutes, which we rarely do. I would like to know what effect reducing the blending time would have on recovery of the organisms.

Williams

We did blending at half-minute intervals for up to about ten minutes. From this we found that thirty seconds was adequate with peak recovery occurring at maybe two minutes. After five minutes, the recovery was reduced over time, probably due to thermal stress and just chopping up the organisms to bits. Thirty seconds would probably be adequate for most purposes.

Seiler

We looked at flour, corn meal and cereals and found that with stomaching, counts peaked at two minutes with the flour. For the corn meal, it took five minutes and with the cereals it took ten minutes for peak counts to occur. So, depending on the commodity, the optimal blending time can vary.

Ferguson

What holding time do you use between blending and plating your sample?

Pitt

This is a separate topic, and will be discussed later.

Splittstoesser

Does anyone have any idea what exactly it is that we are enumerating here, especially in the dry products?

Pitt

We have found it quite difficult to objectively determine what exactly we are enumerating. Some work that Ms. Hocking is conducting suggests that mycelia die off rather quickly, especially in dry foods, and it is the conidia that we are enumerating. From the participants, I see that I get consensus throughout the audience.

Lenovich

We inoculated cocoa with Penicillia. With blending, counts were significantly lower compared to stomaching or shaking. I think the reason is that foaming occurs during blending and, due to the size of the conidia, some are lost in the foam.

Seiler

What kind of blender did you use?

Lenovich

A Waring Blendor or a household blender.

Seiler

The specifications of the blender are important. The ICC standard calls for a high speed blender.

Mossel

In work done by Professor Ingram many years ago under the auspices of ISO, many types of homogenizers and blenders were tested. A publication resulting from this work deals with the different effects that different apparatus will have on counts. Hand shaking, of course, is the best method.

Koburger

One of the observations we have made is that natural fungal populations in foods will act in a certain way, but when isolated and put back in products in the lab they may act entirely different. I believe these are among things we are talking about at this meeting, that is, how inoculated samples are different from naturally contaminated samples. We have seen tremendous differences.

Pitt

It is important that our recommendations should relate as closely as possible to natural conditions.

Beckers

ISO defines the total number of revolutions of a blender for enumerating bacteria and this should be kept in mind when enumerating fungi.

Pitt

It is critical to distinguish what we are doing here from that which has been done with bacteria. A fungal spore or a yeast cell is different from a bacterial cell. Differences seen with bacterial cells may very well not be the same with fungal cells. Therefore, we have to be careful in relating observations made with bacteria to what is observed with the fungi.

Mossel

As you said, a conidium or a yeast cell is quite a different entity from a bacterial cell, and when I refer to previous work, I refer specifically to mycological work. Sorry for the confusion.

Recommendations

Sample preparation

1. The size of the subsample to be examined should be as large and representative as possible.

2. The ratio of sample to diluent should be 1:10 for most foods and applications.

3. For homogenizing subsamples, the stomacher and the Waring type blender are preferable, but high speed top drive macerators are also acceptable. For the stomacher and for normal speed blenders, a homogenizing time of 2 min is recommended, but the optimal time to produce maximum counts should be checked for each particular application. For top-drive macerators this optimal time will usually be less than 2 min.

    Shaking by hand is a less readily standardized technique and will often produce lower counts with particulate foods. However, in situations where mechanical systems are not available, this can be a satisfactory alternative. For particulate foods, glass beads should be used. It is recommended that a standardized protocol such as that in the Compendium of Methods for the Microbiological Examination of Foods published by the American Public Health Association should be used.

4. Because fungal spores settle rapidly, subsampling should be carried out as soon as possible after homogenizing, certainly within 5 min. If necessary, wide-tipped pipettes should be used for particulate foods. If subsampling is not carried out within 5 min, the homogenized sample should be resuspended.

5. To avoid the toxic effects of heavy metals, distilled or deionized water should be used as the diluent solvent. Data presented at this workshop indicated no significant differences among a variety of diluents. However, the addition of 0.1% peptone, 0.1 M phosphate buffer or 0.05% of a food grade detergent such as polysorbitan 80 (Tween 80) will aid dispersal of hydrophobic fungal conidia and survival of sensitive cells.

    For foods of low a_w containing sensitive cells such as yeasts, concentrated diluents such as a 40% (w/v) glucose solution should be used. Further studies on the efficacy of diluents are recommended, particularly for low a_w foods.

6. For the surface sterilization of particulate foods, the use of ca. 0.4% active chlorine solutions is recommended. A treatment time of 2 min is recommended, followed by a 1 min rinse in sterile distilled or deionized water before direct plating.

    Results presented at this workshop indicate that surface sterilization by chlorine solutions is ineffective against some fungal ascospores. Further work is recommended.

Dilution plating

The panel interprets the value of the dilution plate count as essentially an assay for monitoring the quality of raw materials and processed products. If the fungal propagule units per gram or milliliter show deviation from the normal baseline count, then a more comprehensive approach should be followed to investigate the cause.

Pour plates vs. spread plates

There has been little difference shown in the quantitative recovery of fungi when either spread plates were used or pour plates were prepared with adequately tempered agar. If pour plates are prepared with significant variations in quantity and temperature of agar, then pour plating may not be an accurate technique for a quantitative measurement of fungal populations.

The panel recommends that spread plates be used in preference to pour plates for the following reasons:

1. Spread plates are better for recovery of stressed or injured cells because of increased oxygen availability and less thermal shock. (The panel felt that if appropriate samples had been selected for comparative studies, a significant difference between spread and pour plates would have been obtained).

2. Ease of identification and subsequent isolation of fungi are facilitated by spread plates.

The preparation of spread plates is recommended as follows:

1. Plates should be poured the previous day or at least 4 h prior to use to permit drying at 20-25°C or possibly at room temperature. If media containing rose bengal is being used, the plates must be kept away from light to prevent the photosensitive conversion of rose bengal to a toxic product.

2. Aliquots preferably of 0.1 ml but up to 0.3 ml can be spread adequately.

3. Spreaders should have a small diameter to minimize carry over of inocula from plate to plate. The use of bent sealed Pasteur pipettes or hockey sticks made of similar thickness glass or stainless steel is recommended over thicker bore glass tubing.

4. Spreading can be facilitated by the use of electrical or mechanically driven turntables which minimize the exposure of the surface to be spread to aerial contaminants.

Where government regulations require, or suspected low fungal counts are encountered, there is some justification for the use of pour plates. Pour plates are generally quicker to make, allow larger sample sizes to be studied, and are more convenient if the medium is soft. One panel member suggested that if the fungal count is low enough to require a pour plate, the few remaining fungi may be inhibited or inactivated by thermal shock or low redox potential, invalidating one of the reasons given above.

If pour plates are used, the media must be properly tempered to 45-47°C and 15 ml of molten media added to the plates. The plates must not be stacked until all plates have thoroughly set.

Definition of terms

Dilution vs. direct plating. The dilution plate count is a quantitative procedure which is used to determine the number of colony-forming units in a sample on a weight or volume basis. The direct plate assay is more of a qualitative procedure which is used to determine the percentage of pieces of a particulate food that contain one or more viable propagules.

Direct plating is recommended when information on the distribution of fungi through a product is required and cannot be ascertained by the dilution plate count.

In direct plating, pieces plated should be surface disinfected prior to plating, which enables the internal mycoflora to be described. In cases of potential food spoilage or food mycotoxicosis, a description of the internal mycoflora is recommended.

In the direct plating of living samples such as grains, a sprout inhibitor may be incorporated in the medium to prevent the sprouted grain from lifting the lids off the Petri dishes and permitting contaminants to enter. Although 7.5% salt has been recommended as a sprout inhibitor, it is recommended that the effect of salt and 2:4 diphenoxyacetic acid (J. Inst. Brew, 87, 248-251, 1981), used by the European Brewing Committee, be evaluated for fungal and sprout inhibition and a subsequent recommendation made.

Upright vs. inverted plates during incubation

The panel unanimously agreed that upright incubation was the preferred incubation technique. There were no significant advantages to the inverted method of incubation and since there was no significant difference in terms of enumeration, the upright method is recommended.

The upright method has an advantage in that there is less chance of contamination of the colonies for subsequent isolation work and also less chance of laboratory contamination. Upright incubation also means that plates should be counted in the upright position by marking the base and lid of the plate, aligning the marks and proceeding to spot the lid during counting. Examination of the plates under the stereomicroscope and isolation can then follow with minimal risk of cross contamination of the colonies.

CHAPTER 2

GENERAL PURPOSE ENUMERATION AND ISOLATION MEDIA

This chapter contains results of studies done to evaluate media that may be useful for routine enumeration and isolation of fungi from foods. Prior to the workshop, questionnaires were sent to twenty-five laboratories to determine which media were used for enumeration of fungi in foods. A total of fifteen different media were reported to be routinely used. The media most often used were DRBC, RBC, PDA and OGY. Likewise, recommendations in texts as to the medium of choice for routine culture of foodborne fungi vary widely. Some recommended media are not optimal for enumeration or recovery of fungi from foods, e.g., acidified PDA (APDA). Thus, several scientists were asked to compare the media they routinely used with DRBC, DG18 or OGY. The following papers report the results of these studies and others describing media for general purpose enumeration and isolation of fungi from foods.

▶ Properties of the Ideal Fungal Enumeration Medium

Food mycology is just now emerging from the Dark Ages, from which food bacteriology emerged some 30 years ago. The Dark Ages were when it was believed that a "yeast and mold count" on a single all purpose medium was all that was required for any commodity, from fresh cabbage or yogurt to dried fruit or salted fish. Clearly this is not so: fungi which will grow on fresh cabbage or in yogurt almost invariably will not grow on dried foods.

The ideal enumeration medium

In the first paper in this session, it is appropriate to consider what the ideal enumeration medium should achieve. It should have the following attributes:

1. It should suppress bacterial growth completely, without affecting growth of food fungi (filamentous or yeasts)
2. It should be nutritionally adequate and support the growth of relatively fastidious fungi
3. It should slow the radial growth of fungal colonies to permit counting of a useful range of numbers per plate, without inhibiting spore germination
4. It should suppress, but not totally prevent, growth of spreading molds, so they can be enumerated also
5. It should promote growth of fungi of relevance to the contamination or spoilage of the particular commodity being examined and
6. It should suppress growth of irrelevant fungi.

It must be stated at the outset that the ideal enumeration medium does not exist. All practical media must compromise on one or more of the foregoing requirements, and the fact that media are compromises should not be overlooked. Ordinary bacteriological enumeration media will support growth of most fungi, and have been recommended for fungal enumeration by a number of authors. Indeed, if antibiotics are added to suppress bacteria, if spreading fungi are absent, if numbers to be enumerated are low and if there is no interest in screening out fungi which are not food-related, bacteriological enumeration media can be quite suitable for counting fungi. However, spreading fungi do exist, and fungal growth must be limited if useful numbers per plate are to be enumerated, so media designed with these properties are much more satisfactory in most situations.

Points 1 to 4 above are largely self evident. Bacterial growth has traditionally been suppressed by acid media. However, the more recent approach of antibiotic addition has been much more successful, to the point where bacterial contamination of fungal plates or isolates is no longer a serious problem. Media are readily made nutritionally adequate for fungi by the use of malt extract, yeast extract or peptone. Various combinations of chemicals have been found reasonably successful in controlling fungal growth on Petri dishes.

Statements 5 and 6, however, are often overlooked in media design and formulation, and need emphasis here. In short, points 5 and 6 mean that the mycologist should enumerate fungi with a particular purpose in mind, and media should be formulated and used to achieve that purpose. For example, a general counting medium of high a_w and minimal levels of inhibitory compounds is suitable for enumeration of the fungi in wheat grains at harvest when detection of <u>Alternaria</u>, <u>Curvularia</u>, <u>Drechslera</u> and like fungi is important. However, this medium is not suitable for monitoring the deterioration of that wheat in storage several months later, because it will not select for the fungi which could have grown during storage. On the other hand, if the mycologist were monitoring the decline in numbers of spores of <u>Alternaria</u> during storage, the original medium would still be suitable.

Choice of media for specific purposes

A great variety of purposes exist for which specific formulations may be required, and it is not the purpose of an introductory paper such as this to cover this topic in detail. Some general remarks are appropriate, however.

The most important parameter which should influence selection of a fungal enumeration medium is a_w, i.e., the choice between media suitable for high a_w foods such as vegetables, meat and dairy products, and those suited to the enumeration of fungi in dried foods such as cereals, confectionery and dried fruit. A second important consideration relates to whether the mycologist's interest is primarily in molds, or yeasts, or both. Consideration should also be given to whether a medium is to be used for quality control or for selectively monitoring for specific fungi, such as preservative resistant yeasts or <u>Aspergillus flavus</u>.

In quality control, which involves testing large numbers of samples, limitations of cost and time mean that the use of only one or, at most, two media can be justified. A general purpose medium should be used in this situation to monitor mycological quality. After a large number of samples have been analyzed, baseline levels for acceptable quality can then be set. For quality control, it is not necessary to search for the medium which gives the highest possible count, because changes in contamination level are more important than absolute numbers.

64

If fungi can survive processing or enter containers after processing, end product monitoring may be necessary. Here a general purpose medium is usually ineffective. A medium selective for heat resistant molds or preservative resistant yeasts, for example, may be necessary.

As a final example, the screening of raw materials and foods for mycotoxigenic fungi is becoming of increasing importance. Selective media are available for a limited range of such fungi, as will be discussed in a subsequent session, but in the present state of knowledge, screening for mycotoxigenic fungi will usually have to rely on a general purpose high a_w medium which is somewhat selective against adventitious fungi.

J. I. PITT

Principles for Media Evaluation

In September 1978, in the framework of the International Congress of Microbiology, held in Munich, West Germany, a Working Party on Culture Media (WPCM) was founded by the International Union of Microbiological Societies (IUMS). Its assignment was to elaborate and keep up-to-date guidelines for the selection, use and monitoring of selective and, where required, also non-selective culture media used in bacteriology and analytical mycology, particularly media applied in the examination of foods, drinking water and the food environment.

Colloquia were subsequently held in Mallorca (1979), Dallas (1981) and London (1984). After the mostly used and at the same time most effective media had been selected, a first attempt was made to publish a Pharmacopeia-like guideline for the functioning of media. As a framework for all chapters of this guideline, five basic criteria were recommended to be used in assessing the functioning of such media. They are in press in the International Journal of Food Microbiology, the official bimonthly scientific publication of the International Committee on Food Microbiology and Hygiene of the IUMS. With the permission of the Editors of the Proceedings of the London Meeting, Dr. Janet E. L. Corry and Dr. G. Curtis, these principles are presented below.

The following five criteria were recommended for use in every selective medium evaluation:

1. Productivity for pure cultures

   Definition: N_S^S/N_O^O

   where: N_S^S = colonies of sought type obtained on selective medium

   N_O^O = colonies obtained on control medium (ideally, selective medium without inhibitors)

   Assessment: Inoculating with representative test strains, including robust and fastidious types

   Requirement: $\log_{10} N_O^S - \log_{10} N_S^S < 0.5$

2. Selectivity for pure cultures

   Definition: N_S^i/N_0^i

      where: N_S^i = colonies of interfering types on selective medium

             N_0^i = colonies obtained on control medium

   Assessment: Inoculating with refractory and sensitive interfering
               strains
   Requirement: $\log_{10} N_0^i - \log_{10} N_S^i > 5$

3. Electivity

   Definition: Possibilities for easy recognition of colonies of different
               organisms, based on either intrinsic properties of
               organisms, or specific responses to the medium under study

   Assessment: Ranking of ease of differentiation, avoiding parochial bias

   Interpretation: Rank correlation

4. Performance with natural food samples

   Definition: Recovery of sought organisms and inhibition of interfering
               organisms when the test medium is compared to a different
               one

   Assessment: By direct inspection, microscopy or biochemical
               identification

   Interpretation: Grouping of recoveries of sought and non-sought
                   organisms in series of ranges $10^x - 10^{x+i}$ and analysis
                   by rank correlation methods

5. Taxonomic bias with natural food samples

   Definition: Extent to which a medium selects or suppresses a particular
               species or biotype within a selectively enumerated taxonomic
               group

   Assessment: By microscopy or biochemical examination of a representative
               selection of colonies, or by reinoculation of colonies

   Interpretation: Rank correlation

                                                        D. A. A. MOSSEL

▶ Comparison of Various Media for the Enumeration of
  Yeasts and Molds in Food

   Enumeration of yeasts and molds in food provides assurance that the food
was prepared from good quality raw materials under hygienic conditions and
that it was stored properly. Mossel et al. (1962) described an agar medium
(OGY) containing yeast extract and glucose as growth factors and

oxytetracycline as an antibiotic to inhibit bacterial growth. A shortcoming of this medium is that oxytetracyline is not sufficiently thermostable to be included in the medium before autoclaving. Consequently more heat stable antibiotics such as chloramphenicol and gentamicin have been used (Mossel et al. 1975; Jarvis 1978). To insure bacteria-free plates during enumeration of fungi, Koburger & Rodgers (1978) used a combination of different antibiotics. Overgrowth of slowly growing colonies by rapidly spreading molds may be reduced by the addition of inhibitors such as rose bengal (Overcast & Weakley 1969; Jarvis 1973) and dichloran (King et al. 1979) to the media.

The foregoing considerations have led to a confusing number of media being used. For the purpose of standardization of methods for mycological examinations of foods, a large inter-laboratory trial was carried out in the Netherlands to compare various media. The aim of the study was to select a simple culture medium supporting good recovery of yeasts and molds which could be used for a wide range of foods without disturbing growth of bacteria and spread of molds. As OGY agar (Mossel et al. 1962) is widely used in the Netherlands for the mycological examination of different kinds of food such as salads, pastries, fruit juices, beer, vegetables, fermented dairy products, spices and meal, it was chosen as the reference medium.

Since mold counts may be influenced during preparation of dilutions by fragmentation of hyphae and breaking of mold spore clumps, resulting in a higher number of cfu (Jarvis et al. 1983), repeatability was also taken into consideration.

Materials and methods

Samples. In all, 270 samples of 13 different kinds of food were examined (Table 1). The samples were selected for high (> 10^3/g) mold or yeast counts by screening with OGY. They were re-examined with four or five of the following media:

1. OGY (Mossel et al. 1962)

2. OGGY (Mossel et al. 1975)

Table 1. Foods examined in comparative study

Foods	Number of samples
Meat and meat products	55
Spices	39
Meal	27
Beer	26
Raw cut vegetables	23
Pastries	21
Dairy products	20
Salads	20
Dried fruit and vegetables	13
Poultry products	11
Fish and fish products	8
Tempe	4
Lemonade and syrup	3
Total:	270

3. CGY (ISO/DIS-6611, Anon. 1982)

4. RBCC (Baggerman 1981)

5. DRBC (King et al. 1979)

6. DG18 (Hocking & Pitt 1980)

7. GYS (glucose-yeast extract-sucrose) agar following the formula of Northolt: glucose, 20 g; yeast extract, 5 g; sucrose, 450 g; agar, 20 g and water, 505 g; after sterilization (15 min at 110°C), plates were poured and subsequently dried at 37°C for 3 days; drying should result in an a_w value in the medium of 0.91.

Methods. Each sample was examined according to a carefully standardized procedure: 20 g were mixed with 180 ml of 0.1% peptone saline solution in a Stomacher for 1 min. Large particles were allowed to settle for 30 sec. Further decimal dilutions were also made in 0.1% peptone saline solution. Each new dilution was mixed mechanically (Vortex) for 5 sec (for details see ISO 6887, Anon. 1983). From each dilution, 1.0 ml was pipetted in duplicate into 90-mm Petri dishes and mixed with OGY, OGGY, CGY or RBCC, and cooled to 45°C; or 0.1 ml was pipetted in duplicate onto DRBC, DG18 or GYS and spread over the medium surface. Thus the media were used in accordance with the original descriptions.

All media were incubated for 5 days at 22 ± 1°C. Plates with 10-100 yeast and/or mold colonies were selected for counting. If identification was in doubt, colonies were examined microscopically to discriminate between bacteria and yeasts and between yeasts and molds. Mold colonies with a diameter over 2.5 cm were characterized as "spreaders."

Isolation and identification. From the medium with the highest count, one or two colonies of the predominant yeast or mold flora were isolated and purified. Yeast cultures were identified by the Yeast Division of the Centraalbureau voor Schimmelcultures at Delft (Dr. M. Th. Smith); mold cultures were identified by the Centraalbureau voor Schimmelcultures at Baarn (Dr. R. A. Samson).

To evaluate growth potentials, several isolated pure cultures were streaked on all media. The plates were incubated at 20-25°C. After 5 days, plates were checked macroscopically for visible colonies.

Repeatability. To assay repeatability of methods for enumerating yeasts and molds, four laboratories examined selected samples in replicates of five. In total, seven samples were examined in this way, each sample by only one laboratory. Each sample was well mixed and divided into five subsamples; each subsample was examined as described above using OGY and/or DRBC.

Results and discussion

Results of yeast and mold counts with the seven media evaluated are summarized in Table 2. Counts with OGGY, CGY, RBCC, DRBC, DG18 and GYS were separately compared with counts with OGY, applying the t test to the differences in paired log counts. Results of this comparison are also summarized in Table 2.

Yeast counts from all media except CGY were statistically significantly different from counts with OGY ($P < 0.05$). However, only the means of differences for OGGY and GYS relative to OGY were of practical significance (0.45 and 0.32 log units, respectively). The mean of differences between

Table 2. Fungal counts in different foods with six different media compared to fungal counts with a reference method

Fungi	Medium[a]	Number of samples	Mean of log counts	Standard deviation	Mean of differences	Standard deviation	Standard error	2-Tail probability	Correlation
Yeasts	OGY	251	4.38	1.65	–	–	–	–	–
	OGGY	90	4.52	1.23	0.45	0.55	0.06	0.00	0.90
	CGY	60	4.90	1.39	0.08	0.44	0.06	0.16	0.95
	RBCC	195	4.27	1.82	0.09	0.33	0.02	0.00	0.98
	DRBC	251	4.44	1.69	-0.06	0.45	0.03	0.03	0.97
	DG18	193	4.47	1.88	-0.12	0.39	0.03	0.00	0.98
	GYS	133	3.44	1.62	0.32	1.34	0.12	0.01	0.66
Molds[b]	OGY	116	3.43	0.96	–	–	–	–	–
	RBCC	104	3.60	0.90	-0.04	0.25	0.02	0.08	0.96
	DRBC	116	3.44	0.96	-0.01	0.31	0.03	0.79	0.95
	DG18	104	3.66	0.86	-0.11	0.42	0.04	0.01	0.88
	GYS	106	3.09	1.04	0.29	0.67	0.07	0.00	0.78

aMean value of all differences between log counts for yeasts with OGY (reference medium) and log counts for yeasts with test medium for individual samples

bFor enumeration of molds, the number of samples examined with OGGY and CGY having counts exceeding 2.50 was too low (< 10)

OGY on one hand and RBCC, DRBC and DG18 on the other was only 0.1 log unit and in the same order of magnitude as the mean of differences between OGY and CGY, and therefore acceptable. This is underlined by the good correlation between these media (Table 2).

Mold counts from DG18 and GYS were statistically different from counts with OGY ($P < 0.05$), while mold counts with RBCC and DRBC were not statistically different. The difference between DG18 and OGY is acceptable in view of the correlation between both media (Table 2); the difference between GYS and OGY is unacceptable in this respect.

The results lead to the conclusion that DG18 is quantitatively the best medium for enumeration of yeasts and molds and that OGY and DRBC are second best. For enumeration of molds, RBCC is also useful, but has the disadvantage of inferior performance for enumerating yeasts. GYS would be a rather poor choice for the enumeration of yeasts and molds normally occurring in food. CGY and OGGY have only been tested for yeasts. For this purpose, CGY scored similarly to RBCC, while OGGY was inferior to GYS.

The qualitative aspects of the media were also taken into account. Comments on these aspects are summarized in Table 3. Spreading of mold colonies occurred on each medium. Bacterial growth occurred on OGY and CGY, especially in diluted samples of foods heavily loaded with bacteria, but also on DRBC and DG18. Thus, in this respect, there were no differences among most media. The reduced a_w of DG18 (a_w 0.95) combined with the

Table 3. Summary of comments on different media

Medium	Comments
OGY	Spreading of molds, especially from spices and meal (in 21 out of 66 samples); in three samples, no counts were made due to spreading. Antibiotic supplement was sometimes insufficient to inhibit bacterial growth, especially when foods were heavily loaded with bacteria, e.g., raw meat.
OGGY	No bacterial growth, even if bacterial flora outnumbered yeasts and molds.
CGY	Antibiotic supplement was sometimes insufficient to inhibit bacterial growth.
RBCC	Some spreading of molds (9 out of 66 samples of spices and meal) in two samples, no counts were made due to spreading.
DRBC	Spreading of molds in only three samples of spices; in 1 sample no counts were made due to spreading. Antibiotic supplement was sometimes insufficient to inhibit bacterial growth.
DG18	Spreading of molds in only two samples of spices. Growth, especially of molds, appeared to be retarded, making counting and differentiating between yeast and mold colonies difficult. Antibiotic supplement was sometimes insufficient to inhibit bacterial growth.
GYS	Difficult to prepare; not for routine use.

presence of dichloran retarded growth of mold colonies. Differentiation between yeast and mold colonies was therefore difficult. In conclusion, there was qualitatively a preference for media without artificial reduction of a_w, such as OGY, CGY, RBCC and DRBC.

Results of identification of yeasts and molds isolated from various foods with different media are presented in Tables 4 and 5. Because colonies of the predominant yeast and mold flora were isolated from the medium with the highest count, the greatest number of strains originated from DRBC and DG18. Some yeasts and molds (Debaryomyces hansenii, Saccharomyces cerevisiae, Yarrowia lipolytica, Aspergillus spp., Fusarium sp., Geotrichum candidum, Mucor spp., Penicillium spp.) were isolated from various media; others (Cryptococcus spp., Rhodosporidium spp., Trichosporon pullulans, Verticillium sp.) were only isolated from one medium.

The results of experiments with pure cultures (Table 6) demonstrated that media containing gentamicin, rose bengal and/or dichloran or media with

Table 4. Frequency of isolation of yeasts on different media

Yeast	Media							
	OGY	OGGY	CGY	RBCC	DRBC	DG18	GYS	Total
Apiotrichum humicola	1				4			5
Candida famata						1		1
C. glaebosa					1			1
C. guilliermondii						1	1	2
C. ingeniosa					1			1
C. kefyr				1				1
C. lusitaniae	1		1	1	1	1		5
C. sake			1		2			3
C. tropicalis					1			1
C. zeylanoides	1		1	3	1	1		7
Cryptococcus albidus					2			2
C. laurentii					1			1
C. macerans					3			3
Debaryomyces hansenii	7	3	2	5	10	13		40
Hansenula anomala	1				1	1		3
H. subpelliculosa			1			1		2
Hanseniaspora uvarum	3		3	3	3	2		14
Leucosporidium sp.					1			1
Pichia guilliermondii	1			1				2
P. membranaefaciens	2		3	2	3	1		11
Rhodosporidium bisporidiis					1			1
R. infirmo-miniatum					3			3
Rhodotorula acheniorium					1			1
R. rubra						1		1
Saccharomyces cerevisiae	7		4	4	5	3		23
S. exiguus	1							1
Torulaspora delbrueckii	1	1		1	3	3	1	10
Trichosporon pullulans					3			3
Yarrowia lipolytica	12	6	3	7	11	8		47
Zygosaccharomyces rouxii						3	1	4
Total:	38	10	18	29	62	40	3	200

Table 5. Frequency of isolation of molds on different media

Mold	Media							Total
	OGY	OGGY	CGY	RBCC	DRBC	DG18	GYS	
<u>Absidia</u> <u>corymbifera</u>				1				1
<u>Alternaria</u> <u>alternata</u>						1		1
<u>Aspergillus</u> <u>candidus</u>	7					1		8
<u>A</u>. <u>flavus</u>	2			4	5			11
<u>A</u>. <u>nidulans</u>	1							1
<u>A</u>. <u>niger</u>	3			3	1	7		14
<u>A</u>. <u>ochraceus</u>	1				1	1		3
<u>A</u>. <u>oryzae</u>							1	1
<u>A</u>. <u>penicilloides</u>							1	1
<u>A</u>. <u>phoenicis</u>					1	1		2
<u>A</u>. <u>sydowii</u>							1	1
<u>A</u>. <u>versicolor</u>							1	1
<u>A</u>. <u>wentii</u>				1		1		2
<u>Cladosporium</u> <u>herbarum</u>					1			1
<u>C</u>. <u>macrocarpum</u>					4			4
<u>C</u>. <u>sphaerospermum</u>	4					1		5
<u>C</u>. <u>tenuissimum</u>				2				2
<u>Eurotium</u> <u>herbariorum</u>						1		1
<u>Fusarium</u> sp.	2		2	2	2			8
<u>Geotrichum</u> <u>candidum</u>	4	1	2	4	3			14
<u>Mucor</u> <u>circinelloides</u>	1							1
<u>M</u>. <u>racemosum</u>	1				1			2
<u>Mucor</u> sp.						1		1
<u>Penicillium</u> <u>brevicompactum</u>				1	3	5		9
<u>P</u>. <u>chrysogenum</u>				2	4	2		8
<u>P</u>. <u>citrinum</u>				1	2	1		4
<u>P</u>. <u>corylophilum</u>					1			1
<u>P</u>. <u>nigricans</u>						1		1
<u>P</u>. <u>roqueforti</u>	1		1	1	1	1		5
<u>P</u>. <u>steckii</u>						1		1
<u>P</u>. <u>verrucosum</u> var. <u>cyclopium</u>	3			1	2	7		13
<u>Pencillium</u> sp.				1				1
<u>Phaeotheca</u> sp.					1			1
<u>Phoma</u> <u>exigua</u>				1				1
<u>Rhizopus</u> <u>oryzae</u>	1			2				3
<u>R</u>. <u>stolonifer</u>	1			1				2
<u>Verticillium</u> <u>lecannii</u>	3							3
<u>Wallemia</u> <u>sebi</u>							1	1
Total:	35	1	5	28	33	33	5	140

reduced a_w inhibited growth of some species. <u>Rhizopus</u> <u>stolonifer</u> showed sensitivity to dichloran. In general, the majority of the isolates grew well within 5 days on all media except GYS. Growth of some yeasts was affected adversely on some media (circled entries in Table 6). <u>Apiotrichum</u> <u>humicola</u> and <u>Rhodosporidium</u> <u>infirmo-miniatum</u> exhibited sensitivity to reduced a_w, and <u>Candida</u> <u>zeylanoides</u>, <u>Debaryomyces</u> <u>hansenii</u>, <u>Pichia</u> <u>membranaefaciens</u>, <u>Rhodosporidium</u> <u>infirmo-miniatum</u> and <u>Rhodotorula</u> <u>rubra</u> were sensitive to gentamicin. <u>Torulaspora</u> <u>delbrueckii</u> and <u>Zygosaccharomyces</u> <u>rouxii</u> were sensitive to rose bengal, although the results with the latter

Table 6. Growth of pure cultures of yeasts and molds on different media after 5 days of incubation

Fungus	Media[a]						
	OGY	OGGY	CGY	RBCC	DRBC	DG18	GYS
Yeasts							
Apiotrichum humicola	++	++	++	++	++	⊕	−
Candida guilliermondii	++	++	++	++	++	++	++
C. kefyr	++	n.d.	++	++	++	++	n.d.
C. sake	++	++	++	++	++	++	−
C. tropicalis	++	++	n.d.	++	++	++	n.d.
C. zeylanoides	++	⊖	++	++	++	++	n.d.
Debaryomyces hansenii	++	⊕	++	++	++	++	n.d.
Hansenula anomala	++	++	++	++	++	++	+
Pichia membranaefaciens	++	⊖	n.d.	++	++	++	n.d.
Rhodosporidium infirmo-miniatum	++	⊖	++	++	++	⊕	−
Rhodotorula rubra	++	⊖	++	++	++	++	+
Saccharomyces cerevisiae	++	++	++	++	++	++	+
Torulaspora delbrueckii (a)	++	++	++	⊕	++	++	++
T. delbrueckii (b)	++	n.d.	++	++	++	++	n.d.
Trichosporon pullulans	++	++	++	++	++	++	−
Yarrowia lipolytica	++	++	++	++	++	++	−
Zygosaccharomyces rouxii (a)	++	++	++	⊖	⊖	++	+
Z. rouxii (b)	++	++	++	++	++	++	++
Molds							
Absidia corymbifera	++	⊕	++	⊕	⊕	⊕	+
Aspergillus candidus	++	++	++	++	++	+	+
A. flavus	++	++	++	++	++	++	+
A. nidulans	++	++	++	++	++	++	++
A. niger	++	++	++	++	++	++	+
A. oryzae	++	++	++	++	++	++	+
A. sydowii	++	++	++	++	++	++	+
A. versicolor	++	++	++	++	++	++	++
A. wentii	++	++	++	++	++	++	+
Cladosporium herbarum	++	⊖	n.d.	⊖	++	++	n.d.
C. macrocarpum	++	⊖	++	++	++	++	n.d.
C. sphaerospermum	++	++	++	++	++	++	+
Eurotium herbariorum	++	++	++	++	++	++	++
Mucor circinelloides	++	++	++	++	++	++	+
Penicillium brevicompactum	++	++	++	++	++	++	+
P. citrinum	++	++	++	++	++	++	+
P. corylophilum	++	n.d.	++	++	++	++	n.d.
P. nigricans	++	++	++	++	++	++	+
P. steckii	++	⊕	n.d.	++	++	++	n.d.
P. verrucosum var. cyclopium	++	++	++	++	++	++	+
Rhizopus stolonifer	++	++	++	++	⊖	⊕	++

[a] ++, good growth; ⊕, poor growth; ⊖, no growth; n.d., not done

organism demonstrated that this was strain-dependent.

Growth of molds appeared to be less affected than yeasts by additives such as rose bengal and dichloran. Absidia corymbifera grew well only on OGY and CGY, while Cladosporium herbarum, C. macrocarpum and Penicillium

<u>steckii</u> showed sensitivity to gentamicin. <u>C</u>. <u>herbarum</u> did not grow on RBCC.

In different laboratories, yeast and mold populations in foods were determined with one medium in five subsamples. The standard deviation of each group of five log counts resulting from the subsamples did not exceed 0.10 log units, indicating that the repeatability of the counts was rather good in each laboratory. It is assumed that the results may have been favorably influenced by careful standardization of macerating and diluting procedures. There were no indications that the repeatability was influenced by the medium.

Results of these studies clearly indicate that except for OGGY and GYS media there were no important quantitative differences between the media tested. Also, spreading of mold colonies occurred on all media. This phenomenon is well known and is the reason why ingredients such as rose bengal and dichloran are added to media, e.g., RBCC, DRBC and DG18. Nevertheless, spreading also occurred on these media, although less than on OGY and CGY. Spreading was less than was expected in view of the types of food examined, especially on OGY and CGY. Bacterial growth was observed on OGY, CGY, DRBC and DG18, all containing only one antibiotic. Bacterial growth occurred especially when foods were heavily contaminated with bacteria. This observation has been made in previous studies and is the reason why gentamicin was introduced as a supplementary antibiotic by Mossel et al. (1975); Koburger & Rodgers (1978) also advise of the use of a combination of antibiotics. However, bacterial growth did not occur with the vast majority of other food samples examined.

OGGY medium was only tested for enumeration of yeasts. For that purpose, it did not appear to be satisfactory. This result is in contradiction with earlier results of Dijkmann et al. (1979). They demonstrated that OGGY did not suppress populations of yeasts, even when they were stressed by refrigeration or storage at -20°C. However, there is an essential difference between the present study and that of Dijkmann et al. (1979). They used OGGY as a spread-plate medium, while in the present study it was used in a pour-plate system. When using OGGY for pour plating, there is an accumulation of potential growth inhibiting effects, e.g., antibiotics in combination with a reduction of oxygen availability. Moreover, pouring of warm agar on yeasts may lead to cell damage that might not be repaired in combination with the accumulated inhibiting effects. Loss of propagules was experienced earlier by one of us (K. Dijkmann) when poured plates of OGGY were used instead of spread plates.

Results clearly demonstrated that GYS is not suitable for the enumeration of yeasts and molds in food in general when the incubation period is limited to 5 days. Growth on GYS may be better when the incubation period is prolonged, but the incubation period should be limited for practical reasons. Besides, GYS was never intended for the enumeration of yeasts and molds in general. Because of its reduced a_w, it was originally intended for isolation of xerophilic fungi.

Results from growth experiments with pure cultures demonstrate that while the greatest number of strains was isolated from DRBC and DG18, most strains grew well on all other media. It can be concluded that these strains could also have been isolated from other media. Results indicate that when media containing gentamicin, rose bengal and/or dichloran or when media with reduced a_w are used, some species will be suppressed. However, results also indicate that strains of one species may differ in sensitivity. Although differences between media and even methods (poured or spread plates) may be significant when using pure cultures, the differences may be eliminated when examining food samples in general. In most instances, food will contain a mixture of cultures and therefore a significant effect on one organism may be eliminated by the absence of the same effect on another. This may be one of the most important reasons why there were only minor quantitative differences between various media.

The major purpose of the enumeration of yeasts and molds in food is to provide assurance that the food under examination meets certain hygiene requirements, e.g., choice of good quality raw materials, hygienic conditions during production and proper storage. Then, the number of yeasts and molds in a food may be used as an indicator for poor manufacturing and distributing practices. In this respect, the numbers of yeasts and molds are of more interest than the exact species. Moreover, cultures isolated from media containing inhibitors such as rose bengal and dichloran need subculturing prior to identification because of abnormal pigmentation and growth rate of propagules induced by these media (Jarvis et al. 1983). Therefore, the major requirements of a medium to enumerate yeasts and molds in food is that it supports good recovery of yeasts and molds without growth of bacteria and spread of mold colonies. Although an all-inclusive medium may never exist (Dijkmann 1982) the medium of choice should be the medium that can be used for many kinds of food without supporting growth of bacteria and spread of mold colonies in most instances. From this point of view, only minor differences were noted between most media examined and it can be concluded that a simple basal agar formula containing glucose, yeast extract and one antibiotic fulfills the requirements of the intended medium. Based on the results of this study, oxytetracycline may be preferred but, because of its heat-stability, chloramphenicol might also be used. It should be realized that the selected medium may fail in certain instances (Mossel et al. 1980). For example, it has been demonstrated by Dijkmann (1982) that the addition of gentamicin is necessary for enumeration of yeasts and molds in fresh and fermented meat products. Likewise, the addition of rose bengal or dichloran is advised when spread of mold colonies is suspected. Nevertheless, a rather simple medium can be used in most instances.

References

ANON. 1982 International Organization for Standardization. Milk and Milk-Products - Enumeration of Yeasts and Moulds (Colony Count Technique at 25°C). ISO/DIS-6611.

ANON. 1983 International Organization for Standardization. Microbiology - General Guidance for the Preparation of Dilutions for Microbiological Examination. ISO-6887.

BAGGERMAN, W. I. 1981 A modified rose bengal medium for the enumeration of yeasts and moulds from foods. European Journal of Applied Microbiology and Biotechnology 12, 242-247.

DIJKMANN, K., KOOPMANS, M. & MOSSEL, D. A. A. 1979 The recovery and identification of psychrotrophic yeasts from chilled and frozen comminuted fresh meats. Journal of Applied Bacteriology 47, ix.

DIJKMANN, K. E. 1982 The optimal medium - always the best choice? In Quality Assurance and Quality Control of Microbiological Culture Media. Proceedings of the IUMS-ICFMH Symposium held on 6th and 7th September 1979, in Calas de Mallorca ed. Corry, J. E. L. Darmstadt, West Germany: G. I. T.-Verlag Ernst Giebeler.

HOCKING, A. D. & PITT, J. I. 1980 Dichloran-glycerol medium for enumeration of xerophilic fungi from low moisture foods. Applied and Environmental Microbiology 39, 488-492.

JARVIS, B. 1973 Comparison of an improved rose bengal chlortetracycline agar with other media for selective isolation and enumeration of moulds and yeasts in foods. Journal of Applied Bacteriology 36, 723-727.

JARVIS, B. 1978 Methods for detecting fungi in foods and beverages. In Food and Beverage Mycology ed. Beuchat, L. R. pp. 471-504. Westport, CT: AVI Publ., Inc.

JARVIS, B., SEILER, D. A. L., OULD, A. J. L. & WILLIAMS, A. P. 1983
Observations on the enumeration of moulds in food and feedingstuffs.
Journal of Applied Bacteriology 55, 325-336.

KING, A. D., HOCKING, A. D. & PITT, J. I. 1979 Dichloran-rose bengal medium
for enumeration and isolation of molds from foods. Applied and
Environmental Microbiology 37, 959-964.

KOBURGER, J. A. & RODGERS, M. F. 1978 Single multiple antibiotic amended
media to enumerate yeasts and molds. Journal of Food Protection 41,
367-369.

MOSSEL, D. A. A., VISSER, M. & MENGERINK, W. H. J. 1962 A comparison of
media for the enumeration of moulds and yeasts in foods and beverages.
Laboratory Practice 11, 109-112.

MOSSEL, D. A. A., VEGA, C. L. & PUT, H. M. C. 1975 Further studies on the
suitability of various media containing antibacterial antibiotics for the
enumeration of moulds in food and food environments. Journal of Applied
Bacteriology 39, 15-22.

MOSSEL, D. A. A., DIJKMANN, K. E. & KOOPMANS, M. 1980 Experience with
methods for the enumeration and identification of yeasts occurring in
foods. In Biology and Activities of Yeasts, The Society for Applied
Bacteriology Symposium Series no. 9 ed. Skinner, F. A., Passmore, S. M. &
Davenport, R. R. London: Academic Press.

OVERCAST, W. W. & WEAKLEY, D. J. 1969 An aureomycin rose bengal agar for
enumeration of yeasts and molds in cottage cheese. Journal of Milk and
Food Technology 32, 442-445.

H. J. BECKERS
E. de BOER
K. E. DIJKMANN
B. J. HARTOG
J. A. van KOOIJ
D. KUIK
N. MOL
A. J. NOOITGEDAGT
M. D. NORTHOLT
R. A. SAMSON

▶ Comparison of Media for the Enumeration of Fungi
from Dried Foods

The major function of yeast and mold enumeration media has been
selective inhibition of bacteria without compromising the recovery of
fungi. To this end, acidulants and antibacterial antibiotics have been the
most common selective agents. Koburger (1976), Overcast & Weakley (1969)
and Henson (1981) reviewed the problems encountered with use of acidified
media, including spreading mold colonies, bacterial growth, inhibition of
sublethally injured fungi and precipitation of sample constituents.
Antibiotics, often in combination, suppress bacterial growth (Mossel et al.
1975), while permitting higher recoveries of fungi on a less acidic medium.
A major drawback of antibiotic-amended media is that colonies of some mold
species rapidly spread and sporulate on the surface. Rhizopus and Mucor
spp., as well as other members of the order Mucorales, frequently overgrow
the surfaces of media, which lessens counting precision and isolation of
other fungi on the plates (King et al. 1979).

Some researchers have incorporated antimycotic compounds into
enumeration media to restrict colony diameters of rapidly growing fungi.

Baggerman (1981) tested the inhibitory effect of various concentrations of rose bengal on radial growth of fungi isolated from foods. He concluded that 150 µg of rose bengal per ml of medium affords more accurate enumeration of yeasts and molds. Yet, Burge et al. (1977) observed that rose bengal concentrations of 15-150 µg/ml significantly lowered recoveries of airborne fungi, including sporadic total repression of fungal growth. Recovery of <u>Cladosporium</u> spp. and nonsporulating molds was significantly reduced by rose bengal. These researchers concluded that a medium containing rose bengal should be used only in combination with less inhibitory media. Ottow (1972) reported that light-exposed rose bengal inhibits Actinomycetes. He suggested that fungi could be similarly affected if light degrades rose bengal to toxic compounds.

Griffin et al. (1975) developed a selective medium containing 2 µg of dichloran (2,6-dichloro-4-nitroaniline) and 30 mg of NaCl per ml to recover <u>Aspergillus</u> <u>flavus</u> from soil. Dichloran inhibited undesirable fungi whereas the diameters of <u>A</u>. <u>flavus</u> colonies were only slightly reduced. King et al. (1979) tested 31 antifungal compounds for their ability to restrict Mucorales. The combination of 2 µg of dichloran and 25 µg of rose bengal per ml was recommended to select against rapidly growing fungi without affecting colony counts of food spoilage and toxigenic molds. Hocking & Pitt (1980) developed a low a_w medium containing glycerol and dichloran to selectively isolate xerophilic fungi from dried foods.

Addition of dichloran to PDA for routine enumeration of fungi in foods was proposed by Henson (1981). He reported that 5 µg of dichloran per ml of antibiotic-amended PDA reduced mold colony diameters and also improved counting efficiency by preventing overgrowth by spreading molds. Tested yeasts were unaffected by dichloran.

In a collaborative study, four fungal enumeration media were compared for recovery of fungi from naturally-contaminated and inoculated foods (Henson et al. 1982). Colony diameter reduction aided ease of counting when media contained either dichloran or rose bengal.

Media supplemented with 5 µg of dichloran or 25 µg of rose bengal per ml have been used in our laboratory for selective isolation of non-Mucorales from raw materials which were known to harbor a variety of molds. We have not yet confirmed that media supplemented with these antimycotics can be used for quality control testing.

The objective of this study is to determine if media supplemented with dichloran and rose bengal are suitable for recovery and enumeration of fungi common to dried foods.

Materials and methods

Media. Antibiotic-supplemented PDA (Difco) was chosen as the reference medium to which other media were compared. The sterile, tempered medium was supplemented with 40 µg/ml each of filter-sterilized chlortetracycline hydrochloride and chloramphenicol (Sigma) to inhibit bacteria (CCPDA).

The dichloran-supplemented medium proposed by Henson (1981) was modified to contain 2.5 µg of dichloran per ml (CCPDA + 2.5 D). A 2% stock solution of dichloran (2,6-dichloro-4-nitroaniline, Aldrich) in 95% ethanol was prepared for addition of the antimycotic prior to sterilization of PDA. Antibiotics were added as described earlier.

DRBC medium was prepared as described by King et al. (1979); DG18 medium was prepared according to the method described by Hocking & Pitt (1980).

These media were autoclaved at 121°C for 20 min. Filter-sterilized chlortetracycline was added to the sterile media to give final concentrations of 5 and 10 µg/ml. All media were dispensed into 15 x 100 mm plastic Petri dishes for inoculation by the spread-plate technique.

Commodities. Based on past experience, six groups of dried foods were chosen for sampling so that a wide variety of yeast and mold species would be recovered. The groups consisted of dried legumes, rice, cocoa, flour, nuts and pasta. Five products per food group, each representing a different food processor or processing technique, were purchased from a retail store to maximize the fungal species available for testing. Foods were held at temperatures and humidities suitable for germination and growth of naturally-occurring mycoflora to allow fungi to multiply to detectable levels. Since actively-growing cultures were enumerated, the effect of selective media on sublethally injured cells was not tested.

Sample preparation and plating techniques. Samples (25 g) were blended with 225 ml of Butterfield's diluent (0.3 mM KH_2PO_4 diluent at pH 7.2) for 2 min at low speed in Waring blenders. Serial dilutions were prepared to obtain 10-100 colonies on plates. Triplicate plates of each medium were spread-plated with 0.1-ml portions of the dilutions. Plates were incubated upright for 3 days at 25°C, counted, and then incubated for a further 2 days until a final count was made.

Evaluation of mycoflora. Lactophenol mounts of fungi recovered on CCPDA were examined by bright field microscopy. Isolates were classified to genus according to Barnett & Hunter (1972) and Raper & Fennell (1965). The effect of each medium on recovery, colony diameter and sporulation of every isolate was noted.

Statistical analysis. We hypothesized that the dilution from which countable plates were obtained contained a normally distributed population of cells. Therefore, the statistical analysis was made at the level of the colony counts from three plates of each medium. The triplicate (replicate) colony counts allowed the single classification analysis of variance (ANOVA) to test whether the subpopulations deposited in the plates were similar (Sokal & Rohlf 1973). When the homogeneity of the subpopulations was demonstrated, any apparent differences exceeding the variation between replicates could be attributed to recovery differences between media. Each sample was analyzed separately since the number and type of fungi varied between samples. Means from samples exhibiting significant treatment (media) effects by the ANOVA were subjected to Duncan's multiple range test to determine which means differed significantly.

Results and discussion

Thirty samples were tested of which seven had detectable levels of yeasts and 28 had detectable levels of molds. Sixteen of the 35 sets of data exhibited significant differences ($P < 0.05$) between media as determined by the ANOVA. The differences between the means of the replicate counts were elaborated by Duncan's multiple range test and are reported in Table 1. Means represented by the same rank value were not significantly different ($P < 0.05$). When rank values were summed, the order of the recovery abilities of the media was DRBC = CCPDA + 2.5 D > DG18 > CCPDA.

For each sample, the statistical analysis was supplemented with observations recorded when the plates were counted. The principal reasons for lower recoveries of fungi, as determined by the macroscopic appraisals of media performance, are listed in Table 1. The observed differences

Table 1. Ranked means of fungal counts by medium[a]

Food group	CCPDA Rank value	CCPDA Reason for rank	CCPDA + 2.5 D Rank value	CCPDA + 2.5 D Reason for rank	DRBC Rank value	DRBC Reason for rank	DG18 Rank value	DG18 Reason for rank
Dried legumes	3	overgrowth[b]	1		1		2	spreading[c]
Rice products	2	overgrowth	2	spreading	1		1	
	2	overgrowth	2	spreading	1		1	
	1		1		2		3	inhibition-y[d]
	2	overgrowth	1		1		2	spreading
	2	overgrowth	1.5	spreading	1		1	
	1		1		1.5		2	spreading
Cocoa products	2		2		1		3	inhibition-Y
Flours	2	overgrowth	1		1.5		1	
	1.5		2	spreading	1		1	
	2	overgrowth	1		1		1	
Nuts	2	overgrowth	1		1		1	
	2	overgrowth	1		1		1	
	1.5	spreading	1		2	inhibition-M[e]	1.5	spreading
Pasta	1.5	spreading	1		1.5		2	spreading
	1		1		2	inhibition-Y	2	inhibition-Y
Total[f]	28.5		20.5		20.5		25.5	

[a]The differences were elaborated by a Multiple Range Test. Means within a sample represented by the same rank value were not significantly different (P ≤ 0.05); [b]lower counts due to overgrowth by Mucorales; [c]lateral spreading of molds and colony edges not distinct; [d]reduced recovery of yeasts; [e]reduced recovery of A. glaucus group; [f]A total value of 16 would indicate the best performance.

between media for recovering specific fungal genera are described briefly in Table 2.

Conclusions

The overall evaluation of performance of media was based on statistical analysis of colony counts and on observed differences between media for recovery and growth characteristics of specific fungal genera. CCPDA, which does not contain an antimycotic, performed poorly due to heavy sporulation and lateral spreading by most mold isolates and rapid development of heavily sporulating aerial mycelia of Mucorales. The lower counts on CCPDA were caused by the inability of this medium to prevent spreading colonies from coalescing and to prevent Mucorales from obscuring slower-growing genera.

Mucorales spread rapidly on CCPDA + 2.5 D. However, the Mucorales did not sporulate on this medium so their abundant aerial mycelia did not appear as dense as on CCPDA. Therefore, the Mucorales rarely interfered with counting of other genera when a mixed population was encountered. Most Aspergillus spp. spread moderately on CCPDA + 2.5 D. Aspergillus flavus, Alternaria and Paecilomyces were more difficult to count on crowded plates than other isolates.

DRBC inhibited sporulation of most isolates so genera could not be identified macroscopically unless the plates were incubated longer than 5 days. Microscopic identification was impaired because cells of most isolates had a mutated appearance. Colonies of Aspergillus glaucus group isolates were often only pinpoint in size. Despite these difficulties, DRBC ranked similarly to CCPDA + 2.5 D for recovery of fungi because discrete, non-spreading mold colonies were easily counted.

Most Aspergilli, including A. glaucus group molds, spread rapidly on DG18 and formed colonies with diffuse edges. Yeasts belonging to three genera were inhibited, possibly due to the lower a_w of this medium. The inhibition of yeasts was the primary reason for the lower ranking of DG18 (Table 1).

CCPDA, CCPDA + 2.5 D and DG18 would probably have received lower rankings if plates had not been precounted at 3 days. Fewer cfu would have been detected at a single 5-day count since coalescing, sporulating colonies often made differentiation between colonies difficult. For a single 5-day count, the order of the recovery performance might be DRBC > CCPDA + 2.5 D > DG18 > CCPDA.

Recommendations

If raw materials and finished products historically contain low numbers of fungi, there is no need to prepare an exotic medium. CCPDA will perform satisfactorily for such samples. Use of more complicated media, such as those containing antimycotics, can be justified when isolation of specific molds is desired and for research needs.

Our selection of a yeast and mold enumeration medium is generally product-specific or organism-specific. Table 3 lists attributes which are useful in selection of media. Each of the four media tested in this study is rated by its ability to meet each condition. Clearly, one medium will not exhibit all the functions listed. One could select an enumeration medium based on the attributes listed in Table 3 by determining the objective of the analysis and selecting only those attributes that best define that objective.

Table 2. Effect of enumeration medium on recovery and appearance of fungal genera

Genus	Number of isolations	Appearance of media			
		CCPDA	CCPDA + 2.5 D	DRBC	DG18
Absidia	1	spreading sporulation	inhibited	inhibited	inhibited
Acremonium	1	spreading	spreading	–a	–
Alternaria	1	spreading	spreading	–	–
Aspergillus ochraceus group	2	–	–	distinct margins	–
A. glaucus group	8	–	poor to no sporulation pinpoint colonies	poor/no sporulation	heavy sporulation
A. flavus group	10	spreading	spreading	–	spreading
A. candidus group	2	–	–	–	–
A. niger group	3	spreading	–	–	spreading
A. ornatus group	1	–	–	spreading	–
Mucor	6	spreading heavy sporulation	spreading no sporulation	distinct margins no sporulation	spreading no sporulation
Neurospora	1	spreading heavy sporulation	spreading heavy sporulation	spreading	inhibited
Paecilomyces	3	spreading	spreading	distinct margins	–
Penicillium	14	–		distinct margins	–
Rhizopus	5	spreading heavy sporulation	spreading no sporulation	inhibited	inhibited
Syncephalastrum	2	spreading heavy sporulation	spreading no sporulation	distinct margins	spreading
Candida	3	–	–	pinpoint	inhibited
Kloeckera	1	–	–	small colonies	portion inhibited
Saccharomyces	1	–	–	yeast phase	mold phase
Dimorphic yeast	1	yeast phase	yeast phase		inhibited
Unidentified yeast A	1	–	–		

aIndicates lack of notable advantage or disadvantage

Table 3. Decision table for choice of enumeration medium

Attributes	Rating[a] of media by desirable attributes			
	CCPDA	CCPDA + 2.5 D	DRBC	DG18
Easily prepared	1	1	1	1
Saves time during counting	3	2	1	2
Reduces analyst stress during counting	4	2	1	2
Increases counting precision by restricting colony size	4	2	1	2
Lessens contamination of lab with conidia	4	2	2	2
Aids selective isolation of a species from a mixed culture	4	2	1	2
Cultures on medium readily identified	1	2	3	2
Enumerates yeast when mold is absent	1	1	1	2
Enumerates yeast when heavy mold contamination is present	3	1	1	2
Enumerates xerophilic molds	2	2	2	1
Enumerates non-xerophilic Aspergilli and Penicillia	1	1	1	1
Inhibits aerial mycelia development	4	2	1	2

[a]Best performance is represented by the lowest value on scale of 1 to 4

We are grateful for the assistance of Richard K. Smith in the statistical analysis of the data.

References

BAGGERMAN, W. I. 1981 A modified rose bengal medium for the enumeration of yeasts and moulds from foods. European Journal of Applied Microbiology and Biotechnology 12, 242-247.
BARNETT, H. L. & HUNTER, B. B. 1972 Illustrated Genera of Imperfect Fungi Minneapolis: Burgess Publishing Co.
BURGE, H. P., SOLOMON, W. R. & BOISE, J. R. 1977 Comparative merits of eight popular media in aerometric studies of fungi. Journal of Allergy and Clinical Immunology 60, 199-203.
GRIFFIN, G. J., FORD, R. H. & GARREN, K. H. 1975 Relation of Aspergillus flavus colony growth on three selective media to recovery from naturally infested soil. Phytopathology 65, 704-707.

HENSON, O. E.   1981   Dichloran as an inhibitor of mold spreading in fungal
     plating media: effects on colony diameter and enumeration.   Applied and
     Environmental Microbiology 42, 656–660.
HENSON, O. E., HALL, P. A., ARENDS, R. E., ARNOLD, E. A. JR., KNECHT, R. M.,
     JOHNSON, C. A., PUSCH, D. J. & JOHNSON, M. G.   1982   Comparison of four
     media for the enumeration of fungi in dairy products – a collaborative
     study. Journal of Food Science 47, 930–932.
HOCKING, A. D. & PITT, J. I.   1980   Dichloran–glycerol medium for enumeration
     of xerophilic fungi from low–moisture foods. Applied and Environmental
     Microbiology 39, 488–492.
KING, A. D. JR., HOCKING, A. D. & PITT, J. I.   1979   Dichloran–rose bengal
     medium for enumeration and isolation of molds from foods. Applied and
     Environmental Microbiology 37, 959–964.
KOBURGER, J. A. 1976.   In Compendium of Methods for the Microbiological
     Examination of Foods ed. Speck, M. L.   Ch. 16, pp. 225–229.
     Washington, D.C.: American Public Health Association.
MOSSEL, D. A. A., VEGA, C. L. & PUT, H. M. C.   1975   Further studies on the
     suitability of various media containing antibacterial antibiotics for
     the enumeration of moulds in food and food environments. Journal of
     Applied Bacteriology 39, 15–22.
OTTOW, J. C. G.   1972   Rose bengal as a selective aid in the isolation of
     fungi and actinomycetes from natural sources. Mycologia 64, 304–315.
OVERCAST, W. W. & WEAKLEY, D. J.   1969   An aureomycin–rose bengal agar for
     enumeration of yeast and mold in cottage cheese. Journal of Milk and
     Food Technology 32, 442–445.
RAPER, K. B. & FENNELL, D. I.   1965   The Genus Aspergillus. Baltimore:
     Williams and Wilkins.
SOKAL, R. R. & ROHLF, F. J.   1973   Introduction to Biostatistics. San
     Francisco: W. H. Freeman and Company.

L. M. LENOVICH
J. L. WALTERS
D. M. REED

▶ A Collaborative Exercise Comparing Media for
Enumerating Fungi in Foods

   Although much work has been reported on the development of selective
media for enumerating molds and yeasts in specific foods, there are
relatively few reports comparing a range of selective media for a range of
foods. This communication summarizes the results from a collaborative
exercise involving nine laboratories who examined twenty–two samples of food
and feed using five different mycological media.

Materials and methods

   Each laboratory prepared two dilutions of each sample selected to give a
high and low count of fungi on the plates. Eight dried replicate plates
containing 15 ml of each of the five test media were surface–inoculated with
0.2 ml of the same dilutions of the sample and spread using sterile glass
rods. Counts of mold colonies and yeast colonies and identification of the
groups of species or molds present, were made after incubation at 25°C for 5
days and again after 7 days.

The agar media compared were RBC (Jarvis 1983) with chloramphenicol instead of chlortetracycline, RBC modified by Baggerman (1981), DRBC, OGY and Ohio Agricultural Experimental Station medium (OAES)(Kaufman et al. 1963).

Results and discussion

For each laboratory, the relative productivity of the media was determined by averaging the count on all foods with each medium and expressing this as a percentage of the mean count on all media. These percentage figures were then averaged for all laboratories. The relative productivity of the five media in terms of total colonies, molds and yeasts is given in Table 1. Most laboratories were in agreement that OAES was the least satisfactory medium. Of the other media, DRBC and RBC had a slight advantage over OGY and RBCC.

Upon completion of their tests, scientists in the nine laboratories were asked to complete a questionnaire in which they placed the five media in order of preference (1 = best; 5 = worst) on the basis of ease of counting, inhibition of spreading colonies, ability to distinguish between molds and yeasts and ease of identifying molds. The mean preference ratings for all laboratories on the basis of these four parameters are given in Table 2. OGY and RBC were generally considered to be the least satisfactory, mainly because of poor inhibition of spreading molds. The preference for OAES and, to a lesser extent, RBCC varied considerably. Some laboratories who regularly used OAES rated it highly, whereas others found it unsatisfactory. Overall, DRBC was the medium which gave the best rankings.

Six of the nine laboratories attempted to identify the mold colonies present on the plates. On average, OAES and OGY gave significantly lower numbers of identified groups or species than the other three media. The number of colonies unable to be identified was greatest with DRBC. Some laboratories found difficulty with identification on the media containing rose bengal, whereas others stated that the variety of colony color was helpful.

Conclusions

The results from this exercise indicate that none of the media tested is ideal for enumerating fungi in all foods. However, in terms of recovery and preference rankings, DRBC was slightly superior to the other media tested.

Table 1. Relative productivity of media in collaborative tests involving nine laboratories and 22 food samples

	Relative productivity (%)		
Medium	Total colonies	Molds	Yeasts
RBC	109	106	120
RBCC	99	100	88
DRBC	111	112	114
OGY	100	92	110
OAES	84	87	71

Table 2. Mean preference rankings of media on basis of ease of counting,
          inhibition of spreading colonies, ability to distinguish between
          yeasts and molds and identification of molds

Medium	Average preference ranking[a]
RBC	3.3
RBCC	3.1
DRBC	1.9
OGY	3.6
OAES	2.4

[a]1 = best, 5 = worst

References

BAGGERMAN, W. I. 1981 A modified rose bengal medium for the enumeration of
    yeasts and moulds from foods. European Journal of Applied Biotechnology
    12, 242-247.
JARVIS, B. 1983 Comparison of an improved rose bengal-chlortetracycline
    agar with other media for selective isolation and enumeration of molds
    and yeasts in food. Journal of Applied Bacteriology 36, 723-727.
KAUFMAN, D. D., WILLIAMS, L. E. & SUMMER, C. B. 1963 Effect of plating
    medium and incubation temperature on growth of fungi in soil-dilution
    plates. Canadian Journal of Microbiology 9, 741-751.

D. A. L. SEILER

A Comparison of DRBC, RBC and MEA Media for the Enumeration
of Molds and Yeasts in Pure Culture and in Foods

The main objective of this study was to compare the efficiency of DRBC
with RBC. Common molds and yeasts were examined in pure culture and a
selection of foods was analyzed on up to three occasions. Blakeslee's malt
extract agar (MEA) (pH 5.5) was used as a non-inhibitory control medium for
the pure cultures.

Materials and methods

Mold and yeast cultures were food isolates from the Leatherhead Food R A
culture collection. Food samples were purchased from local shops. Cheese
samples were incubated overnight at 25°C to encourage the yeast population.
Cultures and samples are listed in Table 1.

The diluent was Ringer/Tween (1/4 strength Ringer solution containing
0.05% polyethylene sorbitan monooleate (Tween 80) which was prepared in 90-ml
or 9-ml volumes and sterilized by autoclaving at 121°C for 15 min.

MEA was made according to the instructions of Pitt (1979). RBC and DRBC
were made in accordance with the manufacturer's (Oxoid) instructions. Plates
were dried overnight at 37°C in the dark.

Table 1. Counts obtained for molds, yeasts and food samples on three
occasions

Sample	Sample number	Log$_{10}$ colony count[a] on:		
		MEA	RBC	DRBC
Fungal cultures				
Penicillium expansum	1	6.998	7.085	7.012
	2	8.189	8.166	8.258
	3	8.810	8.365	8.631
Aspergillus flavus	1	6.903	6.922	6.889
	2	8.000	7.939	8.115
	3	8.114	8.149	8.311
Cladosporium herbarum	1	5.260	5.301	NG
	2	NG	5.114	NG
	3	6.380	6.061	6.186
Mucor plumbeus	1	UC	5.502	NG
	2	7.398	7.000	NG
	3	7.214	7.270	NG
Rhizopus stolonifer	1	UC	UC	NG
	2	UC	UC	6.580
	3	UC	UC	NG
Saccharomyces cerevisiae	1	6.772	6.666	NG
	2	7.691	7.491	7.392
	3	7.455	7.260	6.436
Metschnikowia pulcherrima	2	7.761	7.635	7.476
	3	7.703	7.515	7.555
Kluyveromyces lactis	1	7.203	6.966	NG
	2	7.612	7.223	6.782
	3	7.758	7.371	7.260
Rhodotorula glutinis	1	6.538	6.365	6.149
	2	7.489	7.072	7.370
	3	7.016	6.913	6.815
Candida pseudotropicalis	1	7.225	7.113	NG
	2	7.574	7.269	7.536
	3	7.617	7.550	6,772
Foods				
Mixed spice	1	–	3.996	3.355
	2	–	4.784	4.307
	3	–	4.374	5.457
Black pepper	1	–	3.675	3.133
	2	–	3.733	3.443
	3	–	3.477	3.382

(continued)

Table 1. Continued

Sample	Sample number	Log$_{10}$ colony count[a] on:		
		MEA	RBC	DRBC
Mixed corn	1	–	4.837	4.149
	2	–	4.117	4.070
	3	–	4.183	NG
Brie cheese	1	–	7.277	–
	2	–	6.748	6.484
	3	–	7.197	6.963
Cheshire cheese	1	–	5.353	4.922
	2	–	5.433	4.967
	3	–	7.544	7.556

[a]NG = No growth; UC = Uncountable; – = Not tested

Cultures were prepared by suspending a loopful of cells from a fresh slant of the organism in 9 ml of diluent. Food samples (10 g) were stomached for 2 min in 90 ml of diluent. Serial decimal dilutions were prepared in 9 ml of diluent, each tube being shaken on a vortex mixer for 5 sec.

Spread plates (0.1 ml) were made in duplicate for each medium, using an appropriate range of dilutions. Plates were incubated upright in air at 25°C for 5 days. Counts were made by weighted means of countable plates (less than 100 colonies per plate for molds and less than 300 colonies for yeasts).

Table 2. Means and standard deviations for colony counts

Sample		Log$_{10}$ mean counts (standard deviations)		
		MEA	RBC	DRBC
Molds	mean (SD)	7.327(1.027)	6.906(1.167)	7.498(0.932)
	N[a]	10	12	8
Yeasts	mean (SD)	7.387(0.384)	7.172(0.358)	7.053(0.481)
	N	14	14	11
Foods	mean (SD)	Not tested	5.115(1.421)	4.784(1.458)
	N		15	13

[a]Number of samples tested

Results

The counts obtained are shown in Table 1 and the means and standard deviations are summarized in Table 2. Analysis of the performance of the three media, by paired $\underline{t}$ test, is given in Table 3. A comparison has been made in Table 4 of data from food samples which contained both yeasts and molds.

Conclusions

1. Higher counts of molds in pure culture were obtained on DRBC than on RBC (significant at 90% but not 95%).

Table 3. Analysis of performance of the three media by paired $\underline{t}$ test

Sample	MEA vs. RBC	MEA vs. DRBC	RBC vs. DRBC
Molds	Not significant	Not significant	Significant (6 DF[a]) (0.05 < P < 0.1)
Yeasts	Significant (13 DF) (0.001 ≥ P)	Significant (10 DF) (0.001 < P ≤ 0.1)	Not significant
Foods	Not tested	Not tested	Significant (12 DF) (0.05) < P ≤ 0.1)

[a]DF = degrees of freedom

Table 4. Comparison of RBC with DRBC for food samples containing both yeasts and molds

Food	Sample	Log_{10} colony counts on RBC		Log_{10} colony counts on DRBC	
		Yeasts	Molds	Yeasts	Molds
Black pepper	1	3.243	3.465	2.845	3.330
	2	2.477	3.436	2.176	3.357
Mixed corn	1	4.761	3.989	4.038	3.484
	2	2.833	4.052	2.477	4.060
Brie cheese	1	6.498	6.389	6.079	6.389
	2	7.092	6.535	6.061	6.910
Mean		4.484	4.644	3.946	4.588
SD		1.961	1.432	1.763	1.627

Analysis by paired t test
Yeasts: significant (5 DF) (0.001 < P ≤ 0.01)
Molds: not significant
Yeasts + molds: significant (11 DF) (0.02 < P ≤ 0.05)

2. Yeasts in pure culture were strongly inhibited on both media which
    contained rose bengal.
3. Total counts on foods were lower on DRBC than on RBC (significant at
    90% but not 95%).
4. In food samples containing both yeasts and molds, yeasts were strongly
    inhibited by DRBC (with respect to RBC) but molds were unaffected.
5. <u>Mucor plumbeus</u> (three of three tests) and <u>Rhizopus stolonifer</u> (two of
    three tests) were not capable of growing on DRBC.

<u>References</u>

PITT, J. I. 1979 <u>The Genus Penicillium and its Teleomorphic States
    Eupenicillium and Talaromyces</u>. London: Academic Press.

                                                    A. P. WILLIAMS

▶ A Comparison of DRBC, OGY and RBC Media for the
  Enumeration of Yeasts and Molds in Foods

    There are few data available on the effectiveness of DRBC for recovering
yeasts from foods. There was a strong indication, however, from the data
presented in the preceding paper that DRBC is inhibitory to yeasts, both in
pure culture and in foods. It was not, however, significantly more
inhibitory than RBC against yeasts in pure culture. In the following study,
DRBC has been compared with OGY and RBC for the recovery of yeasts and molds
from twenty-nine food samples.

<u>Materials and methods</u>

    Food samples were purchased from local shops. Cheese samples were
incubated overnight at 25°C to encourage the yeast population. The fruit
juice and yogurt were spoiled. The foods analyzed are listed in Tables 1
and 2. The diluent was 0.1% peptone (Bactopeptone) and other methods were
as in the preceding paper except that DRBC and OGY were both made from
individual ingredients, according to instructions from A. D. King (U.S.D.A.,
Albany, CA).

<u>Results and discussion</u>

    The food samples contained predominantly either yeasts or molds. Data
for each category have therefore been analyzed separately. Counts on foods
containing yeasts are given in Table 1 and counts on foods containing molds
are given in Table 2. Table 3 contains mean counts and standard deviations.

<u>Conclusions</u>

1. Analysis of performance of the three test media by the paired $\underline{t}$ test
    indicates that there was no significant difference between counts of
    yeast r molds on any of the three media tested. Likewise, there was no
    significant difference between total yeast and mold counts between media.

Table 1. A comparison of DRBC, OGY and RBC media for enumeration of yeasts in foods

Food	Sample number	Log$_{10}$ colony count[a] on:		
		RBC	DRBC	OGY
Brie cheese	1	7.179	7.312	6.512
	2	4.398	4.602	4.544
	3	5.477	5.477	5.477
Cheshire cheese	1	6.813	6.792	6.869
	2	5.301	5.362	5.290
	3	5.799	5.712	5.653
	4	5.000	5.190	4.954
	5	4.544	4.778	4.544
	6	4.845	5.161	5.000
Yogurt (natural)	1	7.074	7.237	7.201
	2	6.836	6.839	6.980
	3	6.924	6.568	6.799
	4	6.243	6.000	6.580
	5	5.903	5.301	6.204
	6	5.176	5.699	6.021
Orange juice	1	8.045	UC	7.952
	2	7.826	NG	7.833
	3	7.574	UC	7.778
	4	7.477	7.580	7.371

[a]UC = uncountable; NG = no growth

Table 2. A comparison of DRBC, OGY and RBC media for the enumeration of molds in foods

Food	Sample number	Log$_{10}$ colony count on:		
		RBC	DRBC	OGY
Black pepper	1	3.462	3.470	3.398
	2	3.699	3.653	3.544
	3	2.903	2.813	2.845
	4	3.477	3.813	3.813
	5	3.903	3.813	3.740
	6	4.061	3.929	4.176
Mixed corn	1	3.000	3.362	3.362
	2	3.580	3.618	3.813
	3	4.021	4.000	4.161
	4	4.025	3.884	3.556

Table 3. Means and standard deviations for colony counts of yeasts and molds

Sample		Log$_{10}$ mean counts[a] (standard deviations) on:		
		RBC	DRBC	OGY
Yeasts	Mean (SD)	6.233 (1.166)	5.976 (0.951)	6.293 (1.112)
	N	19	16	19
Molds	Mean (SD)	3.613 (0.415)	3.636 (0.353)	3.641 (0.396)
	N	10	10	10
Total Y + M	Mean (SD)	5.330 (1.592)	5.076 (1.391)	5.378 (1.578)
	N	29	26	29

[a] N = number of samples tested

2. In the preceding paper, DRBC was shown to inhibit yeasts. However, neither the samples nor the methods are exactly comparable with this experiment. In the previous experiment, DRBC was supplied in powder form whereas in this experiment it was made from individual ingredients. It may, therefore, be minor differences in ingredients (e.g., dichloran) which account for the discrepancies. This should be further investigated.

A. P. WILLIAMS

A Comparison of DG18 with RBC Medium for the Enumeration of Molds in Pure Culture and Molds and Yeasts in Foods

This study is in parallel with that reported in a preceding paper entitled "A comparison of DRBC, RBC and MEA media for the enumeration of molds and yeasts in pure culture and in foods." RBC is compared with DG18.

Materials and methods

Materials and methods are the same as described in "A comparison of DRBC with RBC and MEA for the enumeration of molds and yeasts in pure culture and foods," except that the two media evaluated here were RBC and DG18. RBC (Oxoid CM549) and DG18 (supplied by Oxoid) were made according to the manufacturer's instructions.

Results and discussion

Comparisons of counts are shown for molds in pure culture (Table 1), molds from foods (Table 2) and yeasts from foods (Table 3).

Table 1. A comparison of DG18 with RBC medium for enumeration of molds in pure culture

| Mold[a] | Sample number | Log$_{10}$ colony count[b] on: | |
		RBC	DG18
Penicillium expansum	1	7.085	7.082
	2	8.166	8.291
	3	8.365	8.466
Aspergillus flavus	1	6.922	7.056
	2	7.939	8.105
	3	8.149	8.163
Cladosporium herbarum	1	5.301	5.602
	2	5.114	6.850
	3	6.061	6.121
Mucor plumbeus	1	5.502	6.088
	2	7.000	6.936
	3	7.270	7.374
Mean		6.906	7.180
S.D.		1.167	0.945

[a]Colony counts for Rhizopus stolonifer on DG18 were 6.268, 7.259 and 7.266. This mold overgrew plates of RBC and could not be counted

[b]Analysis by paired t test: Significant (11 DF) (0.05 < P < 0.1)

Conclusions

1. Counts of molds in pure culture were significantly higher (90% but not 95% level of significance) on DG18 than on RBC, although the mean difference was only 0.27 log cycles.
2. DG18 was the only medium used in the present study which permitted counts of Rhizopus stolonifer to be made.
3. There was no significant difference in count between DG18 and RBC for either molds or yeasts in foods.

Table 2. A comparison of DG18 with RBC medium for enumeration of molds in foods

| Food sample | Sample number | Log$_{10}$ colony count[a] on: | |
		RBC	DG18
Mixed spice	1	3.996	4.827
	2	4.784	4.711
	3	4.374	3.929
Black pepper	1	3.675	3.666
	2	3.465	3.412
	3	3.436	3.450
Mixed corn	1	3.989	3.457
	2	4.052	3.464
	3	3.320	3.758
Brie cheese	1	6.389	6.431
	2	6.535	7.020
Mean		4.365	4.375
SD		1.122	1.268

[a]Analysis by paired t test: not significant

Table 3. A comparison of DG18 with RBC medium for enumeration of yeasts in foods

| Food sample | Sample number | Log$_{10}$ colony count[a] on: | |
		RBC	DG18
Black pepper	1	3.243	3.477
	2	2.477	2.959
Mixed corn	1	4.761	4.244
	2	2.833	3.418
	3	4.119	4.020
Brie cheese	1	6.498	6.568
	2	7.092	7.269
Cheshire cheese	1	5.353	5.297
	2	5.433	5.593
	3	7.544	7.591
Mean		4.935	5.0454
SD		1.775	1.675

[a]Analysis by paired t test: not significant

A. P. WILLIAMS

Comparison of a General Purpose Mycological Medium with a Selective Mycological Medium for Mold Enumeration

A general purpose fungal enumeration medium should support good recovery of mold propagules, inhibit rapidly spreading mold colonies such as <u>Rhizopus</u> and permit some degree of identification of colonies by allowing sporulation (Jarvis et al. 1983). Acidified PDA (APDA), which is used as a standard method (Speck 1976) and has a low pH (ca 3.5-4.5) to inhibit bacteria, has the disadvantage of also inhibiting certain fungi which are not strongly acid tolerant (Jarvis 1973, 1978). This can be avoided by using specific chemicals (e.g., antibiotics) which inhibit bacteria but not yeasts or molds (Koburger & Farhat 1975; Jarvis 1978). PDA plus antibiotics or other general purpose mycological media plus antibiotics can be used. Another problem associated with mold counts in foods, however, is the presence of rapidly growing and spreading <u>Rhizopus</u> and <u>Mucor</u> spp. which overgrow the culture medium and prevent counting or isolation of slower growing fungi (King et al. 1979). This problem has been addressed in a number of studies reviewed by Jarvis (1978). A study by King et al. (1979) reported the development of a selective medium (DRBC) that inhibited growth of bacteria and restricted the spread of rapidly growing fungi. This study was done to compare DRBC with PDA + tetracycline (TPDA) for mold enumeration.

Materials and methods

<u>Culture media</u>. PDA (Difco) + 40 μg/ml tetracycline (TPDA) and DRBC agar were used in this study.

<u>Sample preparation</u>. Samples were prepared for plating by either blending or stomaching in phosphate buffer (Speck 1976). Eleven grams of sample were added to 99 ml of buffer and mixed by either method for 3 min; serial dilutions were made in phosphate buffer and samples were plated within 1 min of mixing (Jarvis et al. 1983).

<u>Plate counts</u>. Mold counts were done using a spread-plate technique on TPDA and DRBC media. A 0.1 ml aliquot was used. If countable plates were not obtained, counts were repeated using a 1.0 ml aliquot. Each dilution was plated in duplicate and analyses were repeated three times on the same sample. Thus, each count represents the mean of six determinations. Plates were incubated in an upright position at 25°C for 5 days and counted for mold colonies.

<u>Samples</u>. Seven commercial food samples (dry gravy mix, chopped walnuts, dry split peas, white flour, yellow corn meal, pecan pieces and "organic" corn meal) were analyzed. In addition, white wheat flour, inoculated with three levels (10^2, 10^4 and 10^6/g) of a mixture of mold spores from nine different organisms (<u>Aspergillus flavus</u>, <u>A. niger</u>, <u>A. ochraceus</u>, <u>Penicillium martensii</u>, <u>P. roqueforti</u>, <u>P. viridicatum</u>, <u>Fusarium graminearum</u>, a <u>Cladosporium</u> sp. and an <u>Alternaria</u> sp.) was used. Inoculum for the flour was produced by growing each mold on bread cubes in quart Mason jars according to the method of Sansing & Ciegler (1973). When growth and sporulation were complete, the cultures were allowed to dry and propagules were harvested by adding 100-200 g of dry flour to each jar and dry blending the fungal mixtures using an electric blender (Oster). The dry fungal preparations were then combined and thoroughly mixed for use as inocula for flour samples. The total number of mold propagules in the mixture was determined using the surface plate count technique. Appropriate amounts of the mixture were used to inoculate flour samples to obtain approximate levels of 10^2, 10^4 and 10^6 propagules/g.

Statistical design and analysis. There were two experimental factors in
this study: 1. method for preparation of samples (blending vs. stomaching);
and 2. culture media (TPDA vs. DRBC). A split plot design was used. The
method for preparation of inoculum was considered as the main plot. Every
food sample was plated in duplicate, and analyses were done in triplicate.
Mean log_{10} values of counts and differences obtained in these counts for
each food sample were analyzed by analysis of variance (Steel & Torrie
1980). The α-value (level of significance) was chosen to be 0.05. The
standard deviations of the mean differences were also determined.

Results and discussion

Counts obtained on TPDA and DRBC media were very similar (Table 1), and
there did not appear to be a clear trend of higher or lower counts on one or
the other medium. In some samples TPDA gave higher counts and in some
samples DRBC gave higher counts. Overgrowth of TPDA plates by rapidly
growing mucoraceous fungi was definitely a problem with some samples. This
made counting difficult and isolation impossible. This also complicated
preliminary identification of mold colonies. However, with the samples
tested not all dilutions were overgrown, so it was usually possible to
isolate colonies from other plates (particularly higher dilutions) though
these may not necessarily contain all of the types appearing on plates from
lower dilutions. DRBC medium had the disadvantage of making it difficult to
recognize certain genera of organisms because of the rose bengal. Again,
blending and stomaching gave similar counts.

The yeast and mold counts obtained from TPDA and DRBC were not significantly different in most food samples. Counts on DRBC from stomached split
peas were significantly lower than from the other counts. Counts from TPDA
were significantly higher than counts obtained from DRBC for analysis of
both blended and stomached bleached flour but not for inoculated flour.

Table 1. Mean log_{10} yeast and mold counts, and difference obtained with a
general purpose medium (PDA + tetracycline) and a selective medium
(DRBC) using blending and stomaching with a surface plate count
technique

Commodity	Blending			Stomaching		
	TPDA	DRBC	Difference	TPDA	DRBC	Difference
Dry gravy mix	2.52	2.50	0.02	2.22	2.23	-0.01
Chopped walnuts	3.08	3.31	-0.23	3.32	3.43	-0.11
Dry split peas	1.32	1.34	-0.02	1.34	1.23	0.11
Bleached flour	2.46	1.84	0.62	2.42	2.17	0.25
Yellow Corn meal	2.63	2.48	0.15	2.45	2.60	-0.15
Pecan pieces	2.10	1.90	0.20	1.94	1.84	0.10
Organic corn meal	5.42	5.41	0.01	5.39	5.36	0.03
Spiked flour (10^6)	5.19	5.09	0.10	5.34	5.33	0.01
Spiked flour (10^4)	3.13	3.16	-0.03	3.34	3.27	0.07
Spiked flour (10^2)	2.27	2.20	0.07	2.36	2.62	-0.26
Mean of difference:			-0.035			0.004
SD of difference:			0.237			0.146

Conclusion

While for the most part TPDA and DRBC gave similar counts on all samples, it was easier to recognize genera, in terms of colony types, on TPDA. However, TPDA had the disadvantage of being rapidly overgrown by spreading mucoraceous fungi.

Recommendation

If samples are being tested which do not contain rapidly growing and spreading fungi, PDA + antibiotics would be the preferable medium. With samples that are highly contaminated with Rhizopus or Mucor, DRBC would be the preferred medium.

References

JARVIS, B. 1973 Comparison of an improved rose bengal chlortetracycline agar with other media for the selective isolation and enumeration of moulds and yeasts in foods. Journal of Applied Bacteriology 36, 723-727.

JARVIS, B. 1978 Methods for detecting fungi in foods and beverages. In Food and Beverage Mycology ed. Beuchat, L. R. pp. 471-504. Westport, CT: AVI Publ., Inc.

JARVIS, B., SEILER, D. A. L., OULD, A. J. L. & WILLIAMS, A. P. 1983 Observations on the enumeration of moulds in food and feedstuffs. Journal of Applied Bacteriology 55, 325-336.

KING, A. DOUGLAS, JR., HOCKING, A. D. & PITT, J. I. 1979 Dichloran-rose bengal medium for enumeration and isolation of molds from foods. Applied and Environmental Microbiology 37, 959-964.

KOBURGER, J. A. & FARHAT, B. Y. 1975 Fungi in foods VI. A comparison of media to enumerate yeasts and molds. Journal of Milk and Food Technology 38, 466-468.

SANSING, G. A. & CIEGLER, A. 1973 Mass propagation of conidia from several Aspergillus and Penicillium species. Applied Microbiology 26, 830-831.

SPECK, M. L. (ed). 1976 Compendium of Methods for the Microbiological Examination of Foods. Washington, D. C.: American Public Health Association.

STEEL, R. G. & TORRIE, J. H. 1980 Principles and Procedures of Statistics, a Biometical Approach. 2nd edn. p. 235. New York: McGraw-Hill Book Co.

J. W. HASTINGS
W. Y. J. TSAI
L. B. BULLERMAN

Published as Paper No. 7611, Journal Series, Agricultural Research Division, Lincoln, NE. Research reported was conducted under Project 16-042.

Comparison of DRBC Medium with PDA Containing Antibiotics for Enumerating Fungi in Dried Food Ingredients

The U. S. Food and Drug Administration (1978) recommends that molds and yeasts be enumerated in foods by using PDA that is either acidified to pH 3.5 with tartaric acid or supplemented with chloramphenicol and chlortetracycline. Generally the antibiotic-supplemented PDA enumerates more fungi from foods than the acidified PDA because it allows for the recovery of stressed cells and fungi that grow slowly at acid pH (Beuchat 1979; Koburger 1970, 1971, 1973; Koburger & Rodgers 1978; Mislivec & Bruce 1976). Antibiotic-supplemented media does not prevent the spreading growth of some fungi.

Smith & Dawson (1944) used rose bengal to reduce the spreading of fungi recovered from soil. Since that time several media containing rose bengal and antibiotics have been evaluated for enumeration of molds and yeasts in foods (Baggerman 1981; Jarvis 1973; Mossel et al. 1975). Dichloran (2,6-dichloro-4-nitroaniline) has been used alone or in combination with either rose bengal and/or antibiotics to prevent the spreading of mucoraceous molds (Henson 1981; King et al. 1979).

It is important to enumerate fungi in dried ingredients used in the food industry to ensure that specifications are maintained, product quality is high and ingredients do not add substantial numbers of fungi to manufactured products. This research was done to compare DRBC with PDA containing chloramphenicol and chlortetracycline (CCPDA) for enumerating fungi in dried ingredients supplied by a local food manufacturer from their normal processing operations.

Materials and methods

Culture media and food samples. DRBC was obtained form Oxoid Ltd. PDA (Difco) was supplemented with 40 µg/ml each of chloramphenicol and chlortetracycline-HCl (CCPDA) (U. S. Food and Drug Administration 1978). Dried food ingredients used in the manufacture of dried and frozen foods were supplied by a local food company from their processing lines over a 6-week period. Fourteen separate samples each of cocoa, onion powder, dried onion and chicken spice were randomly sampled from the processing lines and placed in plastic bags for transport to the laboratory. These ingredients generally encompassed different lots from the same supplier.

Preparation of samples. Subsamples (25 g) of each product were stomached with 225 ml of sterile peptone (0.1%, w/v) diluent for 2 min. Additional dilutions were prepared with the same diluent as needed. Homogenized samples were plated in duplicate on CCPDA and on DRBC. Plates were incubated at 25°C for 5 days, colonies were counted and statistical analyses for degree of difference were calculated (Nie et al. 1975).

Identification of molds. Molds from each product were identified based on cultural and morphological characteristics (Talbot 1971; Barnett & Hunter 1972; Malloch 1981). Johnson's slide culture technique using MEA or PDA was prepared according to Harrigan & McCance (1976). All slides were incubated at 25°C, microscopically examined twice daily and discarded once molds were identified.

Results and discussion

Range distributions of mold counts in dried food ingredients are listed in Table 1. The mold counts were fairly low, less than 10^3/g for all products analyzed. CCPDA and DRBC were similar for enumeration of mold from cocoa, dried onion and chicken spice; however, DRBC recovered a one log higher count than CCPDA for 5 of 14 onion powder samples (Table 1). This difference may be critical if there are stringent specifications established for this type of spice. Mossel & Shennan (1976) suggested that dried foods should not have mold counts greater than 10^2/g. Most of the samples analyzed in this project would comply with these recommendations. Fungal counts reported in the literature for onion products, such as dried onions, concentrates, extracts and powders ranged from 10 to 5×10^3/g (Firstenberg et al. 1974; Heath 1983). Overall there were significant differences noted for the analysis of variance between products, media and batch, but not for sample. Since the products analyzed were diverse and involved various processing methods, these differences would be expected when comparing all dry products.

The company specifications for these dry ingredients were initial fungal counts below 10^3/g. It is not unusual to have mold and yeast counts up to 10^6/g for some spices (Powers et al. 1975; Schwab et al. 1982). In this study, when all samples except cocoa were directly plated, there was some inhibition of mold growth. Some spices and herbs can have antimycotic activity toward fungi. Garlic, allspice, cinnamon, cloves, oregano and mustard may inhibit mycelial growth of Aspergillus and Penicillium spp. (Llewellyn et al. 1981; Azzouz & Bullerman 1982).

In Table 2, the types and percentages of molds isolated from the four products are shown. Aspergillus spp. represented over 70% of the isolates for the spices and dried onion, but Penicillium spp. predominated in cocoa. Species of Rhizopus, Geotrichum and Cladosporium were isolated less frequently from these products than were Aspergilli and Penicillia.

Table 1. Range of mold counts for dry food ingredients plated on CCPDA and DRBC media

| Food | Media | Range ($\log_{10}$/g) | | |
		< 1.00	1.00–2.00	2.00–3.00
Cocoa	CCPDA	7/14[a]	5/14	2/14
	DRBC	9/14	3/14	2/14
Onion powder	CCPDA	5/14	7/14	2/14
	DRBC	5/14	2/14	7/14
Dried onion	CCPDA	1/14	10/14	3/14
	DRBC	2/14	8/14	4/14
Chicken spice	CCPDA	1/14	13/14	
	DRBC		14/14	

[a]Number of samples/total number of samples. Overall mean ($\log_{10}$) for CCPDA was 1.53 and for DRBC was 1.78 with standard deviations of 1.62 and 1.92, respectively, at 0.95 confidence interval

98

Table 2. Types and percentages of molds isolated from dried food ingredients
on CCPDA and DRBC media

		Molds isolated (%)			
Mold	Media	Cocoa	Onion powder	Dried onion	Chicken spice
Alternaria	CCPDA	0	0	0	3.1
	DRBC	1.4	0	1.4	0
Aspergillus	CCPDA	17.1	86.1	77.9	70.4
	DRBC	13.9	81.4	76.3	77.4
Circinella	CCPDA	0	0	0	0
	DRBC	15.3	0	0	0
Cladosporium	CCPDA	7.1	0	0	0
	DRBC	6.9	0	0	0
Fusarium	CCPDA	0	0	0	3.1
	DRBC	1.4	0	0.9	1.3
Geotrichum	CCPDA	2.9	0	12.4	4.7
	DRBC	9.7	0.3	2.8	1.3
Penicillium	CCPDA	68.6	0	0	10.9
	DRBC	51.4	1.0	6.0	12.0
Rhizopus	CCPDA	4.3	13.9	9.7	7.8
	DRBC	0	16.9	12.6	5.3
Tricoderma	CCPDA	0	0	0	0
	DRBC	0	0.3	0	2.7

Murphy (1973) reported that species of Penicillium, Aspergillus and
occasionally Mucor were isolated from dried vegetables. No yeasts were
isolated from any of these dry ingredients on either medium.

DRBC prevented Rhizopus spp. from spreading, unlike that noted for
CCPDA. It was easier to enumerate molds on DRBC because the colonies
remained small and did not merge as they did on some CCPDA plates.

Conclusions and recommendations

Since a one log higher count of molds was observed for one spice
analyzed on DRBC than on CCPDA, DRBC is recommended for the enumeration of
fungi in dry food ingredients. DRBC also retards spreading of colonies,
thus facilitating enumeration.

References

AZZOUZ, M. A. & BULLERMAN, L. B. 1982 Comparative antimycotic effects of
    selected herbs, spices, plant components and commercial antifungal
    agents. Journal of Food Protection 45, 1298-1301.
BAGGERMAN, W. I. 1981 A modified rose bengal medium for the enumeration of
    yeasts and moulds from foods. European Journal of Applied Microbiology
    and Biotechnology 121, 242-247.
BARNETT, H. L. & HUNTER, B. B. 1972 Illustrated Genera of Imperfect Fungi.
    3rd. edn. Minneapolis: Burgess Publishing Company.
BEUCHAT. L. R. 1979 Comparison of acidified and antibiotic-supplemented
    potato dextrose agar from three manufacturers for its capacity to
    recover fungi from foods. Journal of Food Protection 42, 427-428.

FIRSTENBERG, R., MANNHEIM, C. H. & COHEN, A. 1974 Microbial quality of dehydrated onions. Journal of Food Science 39, 685-688.

HARRINGAN, W. F. & McCANCE, M. E. 1976 Laboratory Methods in Food and Dairy Microbiology. New York: Academic Press.

HEATH, H. 1983 The microbiology of onion products. Food, Flavorings, Ingredients, Packaging and Processing 5(1), 22-24, 27.

HENSON, O. E. 1981 Dichloran as an inhibitor of mold spreading in fungal plating media: Effects on colony diameter and enumeration. Applied and Environmental Microbiology 42, 656-660.

JARVIS, B. 1973 Comparison of an improved rose bengal-chlortetracycline agar with other media for the selective isolation and enumeration of moulds and yeasts in foods. Journal of Applied Bacteriology 36, 723-727.

KING, A. D., JR., HOCKING, A. D. & PITT, J. I. 1979 Dichloran-rose bengal medium for enumeration and isolation of molds from foods. Applied and Environmental Microbiology 37, 959-964.

KOBURGER, J. A. 1970 Fungi in foods. I. Effect of inhibitor and incubation temperature on enumeration. Journal of Milk and Food Technology 33, 433-434.

KOBURGER, J. A. 1971 Fungi in foods. II. Some observations of acidulants used to adjust media pH for yeasts and mold counts. Journal of Milk and Food Technology 34, 475-477.

KOBURGER, J. A. 1973 Fungi in foods. V. Response of natural populations to incubation temperatures between 12 and 32 C. Journal of Milk and Food Technology 36, 434-435.

KOBURGER, J. A. & ROGERS, M. F. 1978 Single or multiple antibiotic-amended media to enumerate yeasts and molds. Journal of Food Protection 41, 367-369.

LLEWELLYN, G. C., BURKETT, M. L. & EADIE, T. 1981. Potential mold growth, aflatoxin production and antimycotic activity of selected natural spices and herbs. Journal of Association of Official Analytical Chemists 64, 955-960.

MALLOCH, D. 1981 Moulds. Their Isolation, Cultivation and Identification. Toronto: University of Toronto Press.

MISLIVEC, P. B. & BRUCE, V. R. 1976 Comparison of antibiotic-amended potato dextrose agar and acidified potato dextrose agar as growth substrates for fungi. Journal of Association of Official Analytical Chemists 59, 720-721.

MOSSEL, D. A. A. & SHENNAN, J. L. 1976 Microorganisms in dried foods: Their significance, limitation and enumeration. Journal of Food Technology 11, 205-220.

MOSSEL, D. A. A., VEGA, C. L. & PUT, H. M. C. 1975 Further studies on the suitability of various media containing antibacterial antibiotics for the enumeration of moulds in food and food environments. Journal of Applied Bacteriology 39, 15-22.

MURPHY, R. P. 1973 Microbiological contamination of dried vegetables. Process Biochemistry 8(10), 17-19.

NIE, N. H., HULL, C. H., JENKINS, J. G., STEINBREENER, K. & BENT, D. H. 1975. Statistical Package for Social Sciences. 2nd edn. New York: McGraw-Hill Book Co.

POWERS, E. M., LAWYER, R. & MASUOKA, Y. 1975 Microbiology of processed spices. Journal of Milk and Food Technology 38, 683-687.

SCHWAB, A. H., HARPESTAD, A. D., SWARTZENTRUBER, A., LANIER, J. M., WENTZ, B., A., DURAN, A. P., BARNARD, R. J. & READ, R. B., JR. 1982 Microbiological quality of some spices and herbs in retail markets. Applied and Environmental Microbiology 44, 627-630.

SMITH, N. R. & DAWSON, V. T. 1944. The bacteriostatic action of rose bengal in media used for plate counts of soil fungi. Soil Science 58, 467-471.

TALBOT, P. H. B. 1971 Principles of Fungal Taxonomy. New York: St.
    Martin's Press.
U. S. FOOD AND DRUG ADMINISTRATION. 1978 Bacteriological Analytical Manual
    for Foods. Washington, D. C.: Food and Drug Administration, U. S.
    Department of Health and Human Services.

M. A. COUSIN
H. H. LIN

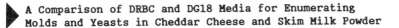

A Comparison of DRBC and DG18 Media for Enumerating Molds and Yeasts in Cheddar Cheese and Skim Milk Powder

DRBC medium was developed for the enumeration of molds and has the
advantage of restricting colony size of spreading molds, especially Rhizopus
and Mucor spp. DRBC medium is also satisfactory for the enumeration of
yeasts from foods. The other medium studied, DG18, was formulated for
enumerating xerophilic fungi from low-moisture foods.

Both media were used in this study for analysis of Cheddar cheese
showing mold spoilage on its surface, and skim milk powder, samples of which
were taken from 'sweepings' following a blow out from the milk drier.

Materials and methods

Media. DRBC and DG18 media were supplied by Oxoid and prepared
according to instructions. The glycerol added to DG18 was BDH Analar grade.

Diluents. Peptone (0.1%, Oxoid L37) in distilled water and 1/4 strength
Ringer solution containing 0.5% (v/v) polyoxyethylene sorbitan monolaurate
(Tween 20, Sigma) were used to prepare homogenates of both cheese and powder.

Examination of samples. Samples (10 g) were stomached in 90 ml of
diluent and 0.1-ml aliquots of the appropriate dilutions were spread over
the medium surface. Duplicate plates were made for each dilution. The
plates were incubated upright at 25°C for 5 days and mold and yeast colonies
were counted.

Statistical analysis. The hypothesis was tested, using Student's 't',
that the mean difference between colony counts, after logarithmic transformation on test media, was not significantly different from zero.

Results and discussion

Data from the enumeration of mold propagules and yeasts are summarized
in Tables 1 and 2, respectively. The mean difference in the propagule counts
on cheese and milk powder did not differ statistically from zero ($P > 0.05$).
Yeasts were detected in only twelve cheese and five milk powder samples.
For both foods, the colony counts of yeasts on DG18 were greater than those
on DRBC. The mean difference in log count differed significantly from zero
($P \leq 0.01$) for the cheese samples but not for the powders. However, had a
greater number of powder samples been contaminated by yeasts, it is probable

Table 1. Comparison of media for enumeration of mold propagules

Sample type	No. of tests	Mean and (range) of log10 propagule count on:		Mean and (range) of difference in log count (DG18-DRBC)	S. D. difference	No. of tests where:	
		DG18	DRBC			DG18 >DRBC	DG18 <DRBC
Cheese	20	6.57 (4.9–8.3)	6.60 (5.0–8.6)	+ 0.047 (−0.46 − +1.26)	0.465	8	10
Milk powder	16	2.65 (2.0–3.9)	2.66 (2.0–3.9)	+ 0.142 (−0.79 − +1.85)	0.613	10	5

Table 2. Comparison of media for enumeration of yeasts

Sample type	No. of tests	Mean and (range) of log10 propagule count on:		Mean and (range) of difference in log count (DG18-DRBC)	S. D. difference	No. of tests where:	
		DG18	DRBC			DG18 >DRBC	DG18 <DRBC
Cheese	12	6.53 (4.85–8.19)	6.56 (4.45–8.08)	0.336[a] (−0.18 − +0.95)	0.300	11	1
Milk powder	5	3.76 (3.03–4.64)	3.36 (2.39–3.85)	0.395 (−0.02 − +0.89)	0.441	4	1

[a] $p < 0.01$

that DG18 counts would have been significantly greater than counts on DRBC. Subjective assessment of mold colonies indicated marked differences in the nature of propagules cultured on the two media. Unfortunately, time constraints have prevented detailed analysis and identification of the propagules.

B. JARVIS
N. SHAPTON

▶ Comparative Study of Media for the Detection and
Enumeration of Molds and Yeasts

This study was undertaken to evaluate several mycological media for optimum recovery and enumeration of molds and yeasts from a variety of food samples. The overall objective was to obtain data which could be used to establish standard procedures for mycological analysis of foods.

Materials and methods

Media. Five media were evaluated in this study. Acidified PDA (APDA) and Sabouraud dextrose agar (SAB) were obtained from Difco Laboratories (Detroit, MI), and prepared and sterilized according to the directions provided by the manufacturer. The other three media were DRBC, OGY and DG18. DRBC and DG18 media were provided as manufacturer's samples by Oxoid Ltd.

Samples and cultural conditions. Samples of flour, wheat, bran, spice, breading mix, pasta and oatmeal were analyzed for mold and yeast populations. Samples (11 g) were homogenized with 99 ml of 0.1% peptone in a laboratory blender. One milliliter quantities of each sample at the 10^{-1} and 10^{-2} dilutions were plated in duplicate using both the spread- and pour-plate technique with the specified media. Data reported here are from spread plates. All plates were incubated in an upright position at 25°C for 5 days. Colonies were counted manually.

Results

Mold recovery and media evaluation. Table 1 lists the mean mold count data for the five test media used for mold recovery from the ten food samples. In terms of overall recovery of the mold flora from the food samples, DRBC and OGY media were most suitable.

The first and second medium preferences for each sample tested based on the mean mold count data are listed in Table 2. Although no clear pattern of preference is noted, it is obvious that SAB is clearly less suitable for mold recovery under these conditions ($P \leq 0.0003$). This is also evident in the overall mold recovery data. It was clear that mold colony resolution and counting were significantly easier with the DRBC medium. There was no attempt made to identify recovered molds.

Yeast recovery and media evaluation. Table 3 exhibits the mean yeast count data for three test media used for yeast recovery from the ten food samples. SAB and OGY were also evaluated in this phase of the study, but under the 5-day incubation conditions, colonies were not countable due to

Table 1. Mold recovery by test media

| | Mean ($\log_{10}$) mold count on various media | | | | | | |
Sample	APDA	SAB	DRBC	DG18	OGY	$\bar{x}$ (all)	SD (all)
Flour	2.778	2.756	2.954	2.820	2.881	2.838	0.081
Wheat	3.041	2.602	2.778	3.041	3.041	2.901	0.202
Bran	2.778	1.602	1.778	2.079	2.176	2.083	0.452
Spice	2.669	2.000	2.000	2.301	2.000	2.200	0.308
Breading mix	2.000	1.000	1.778	2.114	2.041	1.787	0.457
Pasta	2.000	2.000	2.477	2.000	2.415	2.178	0.245
Oatmeal	2.000	2.041	2.531	2.724	2.477	2.355	0.319
Breading mix	2.301	2.699	2.954	2.602	2.903	2.692	0.262
Flour	1.602	1.301	1.954	1.699	2.000	1.711	0.284
Flour	2.000	2.301	2.477	2.000	2.477	2.251	0.240
$\bar{x}$ (all)	2.320	2.030	2.368	2.338	2.441	2.299	
SD (all)	0.472	0.591	0.460	0.435	0.393		

spreading and overgrowth by molds. In terms of overall recovery of the intrinsic yeast flora from the food samples using the spread-plate technique, APDA, DRBC and DG18 were essentially similar in their performance ($P \leq 0.0166$).

In Table 4, the first and second medium preferences for each sample tested are listed, based on the mean yeast count data. There is no clear pattern of preference noted, although DRBC and DG18 have an edge over APDA as the first preference medium. It was also noted that yeast colony

Table 2.   Medium preference for mold recovery

| | Medium preference | |
Sample	1	2
Flour	DRBC	OGY
Flour	OGY	DRBC
Flour	DRBC/OGY	SAB
Wheat	APDA/DG18/OGY	DRBC
Bran	APDA	OGY
Spice	APDA	DG18
Breading mix	DG18	OGY
Breading mix	DRBC	OGY
Pasta	DRBC	OGY
Oatmeal	DG18	DRBC

Table 3. Yeast recovery by test media

| Sample | Mean ($\log_{10}$) yeast count[a] on various media | | | | | | |
	APDA	SAB	DRBC	DG18	OGY	$\bar{x}$ (all)	SD (all)
Flour	3.000	NC	2.845	2.964	NC	2.936	0.081
Wheat	3.699	NC	3.041	2.845	NC	3.195	0.447
Bran	2.778	NC	3.447	3.519	NC	3.322	0.409
Spice	4.114	NC	3.633	4.146	NC	3.965	0.287
Breading mix	2.732	NC	2.740	2.477	NC	2.650	0.149
Pasta	2.903	NC	3.000	3.041	NC	2.981	0.071
Oatmeal	2.322	NC	2.279	2.380	NC	2.327	0.051
Breading mix	2.778	NC	2.845	2.699	NC	2.774	0.073
Flour	2.740	NC	2.778	2.477	NC	2.665	0.164
Flour	2.699	NC	3.000	2.301	NC	2.667	0.351
$\bar{x}$ (all)	2.976		2.961	2.885		2.948	
SD (all)	0.529		0.376	0.576			

[a]NC = not countable

resolution and counting was significantly easier on DRBC medium.
Identification of recovered yeasts was not performed.

Discussion

Three of the test media used in this study, DRBC, DG18 and OGY, were
specifically designed for evaluation. DRBC utilizes dichloran and rose
bengal to inhibit rapidly spreading molds and to restrict colony size.
Chloramphenicol inhibits bacterial growth. DG18 contains 18% glycerol to
provide a reduced a_w and encourage the growth of xerophilic fungi, in
addition to dichloran for restriction of spreading growth. OGY contains

Table 4. Medium preference for yeast recovery

| Sample | Medium preference | |
	1	2
Flour (2 samples)	DRBC	APDA
Flour	APDA	DG18
Wheat	APDA	DRBC
Bran	DG18	DRBC
Spice	DG18	APDA
Breading mix (2 samples)	DRBC	APDA
Pasta	DG18	DRBC
Oatmeal	DG18	APDA

oxytetracycline for inhibition of bacterial growth. APDA is a U. S.
standard medium for recovery of molds and yeasts from food samples. SAB is
a classical medium for mycology and frequently has been used for
differentiation of molds.

For overall recovery of molds from the food samples, DRBC and OGY media
exhibited the best performance. From the technician's viewpoint, throughout
this study DRBC was preferred for overall ease of colony detection and
counting. SAB was also clearly less efficient for mold recovery in this
study.

For recovery of yeasts from the foods, APDA, DRBC and DG18 were
essentially similar in their overall performance, although DRBC was
preferred in terms of ease of colony counting. SAB and OGY media were
utilized but colony resolution and accurate enumeration were not possible
due to spreading and overgrowth.

L. A. KELLEN
T. M. SMITH
C. B. HANNON
K. E. OLSON

Comparison of DG18 with PDA Medium Containing
Antibiotics for Enumerating Fungi in Pasta Products

Generally, media used to enumerate fungi have had high a_w; however,
Hocking & Pitt (1980) reported that these media are not acceptable for
enumerating fungi from dried foods. To reduce the a_w of media, salt,
sucrose, glucose or other sugars have been added to media (Pitt & Christian
1968; Pitt 1975; Pitt & Hocking 1977; Hocking & Pitt 1980).

Since glycerol is acceptable for cultivating several xerophilic fungi
from foods, it has been incorporated into media to reduce the a_w (Hocking
& Pitt 1980). This humectant was used to develop a low a_w medium that
contained dichloran to prevent the spreading of mold colonies.
Statistically significant higher counts were obtained on DG18 than on malt
salt agar when several dried foods were analyzed (Hocking & Pitt 1980).

Rayman et al. (1979) observed that some molds and yeasts were stable in
dry pasta and had D-values greater than $\underline{S}$. $\underline{aureus}$. Since there is interest
in establishing microbiological criteria for pasta products (Rayman et al.
1981; Swartzentruber et al. 1982), media for enumerating fungi in these
foods should be evaluated. This study compares the enumeration of molds and
yeasts in pasta products by using DG18 and CCPDA.

Materials and methods

Culture media. DG18 medium was obtained from Oxoid Ltd. CCPDA was as
described in a preceding paper in this chapter by Cousin & Lin.

Food Samples. Pasta samples were purchased from retail food stores.
Fifteen separate packages each of egg noodles, spaghetti and macaroni were
analyzed.

Preparation of samples and fungal identification. Fifty-gram subsamples
of each pasta product were blended with 450 ml of sterile peptone (0.1%, w/v)
diluent for 2 min. Additional dilutions were made as described previously as
were identification procedures for molds. Nonspreading colonies were
stained with methylene blue (0.1%, w/v) and examined microscopically to
ensure that they were yeasts.

Results and discussion

The range distributions of mold and yeast counts in spaghetti, macaroni
and noodles are shown in Table 1. Spaghetti and macaroni had comparable
counts that were less than 10^3/g. These observations are in agreement with
those reported by Rayman et al. (1981) and Swartzentruber et al. (1982). Egg
noodles generally gave higher counts for molds and yeasts than the other
pasta products and one sample had a count of > 10^4/g. Although the surveys
of pasta products in both the United States and Canada revealed counts higher
than 4 logs for a few samples of noodles, the majority of samples had counts
of 1-4 logs (Rayman et al. 1981; Swartzentruber et al. 1982). Since the
research reported here involved only fifteen samples of noodles, the
distribution of fungi would not be as large as the surveys that included
over one thousand samples.

There were no significant differences noted in the media used to
enumerate molds and yeasts in the three pasta products. Both CCPDA and DG18
media recovered essentially the same number of fungi from each sample. The
only significant differences noted were for product and brand of pasta.
This would be expected since the processing, ingredient quality and
environmental conditions would vary for each product and for each
manufacturer.

The types and percentages of fungi recovered from each product and on
each medium were comparable (Table 2). Penicillium, Geotrichum, Aspergillus
and Sporotrichum spp. were most often isolated from the pasta products.
There were slight variations in the percentages of each mold isolated from

Table 1. Range of mold and yeast counts for pasta plated on CCPDA and DG18
media

| Product | Media | Range ($\log_{10}$/g) | | | | |
		< 1.00	1.00–2.00	2.00–3.00	3.00–4.00	4.00–5.00
Spaghetti	CCPDA	7/15[a]	7/15	1/15		
	DG18	7/15	7/15	1/15		
Macaroni	CCPDA	2/15	9/15	4/15		
	DG18	1/15	10/15	4/15		
Noodles	CCPDA		1/15	12/15	1/15	1/15
	DG18		1/15	12/15	1/15	1/15

[a]Number of samples/total number of samples. Overall mean ($\log_{10}$)
regardless of product for CCPDA was 2.79 and for DG18 was 2.76 with standard
deviations of 3.45 and 3.39, respectively, at 0.95 confidence interval

Table 2. Types and percentages of fungi isolated from pasta products on
CCPDA and DG18 media

Fungi	Media	Product			Total
		Spaghetti	Noodles	Macaroni	
Alternaria	CCPDA	0	0.1	0	0.1
	DG18	0	0	0	0
Aspergillus	CCPDA	11.4	28.0	5.8	24.7
	DG18	13.6	31.6	5.2	27.2
Candida	CCPDA	0	0.8	0	0.7
	DG18	0	0	0	0
Cladosporium	CCPDA	0	1.2	0.6	1.1
	DG18	0	0.6	0	0.5
Fusarium	CCPDA	8.6	0	0	0.4
	DG18	0	0	0	0
Geotrichum	CCPDA	28.6	37.9	27.0	36.3
	DG18	27.2	34.2	40.4	34.6
Mucor	CCPDA	0	0	0	0
	DG18	0	0	0.5	0.1
Penicillium	CCPDA	30.0	18.2	14.4	19.5
	DG18	39.5	18.9	14.5	19.5
Rhizopus	CCPDA	2.9	1.6	1.7	1.6
	DG18	11.1	1.5	1.6	2.0
Sporotrichum	CCPDA	2.9	8.5	32.8	11.0
	DG18	0	10.4	22.3	11.3
Yeasts	CCPDA	15.7	3.8	17.8	6.0
	DG18	8.6	2.9	15.5	4.8

the three types of pasta; however, this may reflect the processing
conditions of the products. Although there were no significant differences
noted between counts on the two media, DG18 medium was more effective for
the recovery of Rhizopus, especially in macaroni, since it prevented the
spreading of the colonies. On CCPDA, Rhizopus colonies generally covered
the entire plate rapidly and obscured enumeration. Rose bengal and/or
dichloran have been incorporated into media to prevent the spreading of
rapidly growing colonies (King et al. 1979; Hocking & Pitt 1980; Henson
1981). Besides preventing spreading of colonies on media, these two
additives reduced the colony size of most molds and facilitated enumeration
and isolation.

Conclusions and recommendations

No significant differences were noted between CCPDA and DG18 media for
the enumeration of fungi in pasta products. However, since DG18 prevented
spreading of Rhizopus spp. and reduced the colony size of most molds, this
medium is recommended for routine enumeration of fungi in pasta products.

References

HENSON, O. E. 1981 Dichloran as an inhibitor of mold spreading in fungal
plating media: Effects on colony diameter and enumeration. Applied and
Environmental Microbiology 42, 656–660.

HOCKING, A. D. & PITT, J. I. 1980 Dichloran-glycerol medium for enumeration of xerophilic fungi from low-moisture foods. *Applied and Environmental Microbiology* 39, 488-492.

KING, A. D., JR., HOCKING, A. D. & PITT, J. I. 1979 Dichloran-rose bengal medium for enumeration and isolation of molds from foods. *Applied and Environmental Microbiology* 37, 959-964.

PITT, J. I. 1975 Xerophilic fungi and the spoilage of foods of plant origin. In *Water Relations of Foods*. ed. Duckworth, R. B. New York: Academic Press.

PITT, J. I., CHRISTIAN, J. H. B. 1968 Water relations of xerophilic fungi isolated from prunes. *Applied Microbiology* 16, 1853-1858.

PITT, J. I. & HOCKING, A. D. 1977 Influence of solute and hydrogen ion concentration on the water relations of some xerophilic fungi. *Journal of General Microbiology* 101, 35-40.

RAYMAN, M. K., D'AOUST, J.-Y., ARIS, B., MAISHMENT, C. & WASIK, R. 1979 Survival of microorganisms in stored pasta. *Journal of Food Protection* 42, 330-334.

RAYMAN, M. K., WEISS, K. F., RIEDEL, G. W., CHARBONNEAU, S. & JARVIS, G. A. 1981 Microbiological quality of pasta products sold in Canada. *Journal of Food Protection* 44, 746-749.

SWARTZENTRUBER, A., PAYNE, W. L., WENTZ, B. A., BARNARD, R. J. & READ, R. B., JR. 1982 Microbiological quality of macaroni and noodle products obtained at retail markets. *Applied and Environmental Microbiology* 44, 540-543.

M. A. COUSIN
H. H. LIN

▶ Comparison of Five Media for the Enumeration of Molds in Cereals and Cereal Products

Media commonly used in the enumeration of molds in foods and feeds vary in their a_w values, composition and conditions of incubation. For mycological examination of dried products such as stored cereals which are commonly contaminated by xerophilic fungi, the most widely used medium is probably malt salt agar (MSA) (Christensen 1946) but more recently a new medium (DG18) with a_w reduced with 18% glycerol has been proposed by Hocking & Pitt (1980). For practical reasons, probably no more than two media should be used in the mycological examination of food products and in order to select the best, a comparison between five possible media has been carried out using five products, i.e., wheat, wheat flour, commercial bran, maize semolina and a mixed feed.

Materials and methods

Samples and media. For each product and each medium, five replicates were performed using subsamples for each commodity. Media compared were 2% malt agar with 100 mg of chloramphenicol/liter (MALT), OGY, DRBC, MSA (malt salt agar, with 75 g of NaCl and 100 mg of chloramphenicol/liter) and DG18.

Methods. Samples of 20 g of each product were suspended in 180 ml of diluent (0.1% peptone, 0.85% NaCl and 0.03 g/liter of polyethylene sorbitan monooleate) and homogenized by shaking prior to preparation of dilutions, except for wheat grains (100 g in 400 ml of diluent) which were treated by grinding in a Waring Blendor for 90 sec, after 25 min of soaking. Serial dilutions were pour-plated in two Petri dishes (plastic, 90 mm dia) for each dilution. Plates are incubated upright at 25°C for 5 days.

Results

All results are given in Table 1 as mean (range) and standard deviation of $\log_{10}$ cfu/g.

Discussion and conclusions

Except for the mixed feed, MSA always gave the lowest counts and strongly inhibited field fungi. Higher counts were obtained with MALT and OGY but on these two media field fungi (<u>Alternaria</u>, <u>Mucor</u>, <u>Rhizopus</u> and <u>Aureobasidium</u>) spread rapidly and overgrew the more slowly developing colonies.

Results were more heterogeneous with DRBC. High counts were obtained in some cases (wheat flour, bran, semolina) but not in others (wheat, mixed feed). The spreading growth of mucoraceous fungi was controlled but direct identification of species was difficult.

DG18 was clearly the best medium for cereal examination, giving in most cases significantly ($P \leq 0.05$) higher counts than other media. Spreading of Mucorales and other field fungi was well controlled without excessive

Table 1. Comparison of colony counts on five cereal products determined with five different culture media

Media	Food/feed sample				
	Wheat	Wheat flour	Bran	Semolina	Mixed feed
MALT	3.07 (2.92-3.27) S.D. = 0.20	3.41 (3.31-3.47) S.D. = 0.09	3.22 (3.20-3.30) S.D. = 0.06	6.98 (6.69-7.14) S.D. = 0.14	4.20 (3.95-4.30) S.D. = 0.21
OGY	3.20 (2.64-3.37) S.D. = 0.22	3.39 (3.30-3.55) S.D. = 0.11	3.18 (3.11-3.22) S.D. = 0.06	7.33 (7.04-7.44) S.D. = 0.13	4.04 (3.84-4.17) S.D. = 0.20
DRBC	2.84 (2.30-3.21) S.D. = 0.28	3.45 (3.34-3.57) S.D. = 0.08	3.36 (3.00-3.56) S.D. = 0.19	7.29 (7.23-7.41) S.D. = 0.09	3.92 (3.77-4.11) S.D. = 0.18
MSA	1.62 (1.30-1.83) S.D. = 0.19	3.31 (2.60-3.51) S.D. = 0.28	3.09 (2.69-3.38) S.D. = 0.23	7.17 (7.00-7.36) S.D. = 0.18	4.50 (4.27-4.70) S.D. = 0.12
DG18	3.13 (2.8-3.27) S.D. = 0.11	3.45 (3.20-3.61) S.D. = 0.12	3.29 (3.08-3.43) S.D. = 0.14	7.28 (7.20-7.41) S.D. = 0.11	4.04 (3.84-4.17) S.D. = 0.16

inhibition of these species. Qualitatively, this medium gave probably the most equilibrated view of "total" mycoflora (yeasts were not taken into account in this study).

The choice of a medium with high a_w is more difficult. It would be of interest to try the addition of dichloran to MALT or OGY.

Reference

CHRISTENSEN, C. M. 1946 The quantitative determination of molds in flour. Cereal Chemistry 23: 322–329.
HOCKING, A. D. & PITT, J. I. 1980 Dichloran-glycerol medium for enumeration of xerophilic fungi from low-moisture foods. Applied and Environmental Microbiology 39, 488–492.

D. RICHARD-MOLARD

▶ Comparison of Media for Enumerating Yeasts and Molds
in Dry Seed-based Foods

The presence of viable fungal propagules on cereal grains and legumes as well as products produced from them is to be expected. Their presence does not represent a public health hazard unless moisture, temperature and atmospheric gas conditions are such that growth and subsequent mycotoxin production might occur. There has been a recent surge in the number of retail grocers offering bulk-stored dried foods for sale. The customer removes from bins, bags, boxes or other containers the amount of product he/she wishes to purchase and generally deposits it in a plastic bag which is then transported to the home. There has been some concern among public health officials regarding the lack of sanitation inherently associated with this type of marketing practice. This study was designed to determine mycological quality and to evaluate six mycological media for their suitability to support colony development by fungi naturally present in bulk-stored cereal- and legume-based foods.

Materials and methods

A total of 109 bulk-stored dry food samples were purchased from eight retail grocers in the Atlanta, GA area. Foods examined included barley, beans, coconut, cereal and legume flours, granola, lentils, dehydrated milk, nuts, oats, peas, popcorn, pumpkin seed, rice, sesame seed, sunflower seed and wheat.

Samples (50 g) were combined with 450 ml of 0.1 M potassium phosphate buffer (pH 7.0) containing 0.01% polyethylene sorbitan monooleate and vigorously shaken for 2 min. Serial dilutions (0.1 ml) of samples were plated in quadruplicate on plates of "dried" test media.

Five media were evaluated for their suitability to support fungal colony formation: DRBC, pH 5.6 as reported by King et al. (1979) and modified by Hocking (1981) and Pitt (1983) to contain chloramphenicol in place of chlortetracycline; OGY, pH 6.5 (Mossel et al., 1970); plate count agar (PCA, pH 7.0) containing 100 µg/ml each of chloramphenicol and chlortetracycline-

Table 1.   Comparison of media for enumerating colony-forming (cfu) in a
           composite of 109 dry foods

Recovery medium	Total cfu/g[a]
DRBC	4270 a
OGY	4228 a
PCA	3860 a
APDA	1491 b
RBC	4965 a

[a]Mean values followed by the same letter are not significantly different
(P < 0.05).

HC1; APDA (acidified to pH 3.5 with tartaric acid); and RBC, pH 7.1 (Jarvis
1973).

    Duplicate plates of all dilutions were incubated for 5 days at 25°C
before yeast and mold colonies were enumerated. Statistically significant
(P < 0.05) differences between mean values were determined subjecting data
to analysis of variance and Duncan's multiple range test (Statistical
Analysis System, 1979).

Results and discussion

    Mean values of populations of fungi enumerated were calculated and
significant differences between means for various media were determined.
Data for composites of mean values of fungal populations detected on test
media are listed in Table 1. There were no significant differences between
DRBC, OGY, PCA and RBC. However, APDA was clearly inferior to DRBC, OGY,
PCA and RBC for supporting colony development.

References

HOCKING, A. D. 1981 Improved media for enumeration of fungi from foods.
    CSIRO Food Research Quarterly 41, 7-11.
JARVIS, B. 1973 Comparison of an improved rose bengal-chlortetracycline
    agar with other media for selective isolation and enumeration of moulds
    in yeasts in foods. Journal of Applied Bacteriology 36, 723-727.
KING, A. D., HOCKING, A. D. and PITT, J. I. 1979 Dichloran-rose bengal
    medium for enumeration and isolation of molds from foods. Applied and
    Environmental Microbiology 37, 959-964.
MOSSEL, D. A. A., KLEYNEN, SEMMELING, A. M. C., VINCENTIE, H. M., BEERENS, H.
    & CATSARAS, M. 1970 Oxytetracycline-glucose-yeast extract agar for
    selective enumeration of molds and yeasts in foods and clinical
    material. Journal of Applied Bacteriology 33, 454-457.
PITT, J. I. 1983 Personal communication.
STATISTICAL ANALYSIS SYSTEM 1979 SAS User's Guide. 494 pp. SAS Institute,
    Inc., Cary, NC. 494 pp.

                                                        L. R. BEUCHAT

Comparison of Media for the Enumeration of Molds in Flour

To standardize methods for routine control of the microbiological quality of flour, a comparative study was done to evaluate the sensitivity and precision of media and to analyze the effect of sample preparation.

Investigations were performed in two series. In the first, two media were used with different methods of suspension preparation. In the second series, two additional media were compared, allowing different times for settling the suspensions.

The effects of sample preparation are discussed separately. Results concerning the media are presented in this paper.

Materials and methods

Eleven laboratories took part in the first series of investigation and twelve took part in the second. Two samples of flour were investigated in duplicate in each study. In order to obtain a higher mold count, wheat meal was used in both series. The main suspension was prepared by adding 10 g of sample to 90 ml of peptone-salt diluent.

Media used were as follows: chloramphenicol glucose tryptone agar (CGT) Sabouraud glucose agar (SGA), OGY and RBC (Oxoid CM 549). However, RBC was not available in all laboratories; hence, each laboratory prepared a medium of similar composition, differing from the Oxoid formula in that tryptone was used instead of mycological peptone. Poured plates were used in both series of experiments. Incubation was made at 25°C for 4 days.

Results and discussion

Results were evaluated by analysis of variance using $\log_{10}$ transformed data. Data were first checked for homogeneity of their mean values and standard deviations by Dixon's test and Bartlett's test, respectively. In

Table 1. Comparison of CGT and SGA media for the enumeration of cfu of molds in wheat meal and flour

	$\log_{10}$ cfu[a]					
	CGT			SGA		
Sample	Mean	SD_0	SD_t	Mean	SD_0	SD_t
Series 1 (wheat flour)	2.61	.0271	.0828	2.48	.0278	.1368
Series 2 (wheat meal)	4.61	.0219	.0711	4.51	.0182	.1455

[a]Data represent means of 66 investigations (11 laboratories, 3 different suspensions in duplicate). SD_0 = intra-laboratory standard deviation; SD_t = inter-laboratory standard deviation

Table 2. Comparison of OGY and RBC media for the enumeration of colony
forming units of molds in wheat flour

Laboratory number	Log_{10} cfu[a]		
	OGY	RBC (Oxoid)	RBC (home-made)
1	3.47	–	3.47
2	3.46	–	3.47
3	3.33	3.41	3.26
4	3.37	–	3.33
5	3.47	3.51	3.53
6	3.53	–	3.53
7	3.51	–	3.57
8	3.51	–	3.40
9	3.47	3.60	3.54
10	3.55	3.51	3.54
Mean	3.46	3.51	3.46

[a]Data represent mean values of 16 analyses, LSD $_{95\%}$ = 0.147

the first series, no outliers were found. However, in the second series,
two laboratories were selected for extremely low mean values.

Data obtained in the first series are summarized in Table 1. The mean
values of cfu on SGA agar were lower while the standard deviations were
higher than those obtained on CGT medium. These differences, however, were
not significant at the 95% level of probability. No difference was found in
relation to the level of mold contamination either.

Data from series 2 are summarized in Table 2. Here again no significant
difference was observed at the 95% level of probability in the performance
of media. No difference was greater than the calculated value of least
significant difference (LSD = 0.147). It is of interest to note that RBC as
prepared by the laboratories was equal to that produced by Oxoid.

T. DEÁ
V. TABAJDI-PINTE
I. FABR

Comparison of General Purpose Media for the Enumeration
of Molds and Yeasts

Investigations were made to compare two reference media prepared as
single homogeneous batches by Oxoid Ltd. with media prepared in two research
laboratories for routine use.

Media. Media studied were DRBC (Oxoid), DG18 (Oxoid), OGY prepared in laboratories, chloramphenicol glucose yeast extract agar (CGY, the same as OGY but oxytetracycline replaced by 100 µg/ml chloramphenicol; prepared in the laboratory according to ISO (1983) and acidified glucose yeast extract agar (AGY), prepared in the laboratory (basic composition was the same as OGY but without oxytetracycline; pH adjusted to 3.5 with tartaric acid).

Samples. Spoiled wine, whole wheat meal and wheat flour were investigated. One milliliter of wine was added to 9 ml of diluent for preparing decimal dilutions. Meal and flour (10-g samples) were added into 90 ml of peptone salt water, shaken by hand for 10 min and allowed to settle for 10 min. This initial suspension was used for inoculation and for further dilutions. Incubation was at 25°C for 5 days.

Evaluation. Each sample was investigated in duplicate, and in one of the laboratories the investigation was repeated. Data were evaluated by analysis of variance after log_{10} transformation.

Results and discussion

Data are summarized in Tables 1, 2 and 3 for wine, whole wheat meal and wheat flour, respectively.

Table 1. Spoiled wine log_{10} yeast counts/ml[a]

Lab no.	Sample no.	DG18		DRBC		OGY		CGY		AGY	
		pp	sp	pp	sp	pp	sp	pp	sp	pp	sp
A	1	5.56	5.67	5.40	5.64	5.40	5.77	5.27	5.68	5.57	5.60
		5.25	5.82	5.42	5.60	5.39	5.48	5.41	5.67	5.43	5.58
	2	5.43	5.76	5.43	5.64	5.62	5.69	5.47	5.68	5.45	5.60
		5.49	5.71	5.44	5.59	5.48	5.64	5.50	5.67	5.46	5.57
B	1	4.83	5.08	4.90	5.11	4.91	5.11	4.94	5.15	–	–
		4.95	5.30	4.96	5.38	5.04	5.30	5.10	5.30		

[a]pp = pour plate; sp = spread plate; $\bar{x}$ 5.275, SD 0.2399

(continued)

Table 1. Continued

1. Comparison of media:
   (A) 3 samples, 2 parallels, 2 inoculations, n = 12
   (B) 2 samples, 2 parallels, 2 inoculations, n = 8

Media	Means (A)	Means (B)
DRBC	5.376	5.520
DG18	5.404	5.585
OGY	5.403	5.559
CGY	5.403	5.544
AGY	–	5.533
$SD_{95\%}$	0.085	0.105

2. Comparison of inoculation (n = 24)

Method	Mean	Standard deviation	Coefficient of variation (%, CV)
Pour–plate	5.270	0.091	23
Spread–plate	5.518***	0.111	29
SD_{95}	0.060		

3. Comparison of laboratories (n = 32)

Lab	Variance	Standard deviation	CV (%)
A	0.008	0.089	23
B	0.015	0.123	33

$F = 2.14 < F_{95} = 2.6$, between–labs standard deviation: 1.52,
Within–lab standard deviation: 1.05

Table 2. Whole wheat meal $\log_{10}$ mold counts/g[a]

Lab No.	Sample No.	DG18		DRBC		OGY		CGY		AGY	
		pp	sp	pp	sp	pp	sp	pp	sp	pp	sp
A	1	3.15	3.24	3.18	2.70	3.22	3.60	3.21	3.41	3.11	3.65
		3.28	3.46	3.19	3.00	3.13	3.28	3.49	3.20	3.15	3.00
	2	3.42	3.00	2.91	3.19	3.50	3.08	3.45	3.23	3.12	3.08
		2.87	2.30	3.27	2.80	2.87	3.00	3.06	3.32	2.94	3.34
B	3	3.25	3.08	3.30	3.11	3.34	3.16	3.20	3.26	–	–
		3.08	2.90	3.48	3.04	3.38	3.00	3.36	3.90		

[a]pp = pour plate; sp = spread plate

1. Comparison of media

Media	Means (n = 2)	Means (n = 8)
DRBC	3.098	3.030
DG18	3.087	3.091
OGY	3.212	3.210
CGY	3.341	3.296*
AGY	–	3.174
SD_{95}	0.189	0.231

2. Comparison of inoculation (n = 24)

Method	Mean	Standard deviation	CV (%)
Pour-plate	3.210	0.212	63
Spread-plate	3.132	0.237	73
SD_{95}	0.134	0.237	

(continued)

Table 2. Continued

3. Comparison of laboratories (n = 32)

Lab	Variance	Standard deviation	CV (%)
A	0.058	0.241	74
B	0.035	0.187	54

$F = 1.67 < F_{95} = 3.2$; between-labs standard error: 0.381;
Within-lab standard error: 0.216

Table 3. Wheat flour ($\log_{10}$ mold counts/g)

Lab no.	Sample no.	DG18	DRBC	OGY	CGY	AGY
A	1	2.48	2.26	2.58	2.61	1.85
		2.40	2.52	2.66	2.40	1.30
B	2	2.65	2.79	2.59	2.48	1.70
		2.49	2.30	2.74	2.91	1.48

Comparison of media (n = 4)

Media	Means
DRBC	2.505
DG18	2.468
OGY	2.643
CGY	2.600
AGY	1.583**
SD_{95}	0.488

   Growth of five molds and six yeast species was compared on RBC, OGY,
gentamicin glucose yeast extract (GGY) and acidified tryptone glucose yeast
extract (ATGY) media (Tables 4 and 5, respectively). It was observed,
according to expectation, that RBC significantly inhibited the growth of
molds. Unexpectedly, however, it stimulated the growth of yeasts; this may
be attributed to the somewhat richer composition of RBC medium.

Conclusions

1. With two exceptions, no significant difference was noted among media.
   For the determination of molds in whole wheat meal, a higher count was
   obtained on CGY compared to DRBC and DG18, while in wheat flour a lower
   count was obtained on AGY than on any other medium. For the
   determination of yeasts, the performance of media was equal.

Table 4.   Growth of molds (colony diameter mm/day, n = 2)

Mold	Medium			
	RBC	OGY	GGY	ATGY
Aspergillus niger	4.63	7.86	5.83	8.40
Penicillium frequentans	4.51	7.17	7.04	10.41
Fusarium graminearum	6.65	4.90	13.06	7.21
Mucor rouxii	7.13	18.40	16.83	10.95
Rhizopus stolonifer	21.86	42.25	38.00	50.50
Mean	8.95**	17.49	16.15	16.12

SD_{99} = 4.40

Table 5.   Growth of yeasts (log_{10} counts in 3 days, n = 2)

Yeast	Medium			
	RBC	OGY	CGY	ATGY
Saccharomyces cerevisiae 1171	9.46	8.47	8.39	8.37
Zygosaccharomyces rouxii	8.09	8.50	8.45	8.57
S. cerevisiae T19	8.87	8.84	8.98	8.88
S. cerevisiae T22	8.53	8.29	7.99	8.08
Kloekera apiculata	8.57	8.10	8.07	8.12
Schizosaccharomyces pombe	7.55	7.95	7.66	7.81
Mean	8.51*	8.35	8.26	8.30

SD_{95} = 0.16

2. Comparing the methods of inoculation, significantly higher counts of yeasts were obtained on spread plates than on pour plates. There was no difference in the standard deviation of the methods.
3. The precision of determination within one laboratory and between two laboratories showed no difference; however, the precision increased with higher counts.

References

ISO 1983 General guidance for detection and enumeration of molds and yeasts. ISO/TC 34/SC 9 N 156.

T. DEÁK
T. TÖRÖK
J. LEHOCZKI
O. REICHART
V. TABAJDI-PINTER
I. FABRI

Merits and Shortcoming of DG18, OGY, OGGY and DRBC
Media for the Examination of Raw Meats

Raw meats are generally heavily contaminated with bacteria and yeasts, especially after chilled or frozen storage, the latter often being followed by a shorter or longer period of more or less chilled storage during further processing or marketing. Besides an assessment of the bacteriological status, the enumeration of yeasts and molds can be used in monitoring the hygienic conditions of the preceding period. For this purpose, a medium should be inhibitory for bacteria but not toxic for yeasts and molds, while the latter organisms should be restricted in their spreading ability. A broad range of media has been proposed with the main inhibitors for bacteria being chloramphenicol, oxytetracycline, gentamicin, a reduced a_w or a combination of them, and rose bengal and/or dichloran as restrictive components against mold spreading. In continuation of the preceding work (Dijkmann et al. 1979; Mossel et al. 1980; Dijkmann 1982), DG18 and DRBC were compared with OGY and OGGY (oxytetracycline gentamicin glucose yeast extract agar).

Materials and methods

The following media were used as both pour and spread plates: DRBC, DG18, OGY and OGGY (Mossel et al. 1975). DG18 and DRBC were provided by Oxoid for studies for this workshop, and OGY and OGGY were prepared in the laboratory. Oxytetracycline (100 µg/ml) and gentamicin (50 µg/ml) were filter-sterilized and added after autoclaving the other ingredients.

Peptone saline (0.5% peptone, 0.85% NaCl) was used for dilution. The 16 samples of minced (ground) meat were bought in butcher shops immediately before enumeration. The minced meat was the Dutch 'half and half' type (50% pork, 50% beef); 20 g of meat was mixed with 180 ml of peptone saline in a Stomacher for 2 min. Large particles were allowed to settle for 30 sec. Decimal dilutions were mixed mechanically (Vortex) for 5 sec. From each dilution, 0.1 ml was spread on the surface of the spread-plate media and 1.0 ml was transferred into Petri dishes (90 mm) and mixed with cooled (47°C) media for pour plates. After 5 days of incubation at 22.5 ± 1°C, suitable dilutions were selected for counting and calculation of the colony count. Because of the very small differences between the duplicates, only the average values are given in the table.

In order to get an impression of the bacterial load, plate count agar (PCA; Difco) was used as pour plates and also incubated for 5 days at 22.5°C.

Results and discussion

The results are given in Table 1 together with the differences between the spread and pour plates of each medium and sample. In Table 2, the differences between the highest and the lowest count of each sample are given for spread and pour plates for all four media and for all plates together. The averages of all data are listed in Table 3. All data represent yeast counts; molds were not detected. On, and very likely also in, DG18 and OGGY not a single colony of bacteria could be detected, while on and in OGY and DRBC the absence of bacteria was an exception.

In individual samples there were quite large differences between the media as well as the methods but the mean values were not so alarming, even when they were significantly different if the enumeration is used only to get

Table 1. Yeast populations ($\log_{10}$/g) in minced (ground) beef detected using five media

	PCA	DG18			OGY			OGGY			DRBC		
Sample No.	pour	spread	pour	Δlog	spread	pour	Δlog	spread	pour	Δlog	spread	pour	Δlog
1	9.14	6.25	6.16	0.09	6.42[b]	6.20[d]	0.22	6.32	5.71	0.61	5.79[b]	5.66[d]	0.13
2	8.95	6.18	6.09	0.09	6.40[b]	6.14[d]	0.26	6.31	5.54	0.77	5.67[b]	5.54[d]	0.13
3	8.25	6.15	6.10	0.05	6.10[c]	5.76[d]	0.34	6.06	5.30	0.76	5.39[a]	5.34[d]	0.05
4	8.11	6.13	6.00	0.13	6.09[c]	5.62[d]	0.47	6.05	5.24	0.81	5.37[a]	5.26[d]	0.11
5	7.95	5.86	5.86	0.00	5.89[b]	5.89[d]	0.00	4.98	4.86	0.12	5.86[b]	5.88[d]	0.02
6	7.92	5.74	5.73	0.01	5.79[b]	5.88[d]	0.09	4.98	4.78	0.20	5.86	5.75[d]	0.11
7	8.59	5.42	5.11	0.31	5.28[c]	4.82[c]	0.46	5.19	4.37	0.82	4.59[a]	4.42[d]	0.17
8	8.58	5.29	5.04	0.25	5.16[c]	4.70[c]	0.46	5.04	4.30	0.74	4.54[b]	4.28[d]	0.26
9	7.00	5.08	4.90	0.18	5.00[b]	4.94[d]	0.06	4.53	3.81	0.72	4.94	4.91	0.03
10	7.12	5.01	5.05	0.04	5.00[a]	5.00[d]	0.00	4.69	4.65	0.04	4.81	4.98	0.17
11	6.81	4.98	4.86	0.12	4.91[b]	4.93[d]	0.02	4.30	3.76	0.54	4.94	4.91	0.03
12	6.95	4.90	4.92	0.02	4.94[a]	5.00[d]	0.06	4.45	4.59	0.14	4.70	4.91	0.21
13	7.17	4.78	4.77	0.01	4.70[a]	4.91	0.21	4.71	4.46	0.25	4.78[b]	4.73[d]	0.05
14	7.18	4.76	4.72	0.04	4.68[a]	4.88	0.20	4.54	4.26	0.28	4.57[b]	4.64[d]	0.07
15	6.66	4.64	4.64	0.00	4.69[a]	4.63[d]	0.06	4.56	4.21	0.35	4.42[a]	4.31	0.11
16	6.48	4.61	4.64	0.03	4.58	4.46	0.12	4.20	4.11	0.09	4.41	4.29	0.12

Recovery medium

[a] 3–10 bacterial colonies/plate

[b] 11–100 bacterial colonies/plate

[c] > 100 bacterial colonies/plate

[d] Only the bacterial colonies which by chance got on the surface could be differentiated

Table 2. Differences ($\log_{10}$) between highest and lowest counts within samples

Sample number	Type of plate		All plates
	Spread	Pour	
1	0.63	0.54	0.76
2	0.73	0.60	0.86
3	0.76	0.80	0.85
4	0.76	0.76	0.89
5	0.91	1.03	1.03
6	0.88	1.10	1.10
7	0.83	0.74	1.05
8	0.75	0.76	1.01
9	0.55	1.13	1.27
10	0.32	0.40	0.40
11	0.68	1.17	1.22
12	0.50	0.41	0.55
13	0.08	0.45	0.45
14	0.22	0.62	0.62
15	0.27	0.43	0.48
16	0.41	0.53	0.53

an impression of the hygienic conditions of raw meat in the period prior to the moment of enumeration.

Assuming that yeasts are not harmful and that the presence of molds above the detection limit of 100/g should always lead to condemnation of raw meat, it does not matter if some yeast propagules are missed by the method or medium. However, molds should be able to grow, and for that reason DG18 should not be used because the low a_w may suppress some species or genera of molds (Lowry & Gill 1984) which may lead to failure because molds, contrary to yeasts, are likely to be present in meat in a pure culture.

OGY and DRBC media should be used as spread plates only to avoid high counts due to bacterial colonies which look on the surface nearly the same as yeast colonies but can be detected by an experienced eye because they are slightly translucent. However, in the medium, bacterial colonies can only

Table 3. Mean values ($\log_{10}$/ml) of all samples

Type of plate	Recovery medium					$\Delta \log_{10}$	$\Delta \log_{10}$ (all plates)
	DG18	OGY	OGGY	DRBC	PCA		
Spread	5.36	5.35	5.06	5.04	–	0.58 }	
						}	0.82
Pour	5.29	5.24	4.62	4.99	0.72	0.72 }	
$\Delta \log_{10}$	0.09	0.19	0.45	0.11	–	–	

be distinguished from yeast colonies by picking off and microscopic examination, which takes quite a lot of time and labor in view of the variable numbers present. Due to that inaccuracy, the counts in pour plates of OGY and DRBC may be high. The low numbers in the OGGY pour plates are not only due to the absence of bacteria but the accumulation of stress conditions (two antibiotics, temperature of agar and partly anaerobic conditions) suppresses some genera, species or strains of yeasts, especially if cells are already stressed, e.g., by freezing.

Conclusions

1. DG18 gives the best performance for yeasts. It suppresses bacteria sufficiently and is easy to prepare, but may also suppress some molds.
2. OGY is also good for enumerating yeasts as well as molds but is an unreliable pour plate medium, while experience is needed for differentiating bacterial from yeast colonies on spread plates; the sterile antibiotic has to be added after autoclaving.
3. OGGY should not be used for pour plates. It suppresses bacteria sufficiently and may be used as spread plates for enumeration of yeasts and molds in raw meat, poultry and fish that are in general highly contaminated with bacteria; antibiotics have to be added after autoclaving.
4. DRBC should be used for spread plates only. Experience is needed to distinguish between yeasts and bacteria; rose bengal in combination with antibiotics may suppress some genera, species or strains of yeasts, especially if they are already stressed (Dijkmann 1982; Engel 1982).

References

DIJKMANN, K. E., KOOPMANS, M. & MOSSEL, D. A. A. 1979 The recovery and identification of psychrotrophic yeasts from chilled and frozen comminuted fresh meats. Journal of Applied Bacteriology 47, ix.

DIJKMANN, K. E. 1982 The optimal medium – always the best choice? In: Quality Assurance and Quality Control of Microbiological Culture Media. Proceedings of the IUMS-ICFMH Symposium held on 6th and 7th September, 1979, in Calas de Mallorca. ed. Corry, J. E. L. pp. 175–180. Darmstadt, West Germany: G.I.T. – Verlag Ernst Giebeler.

ENGEL, G. 1982 Vergleich verschiedener Nährböden zum quantitativen Nachweis von Hefen und Schimmelpilzen in Milch und Milchprodukten. Milschwissenschaft 37, 727–730.

LOWRY, P. D. & GILL, C. O. 1984 Temperature and water activity minima for the growth of spoilage moulds from meat. Journal of Applied Bacteriology 56, 193–199.

MOSSEL, D. A. A., VEGA, C. L. & PUT, H. M. C. 1975 Further studies on the suitability of various media containing antibacterial antibiotics for the enumeration of moulds in food and food environments. Journal of Applied Bacteriology 39, 15–22.

MOSSEL, D. A. A., DIJKMANN, K. E. & KOOPMANS, M. 1980 Experience with methods for the enumeration and identification of yeasts occurring in foods. In Biology and Activities of Yeasts. The Society for Applied Bacteriology Symposium Series No. 9. ed. Skinner, F. A., Passmore, S. M. & Davenport, R. R. pp. 279–288. London: Academic Press.

K. E. DIJKMANN

Discussion

Koburger

In the <u>Bacteriological Analytical Manual</u> (BAM), the <u>Standard Methods for Milk or Dairy Products</u> and the <u>Compendium of Methods for the Microbiological Examination of Foods</u> (compendium), methods recommend plating on plate count agar plus antibiotics, PDA or malt agar. Antibiotics specified in the compendium are 100 g/ml of chlortetracycline and chloramphenicol, as one antibiotic is sometimes inadequate. In the BAM, 40 g/ml of chlortetracycline is specified. Concentrations in the range of 50-100 g/ml are effective.

Pitt

Dr. Koburger, would you agree with me that the methodology mentioned in the documents that you refer to is the type of antiquated and dark ages methodology that I mentioned earlier? Have you, people you work with, or other people characterized the molds that plate count agar will enumerate and have you shown that these are relevant to the type of foods that we are talking about here?

Koburger

In work that we have done and that Phil Mislivec has done, it has been shown that plate count agar works very well. That is, it works as well as anything that we have available. As for the second question, you made the comment in your paper that numbers were not important in quality control. Today I have seen the figures, and some of the counts are log 6.0 and 6.3, which is a 50% difference. To me, numbers are important. If we are checking a salad dressing or orange juice, numbers are important as well as meaningful. Phil has done some work on identifying these things, but as far as the results we have obtained with plate count agar, we are getting the best recovery. Phil can you comment on that?

Mislivec

Well, I have to comment from a regulatory point of view because I get molds from all over the country that I have to identify, regardless of the source from which they come, and anytime I put my name on a report there is a chance that I may end up in court. At the U. S. Food and Drug Administration, we have to use methods that the court will accept. The BAM method is a long accepted method in the courts of law. No useful alternative method is offered by the Association of Official Analytical Chemists. It should be noted that before any method is accepted it must go through comparative studies, several laboratories working independently that come up with relatively the same results. There may be a medium which is better, but we have to stick with what we know and what is acceptable before a judge.

Koburger

I get the feeling that most of the people here are interested in molds because of the mycotoxin problem they pose to foods. That is not necessarily my interest. I am interested in spoilage as well as mycotoxins. This should be taken into account when recommending methods. We are working with all organisms, not predominately molds.

Pitt

We are interested in spoilage. That is why I consider that plate count agar is completely out of date in terms of enumerating molds from foods other than fresh foods. If we are dealing with yogurt or fruit juices, the medium probably does not matter, but if we are looking at dried samples, for example, the molds which will cause spoilage in those commodities simply will not grow on PCA. The molds that are capable of growing on PCA are not capable of producing spoilage in those types of commodities. For this reason, I feel that PCA is out of date.

Mossel

I would like to reiterate the recommendations that the IUMFS has made. In comparative studies, at least four parameters or criteria are to be taken into account. Of course, productivity -- we have plenty of that. Selectivity is the second parameter. You should not compare apples and oranges. Third, electivity, i.e., which medium is the easiest to read for our technicians? You should avoid any bias. And finally, determine if there is any taxonomic bias. This includes two possibilities. First, that you do want taxonomic bias for ecological reasons such as Dr. Pitt has just mentioned or secondly, that you do not want bias and that you want to enumerate all microorganisms. I think instead of discussing this at length that we should give some thought to recommendations of the London principles of media comparisons. It may be that you might want to add some more criteria to this list, but we should give consideration to these.

Lane

Here is the United States we have to conform to Food and Drug Administration regulations. If BAM, or the American Public Health Association Standard Methods for Milk or Dairy Products or the Compendium of Methods for the Microbiological Examination of Foods contain a specific method, we have to conform to it. As it recommends PCA with antimicrobials, we use it. If another medium is suggested, then we use it. If there is another medium better than PCA, then you must have a comparative study. Then you have the evidence and the FDA will accept that evidence. We have gone through this before, so I recommend that a comparative study be done.

Mislivec

All these nice media that we have seen here today -- what is needed is to compare them with something that is already acceptable.

Mossel

On what basis are you going to make your decision if a medium is better or not? You need an international professional recommendation. This is difficult to do unless you have well thought out, well discussed, international criteria on which to judge your media. What we try to do through Dr. Speck and his compendium is to ask our friends in the FDA to look at it because we as an international organization should not be related to any particular national agency. What we should do through scientific channels is to try to tell them to consider new methodology.

Koburger

With regard to the compendium, one message that should come through loud and clear is that no single method is best for all foods. When you are writing about a general purpose medium and method, it has to be qualified that certain foods require certain treatments. This is what it all boils down to.

Deák

I do not think that we as an international group should make any effort to recommend a single medium. We should recommend that several media could do the same thing.

Mossel

This is in essence the structure of the APHA compendium in its fourth edition. They recommend not one method, but at least one. We have suggested that they call these taxomethods because they are aiming at for example, Enterobacteriaceae, Staphylococcus aureus or some other particular group of organisms, but would have to be refined to the level of becoming an ecomethod applied to the specific ecology of foods.

Pitt

We did not produce a medium containing dichloran because it would give higher counts of anything, but because with dichloran spreading colonies are much less problem. If you do not encounter <u>Rhizopus</u> <u>stolonifer</u> or <u>Mucor</u> species, then media without dichloran are equally acceptable or preferable. Climate is an important factor here. In Europe it appears <u>R</u>. <u>stolonifer</u> is not common, while in Australia it constantly invades enumeration plates.

Recommendations

1. Principles of the IUMS Working Party on Culture Media for media evaluation should be taken into account when choosing a medium for enumeration of yeasts and molds.

2. There is no general purpose medium that will be satisfactory for the examination of all foods.

3. Given that there is no definitive medium for all food products, the medium to be used should be appropriate for the food sample and the requirements of the analysis.

4. The medium which is chosen for total yeast and mold counts should be simple to use, have a simple formulation, be stable, give reproducible results and provide the maximal information on types of organisms present.

5. It has been shown that DRBC, RBC, OGY and PDA are most suitable for general purpose enumeration of yeasts and molds in foods. It should be noted that DRBC and RBC may not be suitable for the recovery of some yeasts, and that OGY and acidified PDA will be ineffective in the presence of spreading fungi such as <u>Rhizopus</u>, <u>Mucor</u> and <u>Neurospora</u>.

CHAPTER 3

SELECTIVE MEDIA AND PROCEDURES

In food bacteriology a great deal of effort has been devoted to the development of media selective for particular kinds of spoilage bacteria or pathogens. By comparison, the development of media selective for particular fungi remains in its infancy. Nevertheless, selective media are essential for certain classes of fungi such as xerophiles which will not grow on general purpose media. The detection of preservative-resistant yeasts is greatly simplified by using selective media. Other situations in which selective media are of great value are in the isolation of toxigenic fungi belonging to Aspergillus, Fusarium or Penicillium, for example, and increasing efforts can be expected in this area in years to come. The development of selective media for spoilage fungi, such as specific xerophiles or molds invading cereals before harvest, will also be important in the future.

This chapter comprises a series of papers dealing with aspects of this topic, and also with procedures suitable for the isolation of heat resistant fungi from heat processed foods and raw materials intended for such products.

▶ A Selective Medium for Rapid Detection of Aspergillus
flavus

The discovery in the early 1960s that Aspergillus flavus and A. parasiticus can produce potent carcinogens, aflatoxins, has focused much attention on these fungi. Methods for the chemical detection of aflatoxins in most commodities have now been well established either by TLC or HPLC techniques. However, there are some situations in which it may be more convenient to test for the presence of aflatoxigenic fungi as a guide to the quality and acceptability of a commodity rather than to test for aflatoxins. In a field situation, or perhaps in some developing countries, simple microbiological procedures may be more accessible than procedures involving solvent extraction and chemical analysis. Knowledge of the types of fungi in a particular sample may be a useful guide to which mycotoxins could also be present.

It is fortunate that only two species of fungi, A. flavus and A. parasiticus, are known to produce aflatoxins. These species are closely related to each other and can only be distinguished microscopically (Hocking 1982). They can be readily detected by standard spread-plating techniques on media normally used for fungal enumeration such as DRBC. On this medium, both species form relatively large colonies with characteristic yellow-green, mop-like heads after incubation at 25°C for 4-5 days.

However, such methods are relatively slow. A need existed for a selective medium which would detect aflatoxigenic fungi as reliably as normal counting methods, but much more rapidly. Bothast & Fennell (1974) developed Aspergillus differential medium (ADM), containing 1.0% yeast extract, 1.5% tryptone and 0.05% ferric citrate, recommending incubation at 28°C for 3 days. Hamsa & Ayres (1977) incorporated streptomycin and dichloran into ADM and recommended incubation at 28°C for 5 days. These media rely on a color reaction between A. flavus (and related species) and ferric citrate, which results in the formation of a bright orange-yellow reverse pigment. The pigment is not water soluble and normally remains associated with the colony, but may diffuse into the medium after prolonged incubation. Assante et al. (1981) showed that the pigment is formed when ferric ions react with aspergillic acid, the colored molecule being a complex of three aspergillic acid molecules with associated ferric ions.

Pitt et al. (1983) further refined the formulation of the above two media, producing Aspergillus flavus and parasiticus agar (AFPA). Incubation at 30°C for 42–48 h gives sufficient color development to enable ready recognition of A. flavus (or A. parasiticus) colonies even before conidial heads begin to develop.

Few other fungal species have been observed to produce the orange-yellow color reaction on AFPA. Pitt et al. (1983) observed that A. niger and A. ochraceus can occasionally produce colonies with an orange-yellow reverse, but these species are readily distinguished from A. flavus after a further 24–48 h incubation. They also showed that of 492 isolates picked from AFPA plates and exhibiting orange-yellow reverses, 485 (98.6%) were A. flavus or A. parasiticus. Of the 1.4% false positives, three were A. niger, one was A. ochraceus and three were from other genera. Of 477 isolates from AFPA plates which did not exhibit orange-yellow reverse colors, most were A. niger and Penicillium spp. Only 14 isolates (2.9%) were A. flavus/A. parasiticus and could be considered as false negatives.

AFPA is recommended for the detection of A. flavus and A. parasiticus in commodities such as spices, nuts, oilseeds, grains and stockfeeds. It can also be used to monitor levels of A. flavus in soils where susceptible crops such as peanuts, corn and cotton are grown. The medium enables reliable detection and enumeration of potentially aflatoxigenic fungi within 48 h.

References

ASSANTE, G., CAMARDA, L., LOCCI, R., MERLINI, L., NASINI, G. & PAPADOPOULOS, E. 1981 Isolation and structure of red pigments from Aspergillus flavus and related species, grown on differential medium. Journal of Agricultural and Food Chemistry 29, 785–787.
BOTHAST, R. J. & FENNELL, D. I. 1974 A medium for rapid identification and enumeration of Aspergillus flavus and related organisms. Mycologia 66, 365–369.
HAMSA, T. A. P. & AYRES, J. C. 1977 A differential medium for the isolation of Aspergillus flavus from cottonseed. Journal of Food Science 42: 449–453.
HOCKING, A. D. 1982 Aflatoxigenic fungi and their detection. Food Technology in Australia 34, 236–238.
PITT, J. I., HOCKING, A. D. & GLENN, D. R. 1983 An improved medium for the detection of Aspergillus flavus and A. parasiticus. Journal of Applied Bacteriology 54, 109–114.

A. D. HOCKING
J. I. PITT

Evaluation of Media for Simultaneously Enumerating Total
Fungi and Aspergillus flavus and A. parasiticus in Peanuts,
Corn Meal and Cowpeas

Several media have been formulated to enumerate populations of
Aspergillus flavus and A. parasiticus present among the microflora of
foods. Bell & Crawford (1967) devised a medium containing rose bengal,
dichloran and streptomycin for isolation of A. flavus from peanuts. Bothast
& Fennell (1974) described Aspergillus differential medium (ADM) for
enumerating members of the A. flavus group in grains. The addition of
dichloran (2,6-dichloro-4-nitroanaline) to ADM was subsequently reported to
improve the detection of A. flavus in cottonseeds (Hamsa & Ayres 1977) but
not in a variety of other dry foods (Beuchat 1979). A selective medium for
enumerating A. flavus and A. parasiticus was formulated by Pitt et al.
(1983). This medium (AFPA) had several advantages over ADM including
reduction of the incubation time from 72 h to 42 h.

The study reported here was designed to compare ADM and AFPA for their
suitability to support the development of colonies which would be
interpreted as belonging to A. flavus or A. parasiticus.

Materials and methods

Ten lots each of raw peanuts, corn meal and cowpeas were obtained from
commercial sources. Four subsamples (20 g) of each lot were separately
combined with 80 ml of 0.1 M potassium phosphate buffer (pH 7.0) containing
0.01% polysorbitan 80 and homogenized for 2 min using a Colworth stomacher.
Serial dilutions (0.1 ml) were surface-plated on ADM and AFPA. Chloramphenicol (100 µg/ml) was added to both media; ADM was at pH 6.5 and AFPA was at
pH 4.5, an acidic formulation differing from that specified by Pitt et al.
(1983).

Duplicate plates of each dilution were incubated upright at 30°C and at
37°C. The total number of colonies of yeasts and molds as well as the
number of colonies showing orange-yellow reverse coloration indicative of
potentially aflatoxigenic Aspergilli were recorded after 42-44 h. A second
set of experiments identical to that described above was conducted, except
incubation was extended to 70-72 h before colonies were counted.

Results and discussion

Results from experiments comparing ADM and AFPA for enumerating total
yeasts and molds in peanuts indicate that significant ($P \leq 0.05$) increases
in counts were obtained in three of ten lots by extending the incubation
period from 42-44 to 70-72 h. One incubation temperature was not clearly
superior to the other for supporting colony development. Overall, ADM and
AFPA performed equally with regard to supporting colony development by
yeasts and molds.

In contrast to data for peanuts, populations of yeasts and molds
detected in five of ten lots of corn meal were significantly influenced by
incubation temperature at 42-44 h; populations recovered from six lots
incubated for 70-72 h were significantly affected by temperature.
Incubation at 30°C enhanced recovery compared to incubation at 37°C.

Populations of yeasts and molds detected in nine lots of cowpeas
incubated for 42-44 h and 70-72 h were significantly influenced by
temperature and/or recovery medium. An incubation temperature of 30°C was

clearly superior to 37°C, regardless of recovery medium. Neither ADM nor AFPA was judged superior, and an extension of incubation time did not result in increased colonies.

Only one lot each of peanuts and cowpeas was positive for A. flavus and A. parasiticus, whereas eight lots of corn meal contained the mold. Detection of A. flavus and A. parasiticus in peanuts was not influenced by time, temperature or recovery medium. Six of eight lots of corn meal in which A. flavus and A. parasiticus were detected showed significantly higher counts if incubated at 37° rather than 30°C when ADM plates were examined after 42-44 h of incubation; differences between temperatures and recovery media were essentially eliminated by extending the incubation time to 70-72 h. However, within temperature and medium, an extension of incubation time resulted in significant increases in populations of A. flavus and A. parasiticus in five of eight lots. The optimum conditions for detection of the old group in corn meal, then, are judged to be 42-44 h of incubation at 37°C using either ADM or AFPA. Observations on populations of A. flavus and A. parasiticus in the one positive lot of cowpeas are contrary to those noted for corn meal. An incubation temperature of 30°C was better than 37°C for detecting the mold.

Table 1. Total yeast and mold populations for composites of ten lots each of peanuts, corn meal and cowpeas detected on ADM and AFPA incubated at 30 and 37°C for 42-44 and 70-72 h

Food	Enumeration medium	Incubation temperature (°C)	Incubation time (h)	Population per g[a]
Peanuts	ADM	30	42-44	913 a
			70-72	685 ab
		37	42-44	446 ab
			70-72	555 ab
	AFPA	30	42-44	677 ab
			70-72	764 ab
		37	42-44	427 ab
			70-72	283 b
Corn meal	ADM	30	42-44	34594 a
			70-72	31748 a
		37	42-44	14062 b
			70-72	15911 b
	AFPA	30	42-44	37996 a
			70-72	30791 a
		37	42-44	16738 b
			70-72	15609 b
Cowpeas	ADM	30	42-44	347 b
			70-72	349 b
		37	42-44	49 c
			70-72	36 c
	AFPA	30	42-44	348 b
			70-72	571 a
		37	42-44	74 c
			70-72	57 c

[a]Mean values within food types which are not followed by the same letter are significantly different ($P \leq 0.05$). From Beuchat (1984).

Table 1 lists data representing composites of total yeast and mold populations within food types. For peanuts, incubation at 30°C on ADM for 42-44 h gave significantly higher counts than incubation at 37°C on AFPA for 70-72 h. In the case of corn meal and cowpeas, incubation of plates at 30°C was clearly superior to 37°C, but the incubation time and plating medium were essentially without affect.

Composite data for A. flavus and A. parasiticus are presented in Table 2. Data for peanuts and cowpeas were derived from one lot each, and thus have limited value with respect to eastablishing the best combination of medium, incubation temperature and time. This was not the case for corn meal, eight lots of which contained A. flavus and A. parasiticus. Incubation of ADM plates at 30°C for 42-44 h retarded development of the mold group compared to incubation at 37°C for 42-44 h; incubation time and temperature did not appear to influence counts obtained on AFPA. A larger number of samples of a wider variety of foods may need to be examined before the optimal temperature and time can be established.

Table 2. Aspergillus flavus populations for composites of lots of peanuts, corn meal and cowpeas detected on ADM and AFPA incubated at 30 and 37°C for 42-44 and 70-72 h

Food	Enumeration medium	Incubation temperature (°C)	Incubation time (h)	Population per g[a]
Peanuts	ADM	30	42-44	16 a
			70-72	19 a
		37	42-44	16 a
			70-72	21 a
	AFPA	30	42-44	15 a
			70-72	21 a
		37	42-44	12 a
			70-72	20 a
Corn meal	ADM	30	42-44	86 c
			70-72	140 a
		37	42-44	160 a
			70-72	150 abc
	AFPA	30	42-44	92 bc
			70-72	138 abc
		37	42-44	146 abc
			70-72	154 ab
Cowpeas	ADM	30	40-44	30 b
			70-72	18 ab
		37	42-44	9 bc
			70-72	2 c
	AFPA	30	42-44	34 a
			70-72	10 bc
		37	42-44	11 bc
			70-72	2 c

[a]Mean values within food types which are not followed by the same letter are significantly different ($P \leq 0.05$). Values were determined only from those lots in which A. flavus was detected.

Conclusion

Both media performed similarly for enumerating total fungal propagules but the use of AFPA for detecting A. flavus and A. parasiticus may have some advantage over ADM for inexperienced workers, since color development in colony reverses is quicker and more intense on AFPA. However, neither medium was designed to enumerate total fungal populations, and cannot be promoted for this purpose.

References

BELL, D. K. & CRAWFORD, J. L. 1967 A Botran-amended medium for isolating Aspergillus flavus from peanuts and soil. Phytopathology 57, 939–941.

BEUCHAT, L. R. 1979 Survival of conidia of Aspergillus flavus in dried foods. Journal of Stored Products Research 15, 25–31.

BEUCHAT, L. R. 1984 Comparison of Aspergillus differential medium and Aspergillus flavus and A. parasiticus agar forenumerating total yeasts and molds and potentially aflatoxigenic Aspergilli in peanuts, corn meal and cowpeas. Journal of Food Protection 47, 512–519.

BOTHAST, R. J. & FENNELL, D. I. 1974 A medium for rapid identification and enumeration of Aspergillus flavus and related organisms. Mycologia 66, 365–369.

HAMSA, T. A. P. & AYRES, J. C. 1977 A differential medium for the isolation of Aspergillus flavus from cottonseed. Journal of Food Science 42, 449–453.

PITT, J. I., HOCKING, A. D. & GLENN, D. R. 1983 An improved medium for the detection of Aspergillus flavus and A. parasiticus. Journal of Applied Bacteriology 54, 109–114.

L. R. BEUCHAT

▶ Selective Medium for Penicillium viridicatum in Cereals

Penicillium viridicatum and P. verrucosum sensu Pitt (1979) are perhaps the most important mycotoxin-producing Penicillium spp. These two species may be divided into three groups according to macromorphology, and physiological and biochemical characteristics (Ciegler et al. 1973; Frisvad 1981). A very important feature of this subdivision is that producers of ochratoxin A and citrinin (groups II and III) can be separated from producers of the other potent nephrotoxins, xanthomegnin and viomellein (group I) (Ciegler et al. 1973, 1981; Stack & Mislivec 1978; Frisvad 1981).

The aim of this study was to develop a selective screening medium for P. viridicatum and P. verrucosum as an aid in the examination of stored cereal products for toxic metabolites.

Materials and methods

Yeast extract sucrose (YES) agar was chosen as the basal medium because it distinguishes P. viridicatum group II by a violet brown to reddish brown reverse color (Frisvad 1981). Chloramphenicol (50 µg/ml) and chlortetracycline (50 µg/ml) were added to inhibit bacterial growth. Fungicides added to YES agar to test for their ability to inhibit fast growing molds were pentachloronitrobenzene (100 µg/ml) plus rose bengal (25

μg/ml) (PRYES agar) and dichloran (2,6–dichloro–4–nitroaniline) (2 μg/ml)
plus rose bengal (25 μg/ml) (DRYES agar) (Frisvad 1983). These two media
were compared with DRBC, malt–salt agar (MSA) (Mislivec et al. 1975)
containing 7.5% NaCl, and DG18 medium. The pH of all media was adjusted to
5.6 before and after autoclaving. The plates were incubated at 20°C.

Microbiological examination. The five media described above were
compared for indicative and selective characteristics, using dilution plating
and direct plating on durum wheat (1 sample), buckwheat (1 sample), barley
containing ochratoxin A (15 samples) and barley containing no ochratoxin A
(2 samples). PRYES medium was tested on 40 samples of barley, 35 of which
contained various amounts of ochratoxin A but were not visibly moldy.

For dilution plating, the grains were milled in a water–cooled mill (IKA
Universalmuhle M 20). Following milling, 40 g of flour and 360 ml of
dilution medium (0.85% NaCl, 0.1% peptone and 0.1% polysorbitan in water)
were homogenized in a Stomacher for 2 min.

To test recovery of molds on these media, a good quality barley was
surface sterilized or used untreated (the moisture content was adjusted to
25%), and inoculated artificially with a conidial suspension of P.
viridicatum followed by incubation at 20°C for 3 weeks. After incubation,
grains were surface sterilized by treating with 1% NaOCl for 1 min, washing
twice in sterile water and then placed on PRYES and DRBC media. In
addition, 50 grains from each sample were plated directly on each medium.

After 1 and 2 weeks, plates were examined for the presence of
fast–growing molds such as Rhizopus, Mucor, Absidia, Trichoderma and Botrytis
spp., Aspergilli and Penicillia. Colonies of all Penicillium spp. were
examined, transferred to Czapek yeast autolysate agar (CYA) and MEA (Pitt
1979) and identified.

Results and discussion

Selective properties of the media. The combination of chloramphenicol
and chlortetracycline effectively inhibited bacteria on the media. Zygomycetes such as Rhizopus, Mucor and Absidia were effectively inhibited on
PRYES (Table 1). Rhizopus overgrowth was observed in several cases on DRBC
medium and on MSA. All media except DRBC were unsuitable for the detection
of field fungi such as Fusarium, Alternaria, Stemphylium and Cladosporium.

Recovery of Penicillia and Aspergilli. The recovery of mesophilic
Penicillia and Aspergilli was good on all media tested. The genus Eurotium
(Aspergillus glaucus group) was recovered very effectively on the YES-based
media, producing easily recognizable cleistothecia and medium sized colonies.
On MSA and DG18 medium, this genus had a tendency to grow too quickly, thus
masking slow growing Penicillia and Aspergilli. The recovery of species
capable of producing nephrotoxin was good on all the media. The percentages
of recovery of P. viridicatum on DRBC and PRYES media were almost equal,
indicating that PRYES was effective for this purpose. The low concentration
of pentachloronitrobenzene did not inhibit potential nephrotoxin producers
more than other Penicillia and Aspergilli tested. Quantitative counts of
the flora in barley samples are not presented here, but the counts were not
significantly different on the media tested.

Indication of important nephrotoxin producers. The morphological
differentiation of Penicillium spp. directly on isolation media is difficult
because of the small conidial structures, small differences in
macromorphology and conidial color (Pitt 1979). Frisvad (1981) noted that
P. viridicatum group II was the only terverticillate Penicillium group

Table 1.   Performance of five media containing chloramphenicol and chlortetracycline for inhibition of Zygomycetes and selection for nephrotoxic Penicillia at 20°C

Test	Performance[a] of medium				
	PRYES	DRYES	MSA	DG18	DRBC
Inhibition of Zygomycetes	++	+	+	++	+
Recovery of:					
Field fungi[b]	+	+	+	+	++
Penicillium spp.	++	++	++	++	++
Aspergillus spp.	++	++	++	++	++
Eurotium spp.	++	++	++	++	+
Differentiation of:					
Penicillium spp.	++	++	−	+	+
Penicillium viridicatum type II	++	++	−	−	−
Xanthomegnin producers	++	++	−	−	−

[a]Excellent performance, ++; generally good performance, +; poor performance, −
[b]Recovery of Fusarium, Cladosporium, Alternaria and Stemphylium spp. compared with infection percentages and counts on V8 juice-dextrose-yeast extract agar

producing a violet brown to brownish red non-diffusible reverse pigment on YES-based media.   This was confirmed by Frisvad (1983), where 187 isolates with a violet brown reverse colony color randomly collected from PRYES plates were all identified as P. viridicatum group II.   The violet brown reverse color was easily recognized on PRYES plates in spite of the rose color of the medium.   The obverse of P. viridicatum group II was greyish white to rose, occasionally turning to light yellow with age.   Genera with reverse colors close to the violet brown reverse of P. viridicatum group II were Cladosporium (pure brown colors, brown obverse) and Eurotium (pure orange or red colors) with a yellow, red or green obverse.

Several of the species investigated in subgenus Penicillium had reverse colors in yellow shades.   Other molds with a yellow reverse on PRYES medium were members of the Aspergillus niger group.   Species with a yellow obverse, after 7-8 days growth were P. aurantiogriseum, P. virdicatum group I, P. expansum, P. hirsutum and members of the Aspergillus niger group, the latter always having a small proportion of easily recognizable black to brown conidial heads.

Isolates with yellow reverse and obverse colors on PRYES were collected randomly from each of 40 samples of poor quality barley (5 isolates from each sample).   All 200 isolates were either identified as P. viridicatum group I or as P. aurantiogriseum.

Conclusion

The results show that PRYES medium is very suitable for screening nephrotoxin producers in cereals.   Bacteria and Zygomycetes are effectively inhibited, no mold species grow particularly fast on the medium at the expense of slower molds, seeds do not germinate and the sporulation of most molds is reduced, thus preventing the development of secondary colonies. The ochratoxin A and citrinin producing P. viridicatum group II is consistently indicated by a violet brown reverse, and xanthomegnin and viomellein producers (P. viridicatum group I and P. aurantiogriseum) are consistently indicated by a yellow reverse and obverse.

References

CIEGLER, A., FENNELL, D. I., SANSING, G. A., DETROY, R. W. & BENNETT, G. A. 1973 Mycotoxin producing strains of Penicillium viridicatum: classification into subgroups. Applied Microbiology 26, 271-278.

CIEGLER, A., LEE, L. S. & DUNN, J. J. 1981 Production of naphthoquinone mycotoxins and taxonomy of Penicillium viridicatum. Applied and Environmental Microbiology 42, 446-449.

FRISVAD, J. C. 1981 Physiological criteria and mycotoxin production as aids in identification of common asymmetric penicillia. Applied and Environmental Microbiology 41, 568-579.

FRISVAD, J. C. 1983 A selective and indicative medium for groups of Penicillium viridicatum producing different mycotoxins in cereals. Journal of Applied Bacteriology 54, 409-416.

MISLIVEC, P. B., DIETER, C. T. & BRUCE, V. R. 1975 Mycotoxin-producing potential of mold flora of dried beans. Applied Microbiology 29, 522-526.

PITT, J. I. 1979 The Genus Penicillium and its Teleomorphic States Eupenicillium and Talaromyces. London: Academic Press.

STACK, M. E. & MISLEVIC, P. B. 1978 Production of xanthomegnin and viomellein by isolates of Aspergillus ochraceus, Penicillium cyclopium and Penicillium viridicatum. Applied and Environmental Microbiology 36, 552-554.

J. C. FRISVAD

▶ Comparison of Selective Media and Direct or Dilution
   Plating for the Isolation of Toxigenic Fungi from Foods

Toxigenic molds are often isolated from foods using either direct plating or dilution plating. Mislivec & Bruce (1977) showed that the former method yielded more different species than the latter method. The number of toxigenic species isolated from a food sample will also depend on the use of surface disinfectants and the media used. The choice of methods depends on the information desired and the food or feedstuff investigated. The ability of media to differentiate toxigenic fungi including species of the same genus is also important, because isolation, identification and testing of toxigenicity of each isolate from a mycological analysis is very laborious. We present data which demonstrate that more than one of the methods mentioned above should be used when examining foods and feedstuffs for toxigenic fungi.

Materials and methods

The materials and methods used have been described elsewhere in this book by Frisvad et al. and Frisvad & Viuf. Samples containing various

mycoflora were examined without surface disinfection using direct and
dilution plating on DRBC, PRYES, DRYES, AFPA and DG18 media. The results
are recorded as percentage infection of kernels by each species (direct
plating) and as percentage of the total mesophilic flora, excluding Mucors
and yeasts for each species (dilution plating).

Results and discussion

Table 1 shows the results of the mycological examination of two samples
of green coffee. The important toxigenic species, Aspergillus ochraceus,
was detected only by direct plating and A. candidus was detected only by
dilution plating. It is interesting that A. fumigatus and A. candidus,
found only sporadically by Mislivec et al. (1983), was an important part of
the mycoflora of green coffee when dilution plated.

Table 1. The average mycoflora of two samples of green coffee beans

Mold	Plating method[a]	Media				
		DRBC	DG18	DRYES	PRYES	AFPA
Eurotium spp.	dir	0	100	_[b]	17	–
	dil	0	1	2	5	0
A. niger	dir	100	97	–	100	–
	dil	31	27	34	34	21
A. flavus	dir	4	53	–	52	–
	dil	0	0	0	0	0
A. ochraceus	dir	84	75	–	58	–
	dil	0	0	0	0	0
A. tamarii	dir	2	88	–	17	–
	dil	0	0	0	0	0
A. fumigatus	dir	0	0	–	0	–
	dil	7	0	13	10	9
A. candidus	dir	0	0	–	0	–
	dil	28	12	10	17	50
A. restrictus	dir	10	36	0	2	8
	dil	28	12	10	17	50
Penicillium spp.	dir	3	14	–	1	–
	dil	17	17	12	23	12
Wallemia sebi	dir	0	5	–	0	–
	dil	0	9	5	0	0

[a]dir = direct plating; data indicate % of infected kernels for each
species; dil = dilution plating; data indicate % of total flora for
each species
[b]Plates overgrown by mucoraceous fungi

Table 2. Mycoflora detected on a sample of barley kept in an airtight
silo for 18 months as determined by direct and dilution plating
on five selective media

Mold	Plating method	Media				
		DRBC	AFPA	DRYES	PRYES	DG18
P. roquefortii	dir	100	0	0	0	99
	dil	17	21	11	19	22
A. candidus	dir	75	50	0	0	97
	dil	14	15	25	28	26
Paecilomyces variotii	dir	0	100	0	0	0
	dil	0	0	0	0	0
Eurotium spp.	dir	0	0	0	0	83
	dil	0	0	0	0	0
Candida	dir	0	0	100	100	0
	dil	68	62	64	53	57

A. restrictus, Penicillium citrinum and Wallemia sebi were also more
regularly encountered in dilution plating. This may be due to the
relatively slow growth rate of these fungi and the fast growth rate of A.
niger and A. flavus. Two toxigenic species, A. flavus and A. tamarii, were
only detected by direct plating. The combination of DG18 medium with either
PRYES or DRBC media appears to be necessary to detect all important species
from the coffee samples.

Table 2 shows the result of a mycological analysis of a sample of barley
kept in an airtight silo for 18 months. If only direct plating was used,
DG18, PRYES and AFPA were necessary to detect all important species in the
barley sample. By using both direct and dilution plating, only two media
need be used.

For the general screening of foods for toxigenic species of filamentous
fungi, direct and dilution plating both have advantages. For direct plating,
DG18 and PRYES media should be used. Cereal seed of good quality will
germinate on AFPA and DRBC medium, and the presence of rose bengal in DRBC
prevents the use of alternating cycles of "black light" and darkness for
inducing sporulation of field fungi. For dilution plating, DG18 and DRBC
media are recommended. PRYES and AFPA should be used if rapid identification
of Penicillia and Aspergilli is desired because of the indicative and
differentiation properties of these media (Frisvad 1983; Pitt et al. 1983).

In conclusion, we recommend the use of both direct and dilution plating
for general examination of foods for toxigenic species of molds. The media
used should depend on the information wanted and the food item under
investigation, but DG18, PRYES, AFPA and DRBC media perform well in most
cases. A medium for enumerating field fungi should be developed. When only
limited resources are available, the combination of direct plating on DG18
medium and dilution plating on DRBC medium may be a good compromise.

References

FRISVAD, J. C. 1983 A selective and indicative medium for groups of
    Penicillium viridicatum producing different mycotoxins in cereals
    Journal of Applied Bacteriology 54, 409-416.
MISLIVEC, P. B. & BRUCE, V. R. 1977 Isolation of toxic and other mold
    species and genera in soybeans. Journal of Food Protection 40, 309-312.
MISLIVEC, P. B., BRUCE, V. R. & GIBSON, R. 1983 Incidence of toxigenic and
    other molds in green coffee beans. Journal of Food Protection 46,
    969-973.
PITT, J. I., HOCKING, A. D. & GLENN, D. R. 1983 An improved medium for
    the detection of Aspergillus flavus and A. parasiticus. Journal of
    Applied Bacteriology 54, 109-114.

J. C. FRISVAD
U. THRANE
O. FILTENBORG

▶ Comparison of Five Media for Enumeration of Mesophilic
Mycoflora of Foods

Several excellent media have been devised for the enumeration of
filamentous fungi in foods, but few of these are able to inhibit fast
growing mucoraceous fungi. DRBC and DG18 media appear to be excellent for
the enumeration and isolation of fungi in foods (Corry 1982). We wanted to
compare these media with PRYES and DRYES media developed by Frisvad (1983)
and AFPA for the general enumeration, isolation and direct identification of
molds in foods. Furthermore we wanted to determine the minimum number of
media that should be used to detect the dominant mesophilic mycoflora other
than yeasts and mucoraceous fungi.

Materials and methods

Media. PRYES and DRYES media were prepared as described by Frisvad
(1983). Pentachloronitrobenzene (100 µg/ml), one of the selective
ingredients in PRYES, is very sparingly soluble in water, and should be
added as a wettable powder (commercially available as Quintozen or
Brassicol, often in half strength) or dissolved in a small amount of toluene
(Gyllang et al. 1981). In the latter case, big crystals of PCNB should be
used as they do not contain talc, which is insoluble in toluene.

Sample preparation. Various food materials (5-20 g) were pummelled in a
Stomacher. Kernels, nuts and spices were milled in a water-cooled mill (IKA
Universalmühle, M 20) before stomaching. The dilution medium consisted of
NaCl (0.85%), peptone (0.1%) and polysorbitan 80 (0.1%). Logarithmic
dilutions and intermediate values were used to ensure 30-100 colonies on the
plates counted. The dilutions were spread-plated in duplicate and incubated
upright for 7 days at 25°C in the dark.

Food materials. A variety of foods were examined: green coffee beans
(imported to Denmark from Java and Colombia) (3 samples), black pepper (3
samples), curry (3 samples), mixed herbs (3 samples from France), camomile,
spearmint, alecost and sage (1 sample each from Greece), almonds (2
samples), walnuts (1 sample, U.S.A.), linseed (1 sample), barley (7 samples),

Table 1. Log$_{10}$ total counts of mesophilic filamentous fungi except Mucorales and yeasts on five selective media[a]

	Media				
Sample	DRBC	AFPA	DRYES	PRYES	DG18
Green coffee	4.81	4.57	4.53	4.35	4.70
Green coffee	4.73	4.62	4.50	4.60	4.82
Green coffee	4.88	4.93	4.69	4.97	5.10
Black pepper	4.57	B[c]	4.58	4.53	4.42
Black pepper	4.00	B	3.94	3.83	4.31
Black pepper	4.33	B	M	4.28	4.45
Curry	4.94	B,M[d]	M	5.16	5.26
Curry	4.49	M	M	4.93	4.84
Curry	3.28	B	M	3.75	3.80
Mixed herbs	4.04	3.95[b]	4.12	4.17	4.24
Mixed herbs	4.50	B,M	M	4.54	4.64
Mixed herbs	5.03	4.83[b]	4.72	5.12	5.03
Camomile	4.11[b]	B,M	M	4.07	4.22
Spearmint	3.60	B	3.26	3.54	3.81
Alecost	4.23	B	M	M	4.15
Sage	3.42	B	3.22	3.72	3.61
Almond	3.65	M	M	3.60	3.86
Walnut	4.27	M	M	4.19	4.28
Almond	4.43	M	4.26	4.24	4.38
Linseed	3.27[b]	B	3.38[b]	3.42[b]	3.96
Barley	6.61	6.83	6.83	7.02	7.07
Barley	3.35	3.31	3.13	3.45	3.06
Barley	4.99	B	4.95	5.00	4.87
Barley	3.06	3.18	3.02	3.27	2.48
Barley	5.46	5.47	5.43	5.36	5.73
Barley	6.02	5.56	5.36	5.53	5.54
Barley	3.20	3.60[b]	3.86	3.75	3.99
Oats	4.63	M	M	4.11	4.06
Oats	3.73	3.71	3.72	3.74	3.82
Mixed cereals	3.39	M	M	3.65	3.72
Rice	3.89	B	3.67	3.85	3.94
Onion	7.19	7.10	7.11	7.13	7.23
Radish	4.16[b]	B	B	B	4.01
Parsley	4.15	B	B	4.17	4.08
Artichoke	4.91	B	4.59[b]	4.76	4.88
Turnip	5.25[b]	B	B	B	5.11
Potato	5.26	B	5.32	5.47	5.39
Strawberry	3.65	3.42	3.64	3.61	3.63

[a]On all media, yeasts and Mucorales were sometimes able to be counted, but are not included
[b]Bacteria present
[c]B indicates bacterial overgrowth
[d]M indicates overgrowth by mucoraceous fungi, predominantly _Rhizopus stolonifer_ and _Mucor circinelloides_

oats (2 samples), rye (1 sample), mixed cereals (1 sample), rice (1 sample, Korea), onions (1 sample), radish (1 sample), parsley (1 sample), Jerusalem artichoke (1 sample), early garden turnip (1 sample), potatoes (1 sample) and strawberries (1 sample).

Identification. The fungi were identified according to Samson et al. (1984), von Arx (1981), Nelson et al. (1983), Domsch et al. (1980), Raper & Fennell (1965), Samson (1979), Pitt (1979) and Frisvad & Filtenborg (1983). Statistical analyses were performed according to Niemelä (1983).

Results

The total mesophilic filamentous populations of molds and yeasts in 39 samples of foods, feeds and other subtrates detected on five selective fungal enumeration media are listed in Table 1. DRBC, PRYES, and DG18 media worked very well in the examination of all samples. A t test on paired observations was performed on $\log_{10}$ counts for each combination of media (Table 2). The differences between these media were not significant. However, this close agreement between total counts may be misleading, because significant differences in quantitative counts and occurrence of different species was observed between media. Table 3 illustrates these differences.

AFPA did not work well when incubated for 7 days because it did not contain a carbohydrate and antibiotics other than chloramphenicol, thus permitting growth of Bacillus species such as B. licheniformis. Furthermore, overgrowth of mucoraceous fungi was often observed. Mucoraceous overgrowth was also observed on DRYES medium.

Wallemia sebi and members of the A. restrictus group were only detected on DG18 medium and occasionally on PRYES. Paecilomyces variotii was only found on AFPA. Colonies of A. niger and A. flavus on PRYES were rather extensive, masking other molds and making counting after 7 days difficult. Results from PRYES medium (and other media) should therefore be recorded after 4-5 days incubation. A significantly better recovery of Eurotium spp. on DG18 medium than the other media was expected (Hocking & Pitt 1980). While PRYES medium appeared to give significantly higher counts for Aspergilli other than those mentioned above, DG18 gave significantly higher Penicillium counts than DRBC and PRYES media. The influence of microbial interaction on plates is very important. Counts of dematiaceous fungi did not differ significantly on the various media, but it was difficult to identify these fungi to genus level because of poor conidiation. DRBC,

Table 2. Differences in $\log_{10}$ of counts between DRBC, PRYES, and DG18 agar as tested by a t %-test of paired observations (5% level)

	D_{av}[a]	f[b]	t	$t_{5\%}$	Significant difference
DRBC - PRYES	-0.027	35	0.733	2.03	No
DRBC - DG18	-0.074	38	1.576	2.02	No
PRYES - DG18	-0.063	35	1.603	2.03	No

[a]Average of differences between logarithmic counts
[b]Degrees of freedom

140

Table 3. Differences in $\log_{10}$ counts between DRBC, PRYES and DG18 media as tested by a $\underline{t}$ test of paired observations (5% level) for different groups of fungi

Mold	Media	D_{av}	f	t	$t_{5\%}$	Significant differences
A. niger	DRBC - PRYES	0.150	10	2.969	2.23	Yes
	DRBC - DG18	0.048	11	0.857	2.20	No
	PRYES - DG18	-0.105	10	1.954	2.23	No
Eurotium spp.	DRBC - PRYES	-0.496	3	3.341	2.31	Yes
	DRBC - DG18	-0.628	9	3.962	2.26	Yes
	PRYES - DG18	-0.269	12	3.996	2.18	Yes
Other Apergilli[a]	DRBC - PRYES	-0.054	19	0.621	2.09	No
	DRBC - DG18	0.131	18	1.195	2.10	No
	PRYES - DG18	0.177	18	2.289	2.10	Yes
Penicillia[b]	DRBC - PRYES	-0.004	21	0.060	2.08	No
	DRBC - DG18	-0.181	18	3.269	2.10	Yes
	PRYES - DG18	-0.253	17	2.962	2.11	Yes
Dematiaceae[c]	DRBC - PRYES	-0.010	5	0.106	2.57	No
	DRBC - DG18	-0.078	8	1.089	2.31	No
	PRYES - DG18	-0.30	5	0.255	2.57	No

[a]A. versicolor, A. fumigatus, A. flavus, A. candidus, A. tamarii and Emericella nidulans
[b]P. citrinum, P. chrysogenum, P. brevicompactum, P. glabrum, P. aurantiogriseum, P. roquefortii, P. crustosum, P. mali, P. hirsutum, P. camembertii, P. viridicatum, P. hirsutum, P. griseofulvum and P. melinii
[c]Cladosporium, Ulocladium, Alternaria, Drechslera and Embellisia

PRYES and DG18 media worked well in restricting the colony diameters of mucoraceous fungi, while AFPA and DRYES performed poorly.

In summary, DRBC, PRYES and DG18 media worked well for determining total populations of mesophilic fungi in 37 samples of foods. Total counts were not significantly different on DRBC, PRYES and DG18 media, but a combination of these media was necessary for the optimal recovery of all dominating species. We recommend the use of both DRBC and DG18 media for determining populations and types of fungi in food and feed samples. The addition of PRYES medium in the plating regime will aid in the identification of several Penicillia and ensure optimal recovery of some Aspergilli and Penicillia.

References

VON ARX, J. A. 1981 The Genera of Fungi Sporulating in Pure Culture 3. ed. Vaduz: J. Cramer Verlag.
CORRY, J. E. L. 1982 Assessment of the selectivity and productivity of media used in analytical mycology. Archiv für Lebensmittelhygiene 33, 160-164.
DOMSCH, K. H., GAMS, W. & ANDERSON, T.-H. 1980 Compendium of Soil Fungi. New York: Academic Press.

FRISVAD, J. C. 1983 A selective and indicative medium for groups of Penicillium viridicatum producing different mycotoxins in cereals. Journal of Applied Bacteriology 54, 409–416.

FRISVAD, J. C. & FILTENBORG, O. 1983 Classification of terverticillate Penicillia based on profiles of mycotoxins and other secondary metabolites. Applied and Environmental Microbiology 46, 1301–1310.

GYLLANG, H., KJELLEN, K., HAIKARA, A. & SIGSGAARD, P. 1981 Evaluation of fungal contaminations on barley and malt. Journal of the Institute of Brewing 87, 248–251.

HOCKING, A. D. & PITT, J. I. 1980 Dichloran-rose bengal medium for enumeration of xerophilic fungi from low-moisture foods. Applied and Environmental Microbiology 39, 488–492.

NELSON, P. E., TOUSSOUN, T. A. & MARASAS, W. F. O. 1983 "Fusarium Species. An Illustrated Manual for Identification". University Park: Pennsylvania State University Press.

NIEMELÄ, S. 1983 Statistical Evaluation of Results from Quantitative Microbiological Examinations. Uppsala Sweden: Nordic Committee on Food Analysis. (NMKL Report no. 1, 2. ed.). p. 21.

PITT, J. I. 1979 The Genus Penicillium and Its Teleomorphic States Eupenicillium and Talaromyces. London: Academic Press.

RAPER, K. B. & FENNELL, D. I. 1965 The Genus Aspergillus. Baltimore: Williams and Wilkins.

SAMSON, R. A. 1979 A compilation of the Aspergilli described since 1965. Studies in Mycology (Baarn) 18, 1–40.

SAMSON, R. A., HOEKSTRA, E. S. & VAN OORSHOT, C. A. N. 1984 Introduction to Food-borne Fungi. 2nd. ed. Baarn: Centraalbureau voor Schimmelcultures.

J. C. FRISVAD
U. THRANE
O. FILTENBORG

▶ A Selective Medium for Fusarium Species

Several experiments were conducted to evaluate incorporating 8-hydroxyquinoline sorbate plus chloramphenicol in malt agar for detecting Fusarium spp. in maize grains by direct plating. This fungicide has been promoted by some French agronomists to be relatively ineffective against Fusaria, providing the possibility of using it as a selective agent. The results obtained in our laboratory did not support this opinion. With 10 µg/ml of 8-hydroxyquinoline sorbate or more, the fungicidal effect is complete. In the range 0–2 µg/ml, Fusaria were perhaps less affected than other genera but the selective effect is not clear and further investigations are needed.

D. RICHARD-MOLARD

Comparison of Enumeration Methods for Molds Growing in
Flour Stored in Equilibrium with Relative Humidities
Between 66 and 68%

Species of <u>Xeromyces</u>, <u>Chrysosporium</u> and <u>Eremascus</u> are not often recovered
from foods because media commonly used for enumerating fungi or isolating
xerophilic molds are not suitable for their growth. Media with reduced a_w
such as malt salt agar (Christensen 1946), M40Y (Harrold 1950) and DG18
(Hocking & Pitt 1980) have been found to be valueless for these fungi;
instead, relatively rich media with further reductions of a_w are required
(Pitt 1975). In addition, these fungi are frequently in competition with
members of the <u>Aspergillus</u> <u>glaucus</u> group which readily overgrow them. It is
also known that when using dilution plating methods, damage by osmotic shock
can prevent growth unless diluents with reduced a_w are used (Pitt & Hocking
1982; Beuchat 1983).

This study was initiated when it was noted that certain reference flour
samples stored in glass containers for 10-12 months had an almost
imperceptible change in flow properties. Close visual examination revealed
nothing abnormal; however, when viewed under the microscope a fine network
of fungal mycelium permeated the flour throughout. With the naked eye, no
fungal growth or sporulation was evident. Water activities of several
samples ranged from 0.66-0.68.

Initial attempts to recover the fungus were unsuccessful using DG18
medium or mycological agar (Difco) supplemented with 6% NaCl. Mold
populations detected using these media were only slightly reduced from
initial values and ranged from 100-300/g. Members of the <u>A</u>. <u>glaucus</u> group
constituted about one-third of the population. Failure to recover the mold
prompted the use of other recovery media and a diluent with reduced a_w.

Materials and methods

Flour samples and preparation. The three flours examined in this study
included two white bread flours and one whole wheat flour. The flours had
been placed in tightly capped glass jars shortly after milling and stored at
room temperature for 10-12 months. Samples (22 g) of flour were uniformily
suspended by gentle hand shaking in either 198 ml of Butterfield's phosphate
buffer or 40% (w/v) dextrose water. Samples were suitably diluted using the
same diluents and conventional diluting procedures.

Media. Mycological agar (MA, a_w 0.98, Difco) was prepared according
to the manufacturer's directions. Chloramphenicol (100 µg/ml) was added
before autoclaving. MA supplemented with 6% NaCl (MAS, a_w 0.96) was
evaluated. Chloramphenicol (100 µg/ml) was added before autoclaving.
DG18 medium (a_w 0.96) was prepared as described by Hocking & Pitt (1980).
Chloramphenicol (100 µg/ml) replaced the chlortetracycline originally
prescribed, and was added before autoclaving. Malt extract yeast extract
40% glucose agar (MY40G, a_w 0.93) and 60% glucose agar (MY60G, a_w 0.87)
were prepared as described by Pitt (1975). Chloramphenicol (60 µg/ml) was
added to both media before steaming.

Incubation and inoculation. Pour-plates were prepared by adding 15-20
ml of medium tempered to 46°C to the inoculum in duplicate plates of each
dilution. Each pair was enclosed in a plastic bag to retard drying and
incubated at 25°C. Periodically, colonies were counted and plates were
returned to the incubator. Mold colonies typical of the flora commonly
encountered in flour were not included in the count. For the most part,
these occurred only in the 10^{-1} and 10^{-2} dilutions. The a_w of flour
and media was determined electronically with a Rotronic Hygroskop DT.

Table 1. Enumeration of molds growing in wheat flour stored in equilibrium with relative humidities between 66% and 68%

| | | | Mold counts/g (days incubated at 25°C)[a] | | | | | |
| | | | 10 days | | 14 days | | 22 days | |
Storage a_w	Flour	Medium	PO$_4$	Dextrose	PO$_4$	Dextrose	PO$_4$	Dextrose
0.66	1. White bread	MA	$<1\times10^3$	$<1\times10^3$	$<1\times10^3$	$<1\times10^3$	$<11\times10^3$	$<11\times10^3$
		MAS	$<1\times10^3$	$<1\times10^3$	$<1\times10^3$	$<1\times10^3$	$<11\times10^3$	$<11\times10^3$
		DG18	$<1\times10^3$	$<1\times10^3$	$<1\times10^3$	$<1\times10^3$	$<11\times10^3$	$<11\times10^3$
		MY40G	$<1\times10^3$	$<1\times10^3$	$<1\times10^3$	$<1\times10^3$	$<11\times10^3$	$<11\times10^3$
		MY60G	$<1\times10^3$	$<1\times10^3$	$<1\times10^3$	$<1\times10^3$	$<11\times10^3$	$<11\times10^3$
0.66	2. White bread	MA	$<1\times10^3$	$<1\times10^3$	$<1\times10^3$	$<1\times10^3$	$<1\times10^3$	$<1\times10^3$
		MAS	$<1\times10^3$	$<1\times10^3$	$<1\times10^3$	$<1\times10^3$	$<1\times10^3$	$<1\times10^3$
		DG18	2×10^3	3×10^3	1×10^4	6×10^3	1×10^4	7×10^3
		MY40G	2×10^4	3×10^4	3×10^4	3×10^4	2×10^4	3×10^4
		MY60G	4×10^4	3×10^4	5×10^3	3×10^4	5×10^4	5×10^4
0.68	3. Whole wheat	MA	$<1\times10^3$	$<1\times10^3$	$<1\times10^3$	$<1\times10^3$	$<11\times10^3$	$<11\times10^3$
		MAS	$<1\times10^3$	$<1\times10^3$	$<1\times10^3$	$<1\times10^3$	$<11\times10^3$	$<11\times10^3$
		DG18	$<1\times10^3$	$<1\times10^3$	$<1\times10^3$	$<1\times10^3$	$<11\times10^3$	$<11\times10^3$
		MY40G	$<1\times10^3$	$<1\times10^3$	$<1\times10^3$	$<1\times10^3$	$<11\times10^3$	$<11\times10^3$
		MY60G	$<1\times10^3$	$<1\times10^3$	$<1\times10^3$	$<1\times10^3$	$<11\times10^3$	$<11\times10^3$

[a]Diluents: PO$_4$ = Butterfield's phosphate buffer; dextrose = 40% dextrose water.

Results and discussion

The data in Table 1 demonstrate the importance of selecting the appropriate diluent, medium and length of incubation time. Counts were generally < 1 x 10^3 except for those on MY40G medium diluted with dextrose diluent in flour number 2 (white bread flour, a_w 0.66) after 6 days incubation.

None of the mold counts listed in Table 1 could have been made if the more commonly used methods for enumeration had been used. Normal flora in lower dilutions of flour would have been enumerated well before the molds in question had formed colonies. Even these counts do not adequately characterize the degree of mold proliferation which was observed microscopically in the samples. Lack of sporulation and inability of hyphal fragments to form colonies may have accounted in part for the discrepancy.

Two distinct colony types were isolated from the two white bread flours. These were transferred to plates of MY40G medium and further characterized. The mold of one colony type, later identified as Xeromyces bisporus, formed colonies that were moist, low, lacy early in development, colorless top and reverse and translucent. Cleistothecia were formed containing free "D" shaped ascospores. No asci were seen and the presence of aleuriospores was variable. Cleistothecia initials developed from finger-like projections surrounding a central cell as described by Fraser (1953) for X. bisporus. The mold from the other colony type, identified as a species of Chrysosporium, formed colonies that were moist, low, compact, colorless top and reverse and opaque. Numerous terminal and lateral aleuriospores were present and they became the predominant cell form with age. Colonies from the whole wheat flour were all of a single type characteristic of Xeromyces.

In summary, Xeromyces and Chrysosporium were isolated from wheat flours using a dilution plating procedure. These rarely detected molds were able to grow during storage at a_w values ranging from 0.66-0.68. The two molds grew to form the predominant flora, thus facilitated recovery when samples were diluted to the extent that competing molds would not be present.

Failure of media commonly used to enumerate xerophilic molds to support growth of Xeromyces and Chrysosporium was aptly demonstrated. Only MY40G and MY60G media were uniformly successful. The effectiveness of using a diluent of reduced a_w was also demonstrated.

Harrold (1950) was the first to employ enrichment techniques to isolate xerophilic molds when he attempted to isolate strains of Eremascus albus from samples of household mustard by storing them over saturated salt solutions. The use of similar enrichment techniques with other food commodites might well give information on the prevalence of these molds in nature. Pitt (1975) noted that the rarity of these fungi might be due more to methodology rather than an actual scarcity in nature.

References

BEUCHAT, L. R. 1983 Influence of water activity on growth, metabolic activities and survival of yeasts and molds. Journal of Food Protection 46, 135-141, 150.

CHRISTIAN, C. M. 1946 The quantitative determination of molds in flour. Cereal Chemistry 23, 322-329.

FRASER, L. 1953 A new genus of the Plectascales. Proceedings of the Linnean Society of New South Wales 78, 241-246.

HARROLD, C. E. 1950 Studies in the genus Eremascus. I. The rediscovery of Eremascus albus Edam and some new observations concerning its life-history and cytology. Annals of Botany 14, 127-148.

HOCKING, A. D. & PITT, J. I. 1980 Dichloran-glycerol medium for enumeration of xerophilic fungi from low-moisture foods. Applied and Environmental Microbiology 39, 488-492.

PITT, J. I. 1975 Xerophilic fungi and the spoilage of foods of plant origin. In Water Relations of Foods. R.B. Duckworth, London: Academic Press. pp. 273-307.

PITT, J. I. & HOCKING, A. D. 1982 Food spoilage fungi. I. Xeromyces bisporus Fraser. CSIRO Food Research Quarterly 42, 1-6.

R. B. FERGUSON

▶ Enumeration of Xerophilic Yeasts in Apple Concentrate

Spoilage of products with high sugar concentration, e.g., fruit concentrates, sugar and fruit syrups and jams, as well as other foods containing these ingredients, e.g., carbonated soft drinks, is mainly due to yeasts which tolerate low a_w. Methods for the enumeration of these yeasts were compared using three different media. Each was used in both solid and liquid form as a pour-plate or MPN method, respectively.

Materials and methods

Samples. Uniform samples of an apple concentrate produced by a Hungarian canning factory were evaluated by 17 laboratories of the Veterinary and Food Control Stations participating in this study.

Media. Compositions of the media used were as follows: acidified glucose yeast extract agar (AGY) contained glucose, 20 g; yeast extract, 5 g; agar, 15 g; in 1 liter of water. After melting, pH was adjusted to 4.5 with sterile 10% tartaric acid. Yeast extract medium with 40% glucose, (GY40) contained yeast extract, 5 g; glucose, 400 g; agar, 15 g; in 1 liter of water. De Whalley agar (DW) consisted of peptone, 2 g; yeast extract, 5 g; soluble starch, 2 g; glycerol, 1 g; NH_4Cl, 2 g; glucose, 400 g; agar 15 g. The same media were made as liquids, omitting the agar.

Enumeration. Log_{10} transformed data were statistically evaluated by analysis of variance.

Results and discussion

A total of 952 data points were collected from 17 laboratories. Statistical evaluation is not presented in detail. Before performing the analysis of variance, the homogeneity of means and standard deviations was checked by Dixon's test and Bartlett's test, respectively, and the outliers were omitted from further calculations.

Table 1 summarizes data from those laboratories which gave acceptable results for statistical evaluation. Evaluation of these data revealed that while there was no significant difference among the results on agar media, the two broths of low a_w gave lower counts. The difference was significant at the 0.05 level of probability. Calculation of standard

Table 1.   Log$_{10}$ counts of xerophilic yeasts on different media[a]

		Agar			Broth		
Lab		AGY	GY40	DW	AGY	GY40	DW
1		3.87	3.67	3.84	3.57	3.36	2.97
2		3.81	3.86	3.78	3.77	2.46	2.53
3		3.85	3.81	3.68	3.72	3.71	3.63
4		3.72	3.81	3.81	3.44	3.60	3.50
5		3.67	3.93	3.72	3.33	3.07	2.54
6		3.87	3.89	3.88	3.64	3.11	3.14
7		3.87	3.87	3.83	3.96	2.86	3.22
8		3.97	3.95	3.96	3.88	3.73	3.85
9		3.89	3.59	3.77	3.80	3.09	3.21
10		3.95	3.94	3.87	3.98	3.66	3.63
11		4.10	4.15	3.97	4.23	3.40	3.94
12		3.92	3.95	3.62	4.17	3.57	3.39
	Mean:	3.90	3.90	3.80	3.75	3.16	3.24

[a]$SD_{95\%}$ = 0.47; counts obtained on the 4th day of incubation; data represent means of duplicate trials

deviations gave values of 0.12 and 0.35 for pour–plate and MPN methods, respectively.

With a prolongation of incubation time, the counts of yeasts increased (Table 2). In broth containing 40% glucose, the counts obtained after 12 days were significantly higher than the respective values at 4 days.

From these results it can be reasonably concluded that xerophilic yeasts that occur in apple concentrate can grow well on media with high a_w (AGY agar and broth) and even on solidified media containing 40% glucose. Taking into account both sensitivity and accuracy, the pour–plate method using AGY agar appears to be suitable for the enumeration of xerophilic yeasts in apple concentrate. The method can be characterized with a standard

Table 2.   Increase of log$_{10}$ counts of xerophilic yeasts with incubation time[a]

Time (days)	Agar			Broth		
	AGY	GY40	DW	AGY	GY40	DW
4	3.90	3.90	3.80	3.75	3.16	3.24
6	3.94	3.96	3.93	3.85	3.29	3.56
12	–	–	–	–	3.44	3.75

[a]$SD_{95\%}$ = 0.18; data represent means of 28 measurements

147

deviation of $s_0 = 0.102$; its repeatability is $r = 0.28$ and reproducibility is $R = 0.38$ (in terms of $\log_{10}$ counts/g).

T. DEÁK
V. TABAJDI-PINTER
I. FABRI

▶ Comparison of Media for Counting Yeasts in High Fructose
Syrup

Production of high-fructose syrup from corn by modern immobilized enzyme technology was introduced in Hungary in 1980. At the beginning there was no standardized method to control the microbiological quality of the product, particularly for monitoring occurence of yeasts. Hence, a comparative study was designed with eleven participating laboratories, four of which belonged to inspection institutes and the others to factories that used the product, to evaluate media for enumerating yeasts in high-fructose syrup.

Materials and methods

High-fructose syrup. Because in a preliminary investigation no viable yeasts could be detected in high-fructose syrup, samples were inoculated with Zygosaccharomyces rouxii KE 15, a strain formerly isolated from concentrated grape must and maintained in the collection of the Department of Microbiology, University of Horticulture, Budapest.

Z. rouxii was first propagated in liquid glucose yeast extract medium at 30°C for 2 days. Then, 3 ml of culture were inoculated into 50 ml of high-fructose syrup which was transferred after 1 day into 500 ml of syrup. After thorough mixing, three samples were distributed to each laboratory for analysis.

Enumeration. Samples were diluted and plate counts were determined on three test media after incubation at 30°C for 3-5 days. Data are expressed as populations in 10 g (dry weight) of high-fructose syrup.

Media. The first medium was OGY; after melting and cooling to 45°C, oxytetracycline was added to give 100 µg/ml. The second was De Whalley medium (pH 7.0) (ICUMSA 1974); as described in the previous paper. The third medium was mycophil agar (Schneider 1979) (phytone, 10 g; glucose, 10 g; and agar, 15 g per liter of water). The medium was modified to pH 7.0 instead of 4.5, and chloramphenicol was added to give 100 µg/ml.

Results were evaluated statistically using $\log_{10}$ transformed data. Data obtained from two of eleven laboratories were selected by Dixon's test for their extreme mean values, and two other laboratories were selected by Bartlett's test for their high standard deviations. Mean values of data collected by the remaining seven laboratories were evaluated by two-factor analysis of variance to compare counts obtained on different media and to evaluate repeatability r and reproducibility R. Repeatability, i.e., intra-laboratory error, was evaluated by the formula

$$r = t \sqrt{2 s_0^2}$$

Table 1. Mean values of yeast counts obtained on three different media
by seven laboratories[a]

		Medium			
Lab		OGY	Mycophil	De Whalley	Mean
1		4.01	4.10	4.34	4.15
2		4.08	4.08	4.04	4.07
3		4.46	4.45	4.36	4.42
4		4.35	4.42	4.37	4.38
5		4.41	4.43	4.46	4.43
6		4.33	4.37	4.38	4.36
7		4.20	4.29	4.19	4.23
	Mean:	4.26	4.31	4.30	

[a]Yeast count, expressed per 10 g (dry weight) of syrup; data represent
mean values of triplicate tests; total data from 63 counts

where t is the value of Student's distribution, used here at the level of
P = 0.05 and at infinite degrees of freedom, and s_o^2 is the residual
variance. Reproducibility, i.e., inter-laboratory error, was calculated by
the formula

$$R = t \sqrt{2 \left(\frac{s_t^2}{n} + \frac{n-1}{n} \cdot s_o^2\right)}$$

where t and s_o^2 are as above, s_t^2 is the variance for different
laboratories, and n is the number of parallels.

Results and discussion

Mean yeast populations from the seven acceptable laboratories are listed
in Table 1. No significant differences were observed between the mean values
of yeast counts on three different test media. For the least significant
difference, a value of SD = 0.07 was calculated (P $\leq$ 0.05). A $\log_{10}$ value
of r = 0.2 was obtained for repeatability and R = 0.45 for reproducibility.

With 95% of probability, there was no significant difference between
viable populations detected on OGY, mycophil and De Whalley media. Z.
rouxii is xerophilic and the strain used was isolated from a product of low
a_w. Nevertheless, the counts obtained on media of high a_w (OGY,
mycophil) did not differ from those obtained on media containing 40% sucrose
(De Whalley).

Investigations in four of eleven laboratories were biased by a
systematic error of unknown character. However, in seven laboratories the
method of enumeration proved to be of acceptable repeatability and
reproducibility. The within-laboratory error did not exceed half of the
between-laboratory value of error.

Based on these observations, for the enumeration of yeasts in
high-fructose syrup, the pour-plate method using OGY is appropriate.

References

ICUMSA 1974 <u>Report of the Proceedings of the Sixteenth Session, Ankara</u>.
Manchester: John Roberts and Sons.
SCHNEIDER, F. (ed.) 1979 <u>Sugar Analysis. ICUMSA Methods</u>. Peterborough:
ICUMSA.

T. DEÁK
V. TABAJDI-PINTER
V. NAGEL
I. FABRI

▶ Techniques for the Enumeration of Heat Resistant Fungi

Heat resistant fungi may cause spoilage in acid foods given a
pasteurizing heat process. Spoilage of this type has been encountered most
commonly in canned and bottled strawberries (Olliver & Rendle 1934; Kavanagh
et al. 1963; Spurgin 1963; Put & Kruiswijk 1964). Other products where heat
resistant fungi have caused appreciable losses include canned blueberries
(Williams et al. 1941), grape juice (King et al. 1969), apple juice (Anon
1967), and fruit-gel baby foods and fruit juices packed in cardboard or
plastic without preservatives (Hocking & Pitt 1984).

Spoilage of acid heat processed foods is due to the fungal ascospore,
which possesses sufficient heat resistance to withstand pasteurizing
temperatures. It follows that only fungal genera which produce ascospores
can cause this type of spoilage. Moreover, as ascospores have long
maturation periods, poor factory hygiene is rarely to blame for spoilage
losses. In nearly all cases, the source of heat resistant fungal ascospores
is soil, and only raw materials which have had contact with soil, directly
or indirectly, are likely to be implicated. Berry fruits and pineapples,
which are readily contaminated by rain splash, and fruits harvested from the
ground, such as apples and passionfruit, are the likely sources of
contamination in a product.

<u>Types of spoilage</u>. In canned or bottled fruits affected by <u>Byssochlamys</u>
spp., the first sign of fungal spoilage is usually a slight softening of the
fruit. This progresses until total disintegration takes place, due to the
production of a powerful pectinase by the fungus (Hull 1939; Beuchat & Rice
1979). Off odors, and a slightly sour taste may develop, and there may be
gas production. It is rare for species other than <u>Byssochlamys</u> to be
responsible for spoilage of canned fruits.

<u>Byssochlamys</u> spp. are capable of growing at extremely low oxygen
tensions, and under these conditions in liquid products, they appear to grow
anaerobically and produce CO_2. A small amount of oxygen contained in the
headspace of a jar or bottle or slow leakage of oxygen through a package
such as a Tetra-Brik can provide sufficient oxygen for these fungi to grow.
The production of gas then causes visible swelling and spoilage of the
product. Other types of heat resistant fungi apparently do not produce
CO_2 during growth. Spoilage of products by these latter fungi is usually
visual, e.g., the presence of visible fungal colonies or clarification and
settling in fruit juices.

150

In solidified products such as fruit gels, heat resistant molds cause spoilage by growing as visible colonies on the surface of the product. Under these circumstances growth is aerobic and CO_2 is not produced.

Spoilage prevention. The best way to ensure that heat resistant molds will not cause spoilage of susceptible products is careful selection and handling of raw materials, in conjunction with an effective screening procedure for heat resistant fungal spores. For example, passionfruit crops consist of mostly smooth, sound fruit, together with a proportion which have wrinkled skins. The smooth fruit will generally have a lower mold spore count than the wrinkled fruit, and are much easier to decontaminate. Smooth fruit should be thoroughly washed, preferably in a hypochlorite solution (approximately 100 µg/ml), then rinsed before processing. Wrinkled fruit should be processed separately, and the juice or pulp should be used in products such as refrigerated or frozen desserts that will not be spoiled by heat resistant molds.

Common heat resistant fungi. It is probably true that most fungal ascospores possess relatively high heat resistance and could in theory cause food spoilage. In practice only a few species have been encountered in our work or reported in the literature. Principal among these are Byssochlamys fulva Olliver & G. Smith (Olliver & Rendle 1934; Beuchat & Rice 1979) and B. nivea Westling (Put & Kruiswijk 1964). Other fungi which less frequently cause spoilage in processed fruit products are Neosartorya fischeri (Wehmer) Malloch & Cain (= Aspergillus fischeri; Kavanagh et. al. 1963), Talaromyces flavus (Klöcker) Stolk & Samson and T. bacillisporus (Swift) C. Benjamin. Species from the genus Eupenicillium also form heat resistant ascospores, but do not commonly cause spoilage. Some Penicillium species form hard sclerotia, really undeveloped cleistothecia similar to those in Eupenicillium. These too appear to be highly heat resistant, and have occasionally caused spoilage (Williams et al. 1941). The taxonomy of heat resistant fungi is outside the scope of the present paper; for further information see Hocking & Pitt (1984).

Screening for heat resistant fungi. A number of methods have been developed for the enumeration of heat resistant fungal spores. All include a heat treatment, ranging from 5 min at 75°C to 35 min at 80°C. Most of these methods are outlined by Beuchat & Rice (1979). One of these,

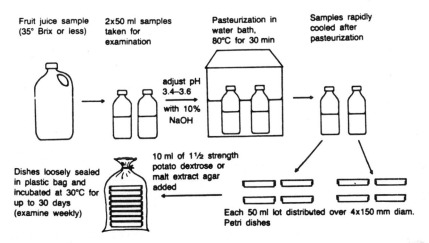

Fig. 1. Procedure for detection and enumeration of heat-resistant mold spores.

developed specifically for screening fruit juices and concentrates by
Murdock & Hatcher (1978), has been adapted (Hocking & Pitt 1984) to enable
the screening of a larger sample (100 ml) of raw material. The method is
described below, and outlined diagrammatically in Figure 1.

If the sample to be tested is more concentrated than 35° Brix, it should
first be diluted 1:1 with 0.1% peptone or similar diluent. For passionfruit
juice, which normally has a pH of about 2.0, the pH should be adjusted to
3.4-3.6 with 10% NaOH. Two 50-ml samples are taken for examination. The two
samples are heated in 200 x 30 mm test tubes or 100-ml medical flats in a
closed water bath at 80°C for 30 min, then rapidly cooled. Each 50-ml
sample is then distributed over four 150-mm Petri dishes and mixed with a
suitable agar (MA or PDA, with 1.5X the usual agar concentration). The
Petri dishes are loosely sealed in a plastic bag to prevent drying, and
incubated at 30°C for up to 30 days. Plates are examined weekly for
growth. Most surviving molds produce visible colonies within 10 days, but
incubation for up 30 days allows for germination and growth of badly heat
damaged spores. This long incubation time also allows most molds to mature
and sporulate, aiding their identification.

The main problem associated with this dilution technique is the
possibility of aerial contamination of the plates with common mold spores
which will give false positive results. The appearance of green _Penicillium_
colonies or colonies of common _Aspergillus_ species such as A. _flavus_ and A.
niger is a clear indication of contamination, as these fungi are not heat
resistant and their spores will not survive the 75 to 80°C heat treatment.
To minimize this problem, plates should be poured in clean, still air or a
laminar flow cabinet if possible. If a product contains large numbers of
heat resistant bacterial spores (e.g., _Bacillus_ spp.), antibiotics can be
added to the agar. The addition of chloramphenicol (100 μg/ml) to media
will prevent the growth of these bacteria.

A more direct method can be used for screening fruit pulps and other
semisolid materials which avoids the problems of aerial contamination.
Place aproximately 30 ml of pulp in a flat bottle such as a 100 ml medical
flat. Heat the bottle in the upright position for 30 min at 80°C and cool,
as described previously. The bottles of pulp can then be incubated
directly, without opening and without the addition of agar. They should be
incubated flat, allowing as large a surface area as possible, at 30°C for up
to 30 days. Any mold colonies which develop should be subcultured onto a
suitable medium for identification. If suitable containers such as Roux
bottles are available, larger samples can be examined by this technique.

References

ANON. 1967 Unusual heat resistance mould in apple juice. _Food
    Industries of South Africa_ 19, 55-56.
BEUCHAT, L. R. & RICE, S. L. 1979 _Byssochlamys_ spp. and their
    importance in processed fruits. _Advances in Food Research_ 25, 237-288.
HOCKING, A. D. & PITT, J. I. 1984 Food spoilage fungi. II. Heat
    resistant fungi. _CSIRO Food Research Quarterly_ 44, 73-82.
HULL, R. 1939 Study of _Byssochlamys fulva_ and control measures in
    processed fruits. _Annals of Applied Biology_ 26, 800-822.
KAVANAGH, J., LARCHET, N. & STUART, M. 1963 Occurence of a
    heat-resistant species of _Aspergillus_ in canned strawberries. _Nature_,
    London 198, 1322.
KING, A. D. Jr., MICHENER, H. D. & ITO, K. A. 1969 Control of
    Byssochlamys and related heat-resistant fungi in grape products.
    Applied Microbiology 18, 166-173.

MURDOCK, D. L. & HATCHER, W. S. 1978 A simple method to screen fruit
    juices and concentrates for heat-resistant mold. Journal of Food
    Protection 41, 254–256.
OLLIVER, M. & RENDLE, T. 1934 A new problem in fruit preservation. Studies
    of Byssochlamys fulva and its effect on the tissues of processed fruit.
    Journal of the Society for Chemical Industry, London 53, 166–172.
PUT, H. M. C. & KRUISWIJK, J. T. 1964 Disintegration and organoleptic
    deterioration of processed strawberries caused by the mould Byssochlamys
    nivea. Journal of Applied Bacteriology 27, 53–58.
SPURGIN, M. M. 1963 Suspected occurrence of Byssochlamys fulva in
    Queensland-grown canned strawberries. Queenland Journal of Agricultural
    and Animal Sciences 21, 247–250.
WILLIAMS, C. C., CAMERON, E. J. & WILLIAMS, O. B. 1941 A facultatively
    anaerobic mold of unusual heat resistance. Food Research 6, 69–73.

J. I. PITT
A. D. HOCKING

▶ Incidence, Properties and Detection of Heat Resistant
  Fungi

Heat resistant fungi are sometimes a problem in canned fruit products as
they can cause spoilage when they survive the normal heat process. The form
of spoilage can be indicated by swollen cans, by decomposed fruit due to
active pectinolytic enzymes or by appearance of fungal growth in the
product. The most frequently reported heat resistant molds are Byssochlamys
fulva and Neosartorya fischeri. Ascospores of these molds are the heat
resistant forms. The other fungal structures are heat sensitive. Conidia,
mycelial fragments and yeasts are inactivated by 1 min at 80°C while
Byssochlamys ascospores can survive for nearly 1 h at 90°C.

Heat processes for fruit products tend to be minimal because the low pH
of the product does not allow growth of spore-forming bacteria. The low pH
also makes microorganisms heat sensitive. Usually only heat sensitive
organisms grow in these products, but the occasional heat resistant mold
contaminant will withstand the mild heat treatment normally used. Mild heat
treatments are desirable because they lessen the thermal degradation of the
product. Apart from heating, ways of preventing spoilage by these molds
include adding preservatives such as benzoic or sorbic acid or sulfur
dioxide (Beuchat 1976) or physically removing the ascospores by filtration
(King et al. 1969). These heat resistant molds can grow in very small
amounts of oxygen, even in the headspace of a vacuum packed can, so removal
of oxygen is not an entirely useful control measure.

Bayne & Michener (1979) surveyed the heat resistance of 25 cultures of
B. fulva and found the maximum ascospore resistance to be 51 min at 90°C in
a defined medium consisting of 16 g glucose plus 0.5 g of tartaric acid per
100 ml water, adjusted to pH 3.6. The heat resistance of other molds has
been summarized by Splittstoesser & King (1984).

Several methods have been developed to isolate B. fulva ascospores from
natural substrates. These address the problem of a low contamination level,
usually by plating large volumes and by reducing the other flora by heat so
detection of B. fulva is possible. However, isolation, enumeration and
determination of heat resistance of molds have not been exhaustively studied.

A discussion of methods to isolate and enumerate heat resistant molds from contaminated products and to measure the heat resistance of these molds has been published (Splittstoesser & King 1984). It is recommended that these procedures be followed. The following details will serve to underline some important points.

When enumerating heat resistant molds, it is important to heat the product or diluted menstruum to inactivate heat sensitive molds and yeasts and to activate the heat resistant ascospores. Heat activation should be 80°C for 30 min or 70°C for 1 h. Plating on unacidified PDA containing rose bengal (8.3 µg/ml) prevents rapid spreading and permits easier counting. Incubation should be at 30°C. Since ascospores are often present in limited numbers (< 1/g of product), large volumes of material should be cultured. Put (1964) described a method that included a heating step before plating on a medium containing either 100 µg/ml chloramphemical or oxytetracycline and a prolonged incubation at 37°C. This procedure was used by Yates (1974) who also developed a centrifuging technique to concentrate spores prior to plating.

Splittstoesser et al. (1970) described a technique wherein 50–100 g of fruit were added to 100 ml of sterile grape juice in a screw-cap blendor jar and mixed for 5 min. The jar was then placed in a plastic bag and heated in a water bath for 2 h at 70°C. This treatment inactivated heat sensitive fungi and activated heat resistant mold ascospores so they would grow. The contents of the blender jar were distributed in approximately 10-ml quantities in 100-mm Petri dishes and an equal volume of double strength acidified PDA was added. The plates were incubated at 32°C and inspected at 48 and 96 h for the typical low tan growth of B. fulva.

A variation of this technique for examining fruit juice concentrates was described by Murdock & Hatcher (1978). They used 8-oz. screw-capped bottles with 50 ml sterile water and added 50 ml juice concentrate. The bottles were heated for 30 min at 77°C, cooled and plated with 2% agar. The plates and bottles were incubated at 30°C for up to 30 days.

Byssochlamys grows on a variety of mycological media so medium selection is not critical. However, for control of bacterial growth it is recommended that an antibiotic be used in the medium rather than acidification. The antibiotic media usually have better recovery rates since low pH tends to inhibit molds, particularly those having sustained some heat injury.

Heat resistance testing, although a related subject to isolation of heat resistant molds, will not be covered here. The subject is complex and a number of heating techniques have been used to determine the heat resistance of isolates. These techniques include testing the prepared spore population in screw-capped test tubes, stirred three-neck flasks, thermal death time tubes or thermal death time cans. The process of calculating heat resistance from the heat resistance data is likewise complex and not well defined. Several methods have been proposed (see Baggerman 1984; Bayne & Michener 1979; King et al. 1979; Michener & King 1974; Splittstoesser et al. 1974; Splittstoesser & King 1984).

It is recommended that the Murdock & Hatcher (1978) or the Splittstoesser (1970) techniques (see also Splittstoesser & King 1984) be used for enumerating and isolating heat resistant fungi but with the qualification that antibiotics rather than acidified agar be used.

154

References

BAGGERMAN, W. I. 1984 Heat resistance of yeast cells and fungal
     spores. In Introduction to food-borne fungi ed. Samson, R. A.,
     Hoekstra, E. S. and Van Oorschot C. A. N. Chapter 5, pp. 227-281.
     Baarn: Centraalbureau voor Schimmelcultures.
BAYNE, H. G. & MICHENER, H. D. 1979 Heat resistance of Byssochlamys
     ascospores. Applied and Environmental Microbiology 37, 449-453.
BEUCHAT, L. R. 1976 Effectiveness of various food preservatives in
     controlling the outgrowth of Byssochlamys nivea ascospores.
     Mycopathologia 59, 175-178.
KING, A. D. Jr., BAYNE, H. G. & ALDERTON, G. 1979 Nonlogarithmic
     death rate calculations for Byssochlamys fulva and other
     microorganisms. Applied and Environmental Microbiology 37, 596-600.
KING, A. D. Jr., MICHENER, H. D. & ITO, K. A. 1969 Control of
     Byssochlamys and related heat-resistant fungi in grape products.
     Applied Microbiology 18, 166-173.
MICHENER, H. D. & KING, A. D. Jr. 1974 Preparation of free
     heat-resistant ascospores from Byssochlamys asci. Applied Microbiology
     27, 671-678.
MURDOCK, D. I. & HATCHER, W. S. Jr. 1978 A simple method to screen
     fruit juices and concentrates for heat-resistant mold. Journal of Food
     Protection 41, 254-256.
PUT, H.M.C. 1964 A selective method for cultivating heat resistant
     moulds, particularly those of the genus Byssochlamys, and their presence
     in Dutch soil. Journal of Applied Bacteriology 27, 59-64.
SPLITTSTOESSER, D. F., EINSET, A., WILKISON, M. & PREZIOSE, J. 1974
     Effect of food ingredients on the heat resistance of Byssochlamys fulva
     ascospores. Proceedings IV International Congress Food Science and
     Technology 3, 79-85 Madrid.
SPLITTSTOESSER, D. F. & KING A. D. Jr. 1984 Enumeration of
     Byssochlamys and other heat resistant molds. In Compendium of Methods
     for the Microbiological Examination of Foods. ed. Vanderzant, C. and
     Gilliland, S. L. Ch. 17, Washington, D.C.: American Public Health
     Association.
SPLITTSTOESSER, D. F., KUSS F. R. & HARRISON, W. 1970 Enumeration
     of Byssochlamys and other heat-resistant molds. Applied Microbiology
     20, 393-397.
YATES, A. R. 1974 The occurrence of Byssochlamys sp. molds in
     Ontario. Canadian Institute of Food Science and Technology 7, 148-150.

A. D. KING JR.

Discussion

Splittstoesser
     When dilution plating is done, what is the most concentrated sample that
can be put onto a plate? Is it a tenth of a gram?

Frisvad
     It is usually one tenth of a milliliter that we plate to give a 10^{-2}
dilution. These are spread-plates.

Seiler
     Dr. Frisvad, I wonder if you could summarize what you would recommend as a
result of your work. It seems to have very widespread applications.

Frisvad

If you have limited resources, I suggest that you dilution plate in DRBC and direct plate onto DG18 medium. This is the best compromise in my estimation.

Pitt

Did you use AFPA at 25°C?

Frisvad

Yes, everything we did was at 25°C.

King

Is there any explanation that you can come up with why Paecilomyces only showed up on AFPA?

Frisvad

I think it is because Paecilomyces grows well on media without much carbohydrate.

Andrews

My experience in plating sorghum and wheat onto different media by direct plating is virtually similar except that I was using DG18, DRBC and my DCPA. The flora developing on different media were remarkedly different. For example, the DG18 gave good growth of Eurotium. In fact, the Eurotia grew too quickly on DG18 and overgrew everything on the plate, so I wouldn't recommend using DG18 for direct plating of grains for that reason. So we found that molds isolated on DG18, DRBC and DCPA were quite different. The only way that we were able to isolate Fusaria and Alternaria was by surface sterilization. In this case, only about 30% of the grains considered to be of good quality would show mold growth.

Frisvad

We need different types of media, depending on the organisms that we are attempting to isolate.

Seiler

I am getting confused here because there are so many types of media and so many types of organisms. Where do we begin?

Hocking

I would like to point out to Mr. Seiler that food bacteriologists have been coping with many different media and many different methods for many years and I don't think that we should restrict ourselves to a single medium. The requirements in mycological testing are so different and that is why we are developing these different selective media.

King

In developing DRBC, we looked at many antifungal ingredients -- PCNB was one of them. But we did not use this compound because it is a reported carcinogen and transportation of it in the U.S. is very difficult. It is a very difficult compound with which to work. Due to its reported carcinogenicity and problems with its safety with regard to laboratory technicians, we did not use this compound in our studies. Nevertheless, PCNB is a good compound for differentiating and selecting fungi.

Pitt

It should be pointed out that PCNB and dichloran are almost exact analogs. PCNB has a benzene ring as its central nucleus whereas dichloran has aniline as its central nucleus. It has always been accepted that aniline compounds are much safer to work with than benzene compounds. It would be sensible to recommend the use of dichloran rather than PCNB in media just for the simple reason that the aniline compound is more likely to be safe. There is very little difference between the two compounds in any of the work we have done.

Recommendations

It is recommended that the use of the term xerophile be maintained for fungi which grow at low a_w. A xerophile as defined by Pitt (1975) is a fungus capable of growth under at least one set of environmental conditions at an a_w of less than 0.85. It is also recommended that the use of an alternative term such as xerotroph be explored. Adequate definitions are needed. The terms osmophile and halophile are rejected for fungi which so far as is presently known exhibit no requirement for high osmotic pressure or high salt concentrations. For fungi previously referred to by these names, the term xerophile should be applied.

For enumeration of fungi from foods of low a_w (< 0.90) such as cereals, confectionery, jams and dried fruit, it is recommended that a reduced a_w medium be used. In some cases, e.g., where xerophilic yeasts are being enumerated, diluents of reduced a_w should also be used. It is recommended that further studies on the applicability of reduced a_w diluents should be carried out.

The evidence presented at this workshop indicates that DG18 medium is an appropriate medium for the enumeration of fungi in reduced a_w foods. Incubation should preferably be at 25°C for 5 days or as long as is necessary to achieve the highest count.

Some fungi such as <u>Xeromyces</u> <u>bisporus</u> and xerophilic <u>Chrysosporium</u> species grow poorly if at all on DG18. For these fungi, media of lower a_w such as MY50G, and prolonged incubation times are necessary. Isolation by direct plating or direct isolation is recommended. For <u>X</u>. <u>bisporus</u> in the presence of <u>Eurotium</u>, the only effective isolation medium is MY70GF.

For the enumeration of heat resistant fungi, the techniques described in this workshop are considered to be adequate and should be followed in conjunction with methods outlined in major papers on the subject. It is recommended that studies in this area be continued.

CHAPTER 4

MEDIA FOR YEASTS AND CONSIDERATION OF INJURED CELLS

A great variety of media are available for enumeration of yeasts and
molds in foods. Many of these have evolved from empirical studies, while
others have been formulated for analysis of specific types of foods, e.g.,
grains and flours, fruits or processed foods. Only a relatively few media
are available specifically for enumerating or at least facilitating the
growth of yeasts at the expense of molds and bacteria.

Certain yeasts are more likely than molds or bacteria to grow or be the
predominant microflora in some foods. For example, spoilage of confectionery
products and fruit syrups is likely to have been caused by yeasts which are
tolerant of or even require low a_w for growth. For these foods and others
which may harbor yeasts, special media for enumeration are desirable.

A proportion of yeast cells as well as mold spores or filaments in any
food being analyzed may be metabolically or structurally injured as a result
of physical or chemical stress. Such cells require optimum conditions for
recovery if subsequent repair and colony formation are to occur on
enumeration media. The following papers give some insight concerning the
appropriateness of specific media for enumerating yeasts and injured fungal
cells, spores or filaments in foods.

▶ Comparison of OGY with DRBC Medium for Enumerating
Yeasts in Dry Food Products

Progression in developing improved media for enumerating yeasts and
molds is well documented (Mossel et al. 1962; Overcast & Weakley 1969;
Mossel et al. 1970; Jarvis 1973; Mossel et al. 1975; Koburger & Rodgers
1978; King et al. 1979; Henson 1981). Much of the early work focused on
overcoming the inhibiting effects of acidified media on yeast and mold
recovery and/or reducing growth of acid tolerant bacteria. Later studies
for the most part were directed at raising pH of media and adding one or
more antibiotics to control bacterial growth. Several studies have shown
that media containing antibiotics are superior to acidified media for
counting fungi in foods (Koburger 1971; Beuchat 1979).

More recently, greater emphasis has been placed on developing media that
control mold overgrowth by species of Rhizopus and Mucor and to slow colony
development in general to favor slower growing fungi. Many inhibitory
compounds have been studied for their effect on retarding mold growth without
reducing fungal counts. Rose bengal and dichloran (2,6-dichloro-4-
nitroaniline) are two chemicals that were shown to be especially effective

and are combined with antibiotics in a number of media in use today (Jarvis 1973; King et al. 1979; Henson 1981).

Materials and methods

Media. In this study, two commonly used media were compared for their ability to enumerate yeasts. OGY (Mossel et al. 1970) has had wide use for many years, especially in European countries. DRBC (King et al. 1979) combines an antibiotic with two inhibitors that restrict colony size of the spreading molds that often prevent an accurate colony count.

Food samples. The foods selected for use in this study were dry food products that normally are contaminated with a wide variety of molds and a lower population of yeasts. Samples were purchased from retail stores specializing in handling bulk quantities. Whole wheat flour, white wheat flour, corn meal, corn flour, raw pecan halves, paprika, cayenne and chili powder were stored under conditions similar to those at point of purchase. Pecans were refrigerated; the remaining samples were held at room temperature.

Enumeration procedure. Five replicates of each food sample were analyzed. Twenty-two gram samples were mixed with 198 ml of 0.1% peptone water for 30 sec using a Colworth Stomacher 400. The mixed samples were suitably diluted further in 10-fold increments using peptone water and conventional diluting procedures. When plating the sample, 0.1 and 0.3 ml quantities of the 10^{-2} dilution were used; 0.1 ml quantities of higher dilutions were plated.

Spread plates of OGY and DRBC were prepared 18-24 h in advance of use to enable partial drying of the surface. Fifteen to 18 ml of medium tempered to 46°C was added to each Petri plate. After the media solidified, plates were inverted and stored at room temperature until used. Plates were inoculated by depositing 0.1 or 0.3 ml quantities of diluted sample on the agar surface. A single glass hockey stick was used to distribute the inoculum over the surface starting at the highest dilution in a series of plates.

Pour plates were prepared by adding 15-18 ml of melted and tempered (46°C) medium to the inoculum in each plate. The inoculum consisted of 0.1 or 0.3 ml quantities of sample dilutions. After mixing and solidifying, the plates were incubated undisturbed, surface side up, for 5 days at 25°C.

Results and discussion

Data comparing populations of yeasts recovered on OGY and DRBC are shown in Table 1. Mean counts from four types of food products are tabulated. Statistical analysis of the data could not be made because of unequal sample numbers and large standard deviation differences between food types. Taken at face value, it appears that OGY medium is more favorable than DRBC medium for recovering yeasts. In six of eight paired comparisons, counts from OGY medium were appreciably higher.

Heavy overgrowth on plates by species of Rhizopus and Mucor occurred much more frequently on OGY compared to DRBC medium. From a total of 80 samples tested, only 66 were countable using OGY medium as compared to 76 with DRBC. In countable plates, these figures represent losses of 18 and 5% respectively.

Not recorded but quite apparent was the reduction in colony size on many of the DRBC plates. This, combined with the effect of rose bengal on colony coloration, made counting an easier task.

160

Table 1. Comparison of OGY and DRBC media for enumerating yeasts in dry food products

Type of product	Method of plating	Evaluation of OGY		Evaluation of DRBC	
		Number of samples	Mean $\log_{10}$ count/g	Number of samples	Mean $\log_{10}$ count/g
Wheat	Spread	8	2.49	10	2.39
Corn	Spread	7	2.96	10	1.94
Nut	Spread	4	1.22	5	0.79
Spice	Spread	14	2.87	14	2.85
		33	$\bar{x} = 2.39$	39	$\bar{x} = 1.95$
Wheat	Pour	10	0.80	10	0.00
Corn	Pour	7	2.61	9	1.75
Nut	Pour	4	0.38	5	0.70
Spice	Pour	12	2.15	13	1.67
		33	$\bar{x} = 1.49$	37	$\bar{x} = 1.03$

Although it is not supported statistically because of abnormal sample numbers and standard deviations, OGY medium generally gave higher recovery of yeasts than DRBC medium. In retrospect, the number of totally overgrown OGY plates would probably have been much less if shorter incubation times and/or lower temperatures had been used. With these food products, 5 days at 25°C is excessive.

The DRBC medium substantially retards overgrowth with _Rhizopus_ and _Mucor_ species. This not only is an advantage in giving greater latitude in choosing countable plates, but also results in better control over contamination in laboratory environments when these organisms are present. However, this study did not demonstrate that DRBC was best for recovering low numbers of yeasts present in dry food products.

References

BEUCHAT, L. R. 1979 Comparison of acidified and antibiotic supplemented potato dextrose agar from three manufacturers for its capacity to recover fungi from foods. _Journal of Food Protection_ 42, 427–428.
HENSON, O. E. 1981 Dichloran as an inhibitor of mold spreading in fungal plating media: effects on colony diameter and enumeration. _Applied and Environmental Microbiology_ 42, 656–660.
JARVIS, B. 1973 Comparison of an improved rose bengal chlortetracycline agar with other media for the selective isolation and enumeration of moulds and yeasts in foods. _Journal of Applied Bacteriology_ 36, 723–727.
KING, A. D., HOCKING, A. D. & PITT, J. I. 1979 Dichloran – rose bengal medium for enumeration and isolation of molds from foods. _Applied and Environmental Microbiology_ 37, 959–964.
KOBURGER, J. A. 1971 Fungi in foods II. Some observations on acidulants used to adjust media pH for yeast and mold counts. _Journal of Milk and Food Technology_ 34, 475–477.

KOBURGER, J. A. & RODGERS, M. F. 1978 Single or multiple antibiotic-amended media to enumerate yeasts and molds. Journal of Food Protection 41, 367–369.

MOSSEL, D. A. A., VISSER, M. & MENGERINK, W. H. J. 1962 A comparison of media for the enumeration of moulds and yeasts in foods and beverages. Laboratory Practice 11, 109–112.

MOSSEL, D. A. A., KLEYNEN-SEMMELING, A. M. C., VINCENTIE, H. M., BEERENS, H. & CATSAVAS, M. 1970 Oxytetracycline-glucose-yeast extract agar for selective enumeration of moulds and yeasts in foods and clinical material.
Journal of Applied Bacteriology 33, 454–457.

MOSSEL, D. A. A., VEGA, C. L. & PUT, H. M. C. 1975 Further studies on the suitability of various media containing antibacterial antibiotics for enumeration of moulds in food and food environments. Journal of Applied Bacteriology 39, 15–22.

OVERCAST, W. W. & WEAKLEY, D. J. 1969 An aureomycin rose bengal agar for enumeration of yeasts and molds in cottage cheese. Journal of Milk and Food Technology 32, 422–445.

R. B. FERGUSON

Effect of Diluent and Medium Water Activity on Recovery of Yeasts from High Sugar Coatings and Fillings

A medium with a reduced a_w is required for enumeration of xerotolerant yeasts and molds in food. Usually media with a_w of about 0.95 are employed. However, such media will permit the growth of non-xerotolerant fungi and occasional bacteria. Where it is desired to enumerate only xerophilic organisms, media which have a reduced a_w of 0.85–0.90 are used but these require a long incubation period to obtain accurate results (Tilbury 1976). Various solutes have been recommended for reducing a_w including sugars such as sucrose, fructose and glucose, or mixtures thereof (Pitt 1975; Tilbury 1980), glycerol (Hocking & Pitt 1980) and sodium chloride (Christensen 1946).

It is often forgotten that the a_w of the diluent must also be reduced when examining foods of low a_w to avoid osmotic shock effects on yeasts (Seiler 1980). This communication describes tests carried out to determine the effects of diluent and medium a_w on recovery of yeasts from jam.

Materials and methods

In two separate tests, populations were determined in samples of jam known to be contaminated by yeasts. Counts were determined using a pour-plate method and various combinations of diluent (0.85% NaCl and 0.1% peptone in distilled water) and agar medium (MEA) containing 0–40% (w/w) anhydrous glucose. The plates were examined after incubation for 5 days at 27°C.

Results and discussion

The effect of glucose concentration in diluent and agar medium on recovery of yeasts from jam in the two tests is shown in Table 1. It is apparent that the number of yeasts detected became progressively higher as the glucose concentration in the diluent and medium was increased. Maximum recovery was obtained when both diluent and medium contained 40% glucose

Table 1. Effect of glucose concentration in diluent and agar medium on recovery of yeasts from jam

Glucose (%, w/w) in		Yeast population per g of jam	
Diluent	Medium	Sample 1	Sample 2
0	0	20	3,000
0	40	65	–
10	40	190	115,000
20	40	1,400	125,000
30	40	9,700	615,000
40	40	22,300	1,000,000
30	30	–	117,000
20	20	–	22,000
10	10	–	11,000

(a_w approximately 0.94). Further tests using mixtures of glucose and fructose to give diluents and media of lower a_w failed to improve recovery over that obtained with 40% glucose.

The results from this work indicate that a misleading indication of the numbers of yeasts present in a high-sugar coating or filling may be obtained if a high a_w diluent or agar medium is used or if the agar medium has a reduced a_w but the diluent does not.

The general implication from this study is that it may always be necessary to adjust the a_w of both diluent and medium in relation to the a_w of the food if the maximum recovery of yeasts is to be achieved. Further work on this important area is required.

References

CHRISTENSEN. C. M. 1946 The quantitative determination of molds in flour. Cereal Chemistry 23, 322–329.
HOCKING, A. D. & PITT, J. I. 1980 Dichloran–glycerol medium for enumeration of xerophilic fungi from low moisture foods. Applied and Environmental Microbiology 39, 488–492.
PITT, J. I. 1975 Xerophilic fungi and the spoilage of foods of plant origin. In Water Relations of Foods ed. Duckworth, R. pp. 273–307. London: Academic Press.
SEILER, D. A. L. 1980 Yeast spoilage of bakery products. In Biology and Activities of Yeasts ed. Skinner, F. A., Passmore, S. M., & Davenport, R. R. pp. 135–152. London: Academic Press.
TILBURY, R. H. 1976 The microbial stability of intermediate moisture foods with respect to yeasts. In Intermediate Moisture Foods ed. Davies, R., Birch, G. G., & Parker, K. J. pp. 138–165. London: Applied Science Publishers.
TILBURY, R. H. 1980 Xerotolerant (osmophilic) yeasts. In Biology and Activities of Yeasts ed. Skinner, F. A., Passmore, S. M., & Davenport, R. R. pp. 153–186. London: Academic Press.

D. A. L. SEILER

Comparison of DRBC, OGY and PDA Media for Enumeration of Yeasts in Beverage Products

The gradual refinement of media used for enumerating fungi in foods has led to the detection of not only a wider variety of fungi, but also sublethally injured cells and spores (Ladiges et al. 1974; Mossel et al. 1975). Media traditionally used were acidified with tartaric acid (e.g., acidified PDA) to preclude the growth of most bacteria while allowing the growth of fungi. Currently, acidified PDA is one of the most widely used medium to enumerate fungi in foods (King et al. 1979).

Recent evidence indicates that while acidified PDA inhibits bacterial growth, it also restricts the detection of many types of fungi as well as sublethally injured cells (Koburger 1971). Recovery of a wider range of fungi and sublethally damaged cells has been accomplished by neutralization of the medium pH and by the addition of various antibiotics (Mossel et al. 1970; Jarvis 1973; Mossel et al. 1975; King et al. 1979).

The choice of an appropriate medium to enumerate fungi is of concern to the food microbiology laboratory. Ideally, a single medium of wide-spread effectiveness would be of the greatest practical value; it is not feasible to employ a multiplicity of media from an economic viewpoint. While a number of studies have been done to evaluate the usefulness of different media for determining yeast and mold populations (Jarvis 1973; Koburger & Farhat 1975), acidified PDA is still widely used (U.S. Food and Drug Administration 1978). In this study, three media were compared for their ability to enumerate yeasts in beverage products and ingredients.

Materials and methods

Media. Media used to enumerate yeasts were acidified PDA (Difco), DRBC (Oxoid) and OGY (Mossel et al. 1970).

Samples. All samples were enumerated undiluted or following serial (1:10) dilutions in sterile 9-ml 0.1% peptone blanks. Pour-plates were prepared by adding 0.1 ml of an undiluted or diluted sample to sterile plastic Petri dishes. The samples were mixed with approximately 18 ml of tempered (50°C) medium and cooled to solidify the agar. Spread-plates were prepared by spreading 0.1 ml of an appropriately diluted sample over the agar surface with a sterile bent glass rod. All plates were incubated upright at 25°C and colony forming units (cfu) were counted after 5 days.

In studies using laboratory inocula, samples of a noncarbonated, high-acid beverage were inoculated with <u>Torulopsis</u> <u>rosei</u>, a yeast previously isolated from a similar contaminated product, to attain a cell concentration of approximately 100/ml. These beverage samples were then enumerated using PDA, DRBC and OGY as previously described.

Results and discussion

Table 1 lists data showing the performance of the three different media for recovering yeasts from high-acid beverages and their ingredients. With twenty naturally-contaminated samples, there was no indication that one medium was significantly more effective than another (P < 0.05). Although DRBC, PDA and OGY recovered the same level of yeasts in a statistical sense, the large sample-to-sample variation accounts for the inability to draw a firm conclusion.

Table 1. Comparison of three media for enumerating yeasts

| | Recovery medium | | |
Sample type	DRBC	PDA	OGY
Naturally contaminated	1.8 (± 1.9)[a]	2.1 (± 2.1)[a]	2.3 (± 2.5)[a]
Artificially contaminated[b]	2.1 (± 1.7)[c]	2.2 (± 1.7)[d]	2.1 (± 1.6)[d]

[a]Mean ± standard deviation (SD) of twenty samples of noncarbonated, high-acid beverage, fruit purees and concentrates
[b]Noncarbonated, high acid beverage receiving an inoculum of 100 cfu/ml of T. rosei
[c]Mean ± SD of three samples
[d]Mean ± SD of six samples

A visual examination of the sample means indicates OGY to be a more effective medium than either DRBC or PDA. In fact, conversion of the data from the log form more strikingly reflects this observation: DRBC (53 microbes/ml), PDA (125/ml) and OGY (200/ml). Additionally, what the numerical data do not reflect is the visual difference in colonial characteristics and morphology. Colonies grew to a larger size on OGY and PDA than on DRBC. Therefore, from a practical standpoint, OGY or PDA would be the media of choice for more rapid enumeration of yeasts in the absence of molds.

Table 1 also includes data summarizing the recovery of T. rosei. The yeast was suspended in the high-acid beverage and then enumerated on all three media. In these experiments, none of the media was significantly better than others in recovering the yeast from a beverage. Mean counts (± standard deviation) from all samples were almost identical. However, as with the naturally contaminated samples, colony size was smaller on DRBC than on PDA or OGY. Therefore, while DRBC may be able to statistically recover the same number of yeasts as the other two media, the small colony size of the yeasts on this medium makes it less desirable.

Recommendations

It is recommended that enumeration of yeasts in noncarbonated, high-acid beverage products or their ingredients be done using either OGY or acidified PDA.

References

JARVIS, B. 1973 Comparison of an improved rose bengal-chlortetracycline agar with other media for the selective isolation and enumeration of molds and yeasts in foods. Journal of Applied Bacteriology 36, 723-727.
KING, A. D., HOCKING, A. D. & PITT, J. I. 1979 Dichloran-rose bengal medium for enumeration and isolation of molds from foods. Applied and Environmental Microbiology 37, 959-964.
KOBURGER, J. A. & FARHAT, B. Y. 1975 Fungi in foods. VI. A comparison of media to enumerate yeasts and molds. Journal of Milk and Food Technology 38, 466-468.

KOBURGER, J. A. 1971 Fungi in foods. II. Some observations on acidulants used to adjust media pH for yeast and mold counts. Journal of Milk and Food Technology 34, 475-477.

LADIGES, W. C., FOSTER, J. F. & JORGENSON, J. J. 1974 Comparison of media for enumerating fungi in precooked frozen convenience foods. Journal of Milk and Food Technology 37, 302-304.

MOSSEL, D. A. A., KLEYNEN-SEMMELING, A. M. C., VINCENTIE, H. M., BEERENS, H. & CATSARAS, M. 1970 Oxytetracycline-glucose yeast extract agar for selective enumeration of moulds and yeasts in foods and clinical material. Journal of Applied Bacteriology 36, 454-457.

MOSSEL, D. A. A., VEGA, C. L. & PUT, H. M. 1975 Further studies on the suitability of various media containing antibacterial antibiotics for the enumeration of moulds in foods and food environments. Journal of Applied Bacteriology 39, 15-22.

U.S. FOOD AND DRUG ADMINISTRATION. 1978 Bacteriological Analytical Manual. 5th edn. Arlington: Association of Official Analytical Chemists.

C. B. ANDERSON
L. J. MOBERG

Sensitivity and Precision of Methods for the Enumeration of Yeasts in Wine

Results of enumeration of yeasts in foods and beverages are influenced by many factors, e.g., sampling technique, preparation of dilutions, composition of recovery media, method of enumeration and incubation time and temperature. In order to evaluate the performance of a microbiological method, one should analyze these variables with regard to their impact upon sensitivity, reliability and efficiency. Reliability manifests itself in the correctness, accuracy and precision of the method, while sensitivity characterizes the usefulness of the method, its specificity and limits in measurement. Efficiency encompasses practical factors (speed, cost, availability of media), while the other two attributes involve variables that can be analyzed in comparative studies.

Besides metabolic characteristics of microorganisms in the sample, their population also strongly influences the sensitivity and reliability of the enumeration method.

The applicability and sensitivity of three enumeration methods, i.e., pour plate, most probable number (MPN) and membrane filtration, were studied for enumerating four populations of yeasts in wine. The influence of the number of viable yeast cells in samples on the precision of enumeration was also evaluated.

Materials and methods

Dry white table wine of type Silvaner green was evaluated. Four samples were pasteurized and then inoculated with an appropriate amount of a suspension of Saccharomyces cerevisiae strain Tokaj 22 to give viable populations of the order of 1, 10, 100 and 1000/ml.

In the first investigation, must agar and acidified glucose yeast extract (AGY) agar were used as enumeration media. Must agar consisted of concentrated must diluted with tap water to give 15-16 (w/w%) sugar, plus 20

166

g agar in 1 liter (pH 5.5); AYEGA was composed of 20 g glucose, 5 g yeast extract and 15 g agar per liter of water; pH was adjusted to 4.5 with 10% tartaric acid. In the second study, AYEGA was used only for pour-plate and membrane filtration methods. A broth having the same composition, omitting agar, was used for MPN determinations. Enumeration by membrane filtration was carried out on samples containing 1 and 10 yeast cells/ml using a Sartorius membrane filter (0.45 μm pore diameter). The pour-plate and most probable number methods were applied to samples with 10, 100 and 1000 cells/ml. Data were evaluated by analysis of variance after $\log_{10}$ transformation.

Results and discussion

Yeast populations in samples were adjusted to differ by one $\log_{10}$ order. This was achieved by determining the yeast count on must agar and AGY agar by the pour-plate method. For samples with the smallest yeast population, membrane filtration was used. Counts obtained closely approached the intended levels, and the results on the two media did not differ significantly from each other (Table 1).

The standard deviations calculated to evaluate the effect of yeast level on the precision of determination are shown in Table 2. Except for sample 2, no significant differences were observed. In this sample the standard deviation of yeast count was higher than in other samples. This may be due to the small number of yeasts in the sample (16 cells/ml) which is lower than the acceptable limit of measurement for accurate statistical analyses by the pour plate method (30 cells/ml).

Table 3 summarizes data from an experiment in which the same sample was investigated by different methods. The MPN method resulted in higher numbers, particularly in samples with low levels of yeasts. Its standard deviation, however, did not differ from that derived from the pour-plate method. The membrane filtration method resulted in the smallest standard deviation.

The MPN and pour-plate methods can be reliably applied for the determination of yeast populations higher than 10 cells/ml. These methods entail a standard deviation about 6 cells/ml, which increases with decreasing cell count.

Table 1. Enumeration of yeasts on must agar and AGY agar from wine
containing a range of populations

$\log_{10}$ inoculum/ml	Mean population per ml[a]	
	Must agar	AGY agar
0.5	0.46 (.175)	0.55 (.205)
1.2	1.25 (.300)	1.22 (.335)
2.4	2.37 (.250)	2.41 (.220)
3.2	3.45 (.205)	3.36 (.220)

[a]Standard deviation shown in parenthesis; 16 replications

Table 2. Precision of enumeration methods to detect various yeast populations in wine

Sample number	Log$_{10}$ inoculum/ml	Method[a]	Standard deviation[b]	
			Within laboratory	Between laboratories
1	0.5	MF	0.11	0.18
2	1.2	PP	0.22	0.35
3	2.4	PP	0.15	0.26
4	3.2	PP	0.11	0.17

[a]MF = membrane filtration; PP = pour-plate
[b]16 replications

Table 3. Comparison of methods for enumerating various yeast populations in wine

Series	Method[a]	Mean population per ml[b]				Mean standard deviation
		Log 0	Log 1	Log 2	Log 3	
1	MPN	0.29	1.17	1.96	2.84	0.37
	PP	0.10	0.83	1.94	2.71	0.38
	MF	0.04	0.98	–	–	0.10
2	MPN	0.24	1.05	2.04	2.49	0.32
	PP	−0.30	0.55	1.61	2.48	0.30
	MF	−0.52	0.67	–	–	0.04

[a]MPN = most probable number; PP = pour plate; MF = membrane filtration
[b]Means of six replications

The precision of the membrane filtration method is high. Its standard deviation is approximately 1 cell/ml and it apparently is independent of the level of population. For the determination of yeast populations in wine, only the membrane filtration method can be recommended.

T. DEÁK
V. NAGEL
I. FABRI

 Consideration of Media for Enumerating Injured Fungi

Damage or injury to fungal cells upon exposure to environmental stress has been demonstrated by researchers in numerous laboratories (Table 1).

Table 1. Selected references reporting injury of yeasts and molds

Environmental condition causing injury	Reference
Refrigeration or subfreezing temperatures	Bank 1973
	Mazur 1966
	Mazur & Schmidt 1968
	Souzu 1967
	Tanaka & Miyatake 1975
Elevated temperature	Adams & Ordal 1976
	Beuchat 1982
	Graumlich & Stevenson 1978
	Hagler & Lewis 1974
	Nash & Sinclair 1968
	Schenberg-Frascino 1972
	Stevenson & Richards 1976
Irradiation	Baldy et al. 1970
	Padwal-Desai et al. 1976
	Sommer et al. 1963

While data on metabolic and structural injury of fungi are not as voluminous as data on bacteria, there is, nevertheless, convincing evidence showing that vegetative fungal cells and spores have increased sensitivity to a wide range of environmental parameters which may otherwise be innocuous to healthy (non-injured) propagules (Table 2).

Understanding of the injury phenomenon in yeasts and molds is in its infancy when compared to our knowledge about injury of bacteria. Especially lacking is information which might enable us to formulate detection and enumeration media which would enhance repair of injured fungal cells. Among the reports in the literature describing modifications of traditional media or development of new formulations, these have not generally been directed toward recovering stressed fungal cells naturally occurring in foods.

Thus there is an urgent need to focus research attention toward more clearly defining conditions under which injury of yeasts and molds will occur. Furthermore, having demonstrated cellular injury, it is essential that the target site or lesion be identified. Having accomplished this, the next step would be to modify existing formulae or develop new ones which will be superior for resuscitating injured cells. Other parameters such as incubation temperature and oxygen supply are also likely to play a role in establishing optimal conditions of fungal cell repair.

It may be desirable or in fact necessary to develop methods for enumerating injured fungi which would be selected depending upon the food type under investigation, even when the same organism is the subject of analysis. For example, enumeration of freeze-injured Saccharomyces cerevisiae in bread dough may require a set of recovery conditions quite different from that for heat-injured S. cerevisiae in fruit juice or osmotically-injured S. cerevisiae in confectionery products. Media for detecting injured spores may differ from those required to detect and resuscitate injured vegetative cells.

Table 2. Selected references reporting increased sensitivity of fungal cells after exposure to stress

Agent parameter to which stressed cells have increased sensitivity	Reference
Acid pH	Graumlich 1981
	Koburger 1970
	Mace & Koburger 1967
	Menegazzi & Ingledew 1980
	Nelson 1972
Water activity	Adams & Ordal 1976
	Beuchat 1982
Sodium chloride	Adams & Ordal 1976
	Beuchat & Jones 1978
	Fries 1969
Food preservatives	Beuchat 1981
	Beuchat & Jones 1978
	Shibasaki & Tsuchido 1973
	Tsuchido et al. 1972
Food antioxidants	Beuchat & Jones 1978
	Eubanks & Beuchat 1982
Essential oils of plants	Conner & Beuchat 1984

If mycological procedures recommended for use in quality assurance and other types of laboratories are to be optimized, a necessary part of future research activities will include studies to determine conditions necessary for recovering injured cells. There is clearly a need for intensifying such research activity in an increased number of food mycology laboratories.

References

ADAMS, G. H. & ORDAL, Z. J. 1976 Effects of thermal stress and reduced water activity on conidia of Aspergillus parasiticus. Journal of Food Science 41, 547-550.

BALDY, B. W., SOMMER, N. F. & BUCKLEY, P. M. 1970 Recovery of viability and radiation resistance by heat-injured conidia of Penicillium expansum Lk ex Thom. Journal of Bacteriology 102, 514-520.

BANK, H. 1973 Visualization of freezing damage. II. Structural alterations during warming. Cryobiology 10, 157-170.

BEUCHAT, L. R. 1981 Influence of potassium sorbate and sodium benzoate on heat inactivation of Aspergillus flavus, Penicillium puberulum and Geotrichum candidum. Journal of Food Protection 44, 450-454.

BEUCHAT, L. R. 1982 Effects of environmental stress in recovery media on colony formation by sublethally heat-treated Saccharomyces cerevisiae. Transactions of the British Mycological Society 78, 536-540.

BEUCHAT, L. R. & JONES, W. K. 1978 Effects of food preservatives and antioxidants on colony formation by heated conidia of Aspergillus flavus. Acta Alimentaria, Hungarian Academy of Science 7, 373-384.

CONNER, D. E. & BEUCHAT, L. R. 1984 Sensitivity of heat-stressed yeasts to essential oils of plants. Applied Environmental Microbiology 47, 229-233.

EUBANKS, V. L. & BEUCHAT, L. R. 1982 Increased sensitivity of heat-stressed <u>Saccharomyces</u> <u>cerevisiae</u> cells to food-grade antioxidants. <u>Applied Environmental Microbiology</u> 44, 604-610.

FRIES, N. 1969 Induced salt sensitivity in fungal cells and its reversal with imidazole derivatives. <u>Journal of Bacteriology</u> 100, 1424-1425.

GRAUMLICH, T. R. 1981 Survival and recovery of thermally stressed yeast in orange juice. <u>Journal of Food Science</u> 46, 1410-1411.

GRAUMLICH, T. R. & STEVENSON, K. E. 1978 Recovery of thermally injured <u>Saccharomyces</u> <u>cerevisiae</u>: effects of media and storage conditions. <u>Journal of Food Science</u> 43, 1865-1870.

HAGLER, A. N. & LEWIS, M. J. 1974 Effect of glucose on thermal injury of yeast that may define the maximum temperature of growth. <u>Journal of General Microbiology</u> 80, 101-109.

KOBURGER, J. A. 1970 Fungi in foods. I. Effects of inhibitor and incubation temperature on enumeration. <u>Journal of Milk and Food Technology</u> 33, 433-434.

MACE F. E. & KOBURGER, J. A. 1967 Effect of pH on recovery of fungi from foods. <u>Proceedings of the West Virginia Academy of Science</u> 39, 102-106.

MAZUR, P. 1966 Physical and chemical basis of injury in single-celled microorganisms subjected to freezing and thawing. In <u>Cryobiology</u> ed. Meryman, H. T. pp. 214-315. London: Academic Press.

MAZUR, P. & SCHMIDT, J. J. 1968 Interactions of cooling velocity, temperature, and warming velocity on the survival of frozen and thawed yeasts. <u>Cryobiology</u> 5, 1-17

MENEGAZZI, G. S. & INGLEDEW, W. M. 1980 Heat processing of spent brewer's yeast. <u>Journal of Food Science</u> 45, 182-186, 196.

NASH, C. H. & SINCLAIR, N. A. 1968 Thermal injury and death in an obligately psychrophilic yeast, <u>Candida</u> <u>nivalis</u>. <u>Canadian Journal of Microbiology</u> 14, 691-697.

NELSON, F. E. 1972 Plating medium pH as a factor in apparent survival of sublethally stressed yeasts. <u>Applied Microbiology</u> 24, 236-239.

PADWAL-DESAI, S. R., GHANEKAR, A. S. & SREENIVASAN, A. 1976 Studies on <u>Aspergillus</u> <u>flavus</u>. I. Factors influencing radiation resistance on non-germinating conidia. <u>Environmental and Experimental Botany</u> 16, 45-51.

SCHENBERG-FRASCINO, A. 1972 Lethal and mutagenic effects of elevated temperature on haploid yeast. II. Recovery from thermolesions. <u>Molecular Genetics</u> 117, 239-253.

SHIBASAKI, I., & TSUCHIDO, T. 1973 Enhancing effect of chemicals on the thermal injury of microorganisms. <u>Acta Alimentaria, Hungarian Academy of Science</u> 2, 327-349.

SOMMER, N. F., CREASEY, M. T., ROMANI, R. J. & MAXIE, E. C. 1963 Recovery of gamma irradiated <u>Rhizopus</u> <u>stolonifer</u> sporangiospores during auto-inhibition of germination. <u>Journal of Cell Physiology</u> 61, 93-98.

SOUZU, H. 1967 Location of polyphosphate and polyphosphatase in yeast cells and damage to the protoplasmic membrane of the cell by freeze-thawing. <u>Archives in Biochemistry and Biophysics</u> 120, 344-351.

STEVENSON, K. E. & RICHARDS, L. J. 1976 Thermal injury and recovery of <u>Saccharomyces</u> <u>cerevisiae</u>. <u>Journal of Food Science</u> 41, 136-137.

TANAKA, Y. & MIYATAKE, M. 1975 Studies on the injury of yeasts in frozen dough. Part I. Effect of prefermentation before freezing on the injury of yeast. <u>Journal of Food Science and Technology</u> (Tokyo) 22, 366-371.

TSUCHIDO, T., OKAZAKI, M. & SHIBASKI, I. 1972 Enhancing effect of chemicals on the thermal injury of microorganisms (II). Mechanism of the enhancing effect on sorbic acid upon the thermal injury of <u>Candida</u> <u>utilis</u>. <u>Journal of Fermentation Technology</u> 50, 341-348.

L. R. BEUCHAT

The "Phoenix Phenomenon" During Resuscitation of
Fungi in Foods

Recognition that an injury process can occur in microbial cells has had
a profound effect on methodology used in the area of food microbiology (APHA
1976; Ray 1979; Read 1979). Studies with bacteria have been numerous and
definitive in their conclusions regarding the many ways that this process
can occur as well as some of the ways in which cellular damage can manifest
itself.

Although bacteria have been extensively investigated, only limited
studies have been conducted on fungal injury. Factors affecting fungal
injury and procedures for recovery have been reviewed by Stevenson &
Graumlich (1978). In order to gain additional insight into the injury
phenomenon with naturally occurring populations of fungi in foods, a study
was undertaken in our laboratory to determine the effect of a pre-incubation
period in a recovery medium prior to plating on the viability of fungi
present in various foods. During these preliminary studies, two modes or
phases of recovery were observed. The first represented the normal growth
of a population enriched in a suitable medium. The second type of response
was that of a rapid initial die-off followed by a period of limited recovery.
This latter growth response has been called the "Phoenix Phenomenon" by
Collee et al. (1961). This report documents our attempts to eliminate or
otherwise alter this initial die-off phase.

Materials and methods

Samples of corn grits, fresh vegetables and ground beef were obtained
from retail food stores in the Gainesville, Florida, area. Dilutions of the
samples were prepared in Butterfield's phosphate buffer, blended at 8000 rpm
for 2 min and incubated at 20°C for up to 48 h. At selected intervals, the
incubated samples were mixed and a subsample was removed for determination
of viable count. Fungal counts were conducted by standard procedures in
antibiotic-supplemented PCA (APHA 1976), using triplicate plates at each
dilution, with incubation at 20°C for 5 days. All media were purchased from
Difco (Detroit, MI), and reagents were of AOAC quality.

Results and discussion

Figure 1 shows the typical "Phoenix Phenomenon" for ten samples of corn
grits incubated in Butterfield's buffer for up to 48 h. Not all food
samples exhibited a measurable reduction in count; for example, the ground
beef and most of the samples of fresh vegetables exhibited logarithmic
growth. Additionally, samples such as grits that contained a high
population of molds, rather than yeasts, exhibited the sharpest initial
decline. Initiation of the decrease, when it occurred, was measurable
within 30 min of dilution of the sample.

When different resuscitation menstrua and plating media were evaluated
in attempts to overcome the die-off, no approach lessened the effect.
Potato dextrose agar, trypticase soy agar and APT agar, in addition to
standard plate count agar, were employed without success. Dilution (and
incubation) menstrua such as plate count broth, Butterfield's buffer, 1%
NaCl, 10% glycerol or sucrose, and 0.1% peptone did not alter the initial
decrease in viable population.

Additional procedures were also investigated to resuscitate the cells.
Sample rehydration ratios (Van Schothorst et al. 1979) of 1:0.05 to 1:2 were

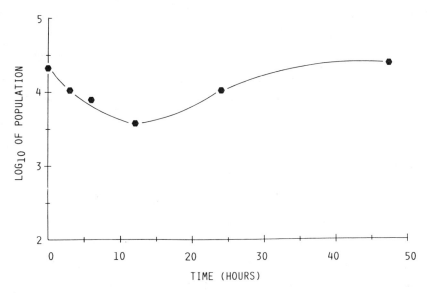

Fig. 1. Typical "Phoenix Phenomenon" of fungal population in corn
grits held in Butterfield's buffer for up to 48 h

used and the addition of NaCl (7%), sucrose (10%), glycerol (5, 10 and 20%)
(Hocking & Pitt 1980), yeast extract (1%), varying dextrose concentrations
(Koburger & Rodgers 1979), calcium (10 and 100 µg/ml) and catalase (50 and
100 units per ml) were added to the diluent and/or agar. Decimal dilutions
of the samples were also prepared and incubated in addition to the initial
1:10 dilution in an attempt to minimize the effect of inhibitory components
which may be present in the samples. Antibiotics (APHA 1976) were also
added to the resuscitation mixture to control growth of bacteria and their
possible production of inhibitory compounds. All of these procedures failed
to eliminate the initial fungal die-off.

Working with Clostridium perfringens, Shoemaker & Pierson (1976)
demonstrated that the "Phoenix Phenomenon" was an injury-recovery process
which could not be eliminated by a nonstress assay medium. Our results in
altering recovery conditions for fungi confirmed this observation, although
Mead (1969) had earlier suggested that this phenomenon may be a result of
the recovery medium used to study the organism.

Previous studies on microbial injury have concentrated on the use of
laboratory cultures to demonstrate injury. Heat, cold and various solutes
have been employed to elicit an expression of injury. Numerous hypotheses
have been suggested to account for the manifestation of injury. Enzyme
inactivation, DNA breakage, substrate effects and membrane damage have all
been shown to contribute to this injury (Postgate & Hunter 1963; Gomez &
Sinskey 1973; APHA 1976; Ray 1979). In dealing with a naturally occurring
mixed population of cells within a complex food system, it is extremely
difficult to design an experiment which would address these hypotheses.
This is particularly true of corn grits which had a microbial population
only in the thousands. However, recognition that such a phenomenon can

occur within a fungal population allows for speculation as to the significance of such an observation. Are spores or hyphal fragments involved? Is the phenomenon species dependent? How may this phenomenon be overcome so as to minimize its effect, and does it have practical implications during the enumeration of fungi? Lastly, these results again point out the limited knowledge we have about the fate of many organisms in foods, particularly fungi, which have received little attention relative to bacteria.

References

AMERICAN PUBLIC HEALTH ASSOCIATION. 1976 In Compendium of Methods for the Microbiological Examination of Foods ed. Speck, M. L. Washington, D.C.: American Public Health Association.

COLLEE, J. G., KNOWLDEN, J. A. & HOBBS, B. C. 1961 Studies on the growth, sporulation and carriage of Clostridium welchii with special reference to food poisoning strains. Journal of Applied Bacteriology 24, 236-339.

GOMEZ, R. F. & SINSKEY, A. J. 1973 Deoxyribonucleic acid breaks in heated Salmonella typhimurium LT-2 after exposure to nutritionally complex media. Journal of Bacteriology 115, 522-528.

HOCKING, A. D. & PITT, J. I. 1980 Dichloran-glycerol medium for enumeration of xerophilic fungi from low-moisture foods. Applied and Environmental Microbiology 39, 488-492.

KOBURGER, J. A. & RODGERS, M. F. 1979 Effect of glucose concentration on recovery of fungi from foods. Journal of Food Protection 42, 249-250.

MEAD, G. C. 1969 Growth and sporulation of Clostridium welchii in breast and leg muscle of poultry. Journal of Applied Bacteriology 32, 86-90.

POSTGATE, J. R. & HUNTER, J. R. 1963 Acceleration of bacterial death by growth substrates. Nature 198, 273.

SHOEMAKER, S. P. & PEIRSON, M. D. 1976 "Phoenix Phenomenon" in the growth of Clostridium perfringens. Applied and Environmental Microbiology 32, 803-807.

STEVENSON, K. E. & GRAUMLICH, T. R. 1978 Injury and recovery of yeasts and molds. Advances in Applied Microbiology 23, 203-217.

RAY, B. 1979 Detection of stressed microorganisms. Journal of Food Protection 42, 346-355.

READ, R. B. 1979. Detection of stressed microorganisms--Implications for regulatory monitoring. Journal of Food Protection 42, 368-369.

VAN SCHOTHORST, M., VAN LEUSDEN, F. M., DE GIER, E., RIJNIERSE, V. F. M. & VEEN, A. J. D. 1979 Influence of reconstitution on isolation of Salmonella from dried milk. Journal of Food Protection 42, 936-937.

J. A. KOBURGER

R. A. DARGAN

Discussion

SEILER

Can I suggest a reason for the "Phoenix Phenomenon?" According to the method you are describing, what you did was to soak the sample, blend it, leave it for a little while and reblend it. Could I suggest damage by blending?

KOBURGER

We still got the same phenomenon with shaking the dilution bottles and not blending it. I don't know if shaking that hard would cause any injury to the cells. I believe that it would not.

SEILER
The pattern looks similar.

KOBURGER
I know, but I still do not know why it comes back up.

SEILER
Probably because it begins to grow again.

KOBURGER
It comes back to the same level and then goes up within thirty minutes, so therefore we see a thirty percent reduction in thirty minutes and forty percent reduction in fifty minutes.

SEILER
You mix the sample while incubating, don't you?

KOBURGER
We mix it once by blending it for two minutes; then we let it sit for thirty minutes followed by thirty seconds of blending just to resuspend it.

SEILER
Well, I believe it is the second blending that is causing the problem.

KOBURGER
That may be, but I believe we have eliminated the possibility of a blender effect when we did the second dilution experiment, where we only blended once.

PITT
The best thing to do to overcome this problem is to plate onto the growth medium as quickly as possible.

KOBURGER
Absolutely. The question that must be asked, however, is this: Is there a population of injured cells that can recover and grow due to resuscitation? Now, there is no argument that the quicker you get the sample onto the medium, the better, but I think there is merit in attempting resuscitation in certain foods to see if additional populations can be recovered. I believe that cells become injured in nature just as we have shown in the lab.

Recommendations

Yeast media

The literature and data presented indicate that rose bengal is inhibitory to some yeasts. The panel recommends that unless mold overgrowth prevents the enumeration of yeast colonies, a general purpose antibiotic-supplemented medium should be used. These would include PDA, MEA, mycological agar, PCA, yeast morphology agar and glucose yeast extract agar.

The pH of the chosen medium should be between 5.8 and 7.0, and supplemented with a suitable antibiotic (40-100 mg/l). When heavy bacterial contamination of the food sample is expected, especially if the flora consists of Gram-negative bacteria, the use of two antiobiotics is recommended. Antibiotics should be added to tempered (46-48°C) agar immediately prior to pouring. Care should be exercised that the antibiotic preparation used possesses the activity claimed by the manufacturer. If mold overgrowth prevents the enumeration of yeasts, the addition of an antispreading agent such as dichloran is recommended. These inhibitors should be used at the minimum concentration necessary.

As a diluent for sample preparation and subsequent dilution, Butterfield's phosphate buffer, 0.1% peptone or 1/4 strength Ringers solution is recommended. The incorporation of a surfactant such as polysorbitan cannot be recommended as evidence for increased recovery of yeasts has not been demonstrated. The primary macerate should be prepared in the recommended diluent by stomaching or blending for 2 min, with surface-plating occurring within 10 min after maceration. Due to insufficient data, no further recommendations on macerate preparation can be made, but the panel has identified this as an area where further study should be undertaken.

The plates should be incubated in an upright position at 20-25°C for 5-7 days. Alternative methodology such as the hydrophobic grid membrane, MPN technique using solid and liquid media and quality presence or absence tests should be investigated.

Special foods may require deviation from the general procedures recommended. For example, the use of special diluents, plating media or incubation temperature for analysis of foods with reduced a_w or acid pH may be necessary for enumeration of a particular group of yeasts (e.g., xerophilic or psychrotrophic organisms).

Recovery of injured cells

Processing of foods may result in sublethal injury of yeast and mold propagules. It has been demonstrated that sublethal injury results in an increased sensitivity to chemical and physical extremes, e.g., pH, a_w and salt concentration. However, procedures for overcoming injury of fungi have not been fully described. The only recommendation that the panel can make is that if sublethal injury is suspected a general purpose plating medium containing minimal amounts of chemical and physical inhibitors should be used.

Due to the paucity of information on this subject, the panel recommends that the areas of diluent, maceration, liquid and solid media, inhibitor and nutrient requirements as well as incubation time and temperature should be studied with respect to resuscitation of sublethally injured fungal cells.

CHAPTER 5

SIGNIFICANCE OF FUNGAL POPULATIONS ON FOODS (BASELINE COUNTS)

"Baseline" counts are defined as the populations of molds and yeasts that normally occur in wholesome food or feed ingredients. Enumeration of fungal propagules is justified for the following reasons:

1. They provide an indication of compliance with Good Manufacturing Practice (GMP) conditions during the growing, processing or storage of the food
2. High numbers of particular groups or species may indicate that mycotoxins are present
3. Even low numbers of propagules of some species may be highly significant and undesirable in some instances (e.g., Zygosaccharomyces bailii in soft drinks or Byssochlamys species in fruit juice).

Maximum fungal populations are often included in specifications for commercial food transactions and occasionally standards are set by governments. The ICMSF (1974) suggested the same limits for molds in cereal by-products (e.g., bran and flour), sun-dried fruits, cocoa, coconut, nuts and spices. These were: $n = 5$; $c = 2$; $m = 10^2$; and $M = 10^4$. Thus, of five samples, none should contain more than 10^4 cfu/g and not more than two should contain more than 10^2 cfu/g. The ICMSF also suggested limits for xerophilic yeasts in dried fruits: $n = 5$; $c = 2$; $m = 10^1$; and $M = 10^3$ cfu/g. It is unlikely that these products can consistently comply with these specifications, or that samples which do not comply are necessarily unsafe to eat.

Consideration must also be given to the effect of storage on the viability of fungal propagules. It may be necessary to cautiously interpret the results of fungal counts, e.g., does a high count in processed food reflect a high count in the raw material or is it the result of poor sanitation during processing or poor storage after production? Identification of certain groups of fungi present may be of assistance in this respect.

The absence of mycotoxigenic molds in a food sample is obviously not sufficient to conclude that mycotoxins are not present. The molds responsible may no longer be viable because the food has been subjected to heat or other lethal processes such as drying or freezing. Alternatively, mycotoxigenic molds may be present but may not have produced mycotoxins in the food. In some products, especially low numbers of fungi may be required, e.g., xerophilic yeasts and molds in bottlers' sugar (ICMSF 1980).

References

ICMSF. 1974 Micro-Organisms in Foods 2. Sampling for Microbiological Analysis: Principles and Specific Applications. International Commission on Microbiological Specifications for Foods, Toronto: University of Toronto Press.

ICMSF. 1980 Microbial Ecology of Foods 2. Food Commodities. International
   Commission on Microbiological Specifications for Foods, New York:
   Academic Press.

▶ Fundamentals of the Rational Use of Target Values
   ("Standards") in the Microbiological Monitoring of
   Foods

   Optimal use of target values (baseline counts, acceptable levels,
specifications, limits, guidelines, standards, etc.) in the microbiological
monitoring of foods has long been marred by a lack of scientific principles.
Following critical papers on this subject by Thatcher (1955), Ingram (1961)
and Levine (1961), a basic philosophy was elaborated which is referred to as
a three-stage approach (Mossel 1980).

   The first stage consists of an assessment of attainable levels of
contamination with organisms of either health or keeping quality significance
when the best possible manufacturing, storage and distribution practices are
followed. Clinical medicine has demonstrated how to achieve this goal.
Surveys are carried out using a predetermined method for the assessment of
numbers of relevant organisms or levels of given metabolites. 'Relevance'
should be derived after thorough epidemiological surveillance and ecological
studies of every specific commodity. The results are plotted in a frequency
distribution diagram (profile) and subsequently tentative reference values
are derived.

   Attainable levels may not always be safe levels, however. Experience
has shown that commodities like raw milk or pre-processed frozen seafoods
are not necessarily safe, even when the best current sanitary practices are
followed. Consequently the second stage will ensure that the product is
henceforth safe, relying on G. S. Wilson's principle of active and validated
intervention (Kayser & Mossel 1984). Adopted reference values will
subsequently be determined by the surveys described earlier. Finally, the
third stage will aim at implementation of the reference values arrived at by
a survey of appropriate criteria. Decisions to be made include setting
acceptance values, expressed as m (roughly the 95th percentile of the
frequency distribution curve) and M, the maximum level that seems
acceptable; determining tolerances, usually a maximum of about 20% of
samples in the count area between m and M; and deciding on the destiny of
consignments found not acceptable (sorting, reprocessing, use in animal
feeds, etc).

References

INGRAM, M. 1961 Microbiological standards for foods. Food Technology,
   Champaign 15(2), 4-16.
KAYSER, A. & MOSSEL, D. A. A. 1984 Intervention sensu Wilson: the only
   valid approach to microbiological safety of food. International Journal
   of Food Microbiology 1, 1-4.
LEVINE, M. 1961 Facts and fancies of bacterial indices in standards for
   water and foods. Food Technology 15(11), 29-38.
MOSSEL, D. A. A. 1980 Assessment and control of microbiological health
   risks presented by foods. Reference values ('Standards'). In Food and
   Health: Science and Technology. ed. Birch, G. G. & Parker, K. J. pp.
   146-150. London: Applied Science Publishers Ltd.

THATCHER, F. S. 1955 Microbiological standards for foods: their function and limitations. Journal of Applied Bacteriology 18, 449–461.

D. A. A. MOSSEL

▶ Baseline Counts of Yeasts and Molds in Foods in The Netherlands

Several surveys have been carried out by the Food Inspection Services in the Netherlands to determine the microbiological quality of various types of food (Hartog et al. 1980; Northolt et al. 1980; Northolt & De Boer 1981; Hartog et al. 1982; Hartog & Van Kooy 1982; Hartog & Kuik 1982; Hartog & Jansen 1984; Northolt et al. 1984; Wiedemeijer & Pateer 1984). In addition, the National Institute of Public Health and Environmental Hygiene has carried out a monitoring program for some years for the same purpose. Fish and fish products, fruits, vegetables, spices, milk and pudding powders, powdered cheese, rye bread and pastries have been examined. Yeast and mold counts were included in these microbiological examinations and are summarized in this paper.

Materials and methods

The majority of samples was collected in retail stores, but samples were also collected in production areas. All foods sampled were ready to be sold to the consumer and were wholesome.

From each analysis, 20 g of sample were mixed with 180 ml of 0.1% peptone saline solution in a Stomacher. Further decimal dilutions were also made in 0.1% peptone saline solution (for details, see ISO-6887, Anon. 1983). From suitable dilutions, 1.0 ml was pipetted in duplicate into 90–mm Petri dishes and mixed with OGY (Mossel et al. 1962). After 5 days of incubation at 20–25°C, the number of yeast and/or mold colonies was counted and the colony count/g of food was calculated. When in doubt, several colonies were examined microscopically to discriminate between bacteria and yeasts and between yeasts and molds.

Results and discussion

Yeast and mold counts were divided into three groups. For yeast counts, the groups were $< 10^4$, 10^4-10^6 and $> 10^6$ cfu/g. For mold counts, the groups were $< 10^3$, 10^3-10^4 and $> 10^4$ cfu/g. Categorized in this way, Table 1 shows the baseline counts in fish and fish products, Table 2 in fruits, vegetables and spices and Table 3 in various other foods. Yeast counts exceeding 10^6/g and mold counts exceeding 10^4/g were rare (in all, 2% and 0.75%, respectively). The vast majority of the samples examined had yeast counts less than 10^4/g and mold counts less than 10^3/g. Some food items had a considerable percentage of samples with yeast or mold counts in the middle categories. For yeast counts, these included smoked fish, shrimp, raw cut vegetables, powdered cheese and pastries with $a_w \geq 0.90$ (in all, 15% of the samples), and for mold counts, semi-preserved fish, frozen cauliflower and kale, dried vegetables, spices, powdered cheese and pastries with $a_w < 0.90$ (in all, 5% of the samples).

Table 1. Baseline counts (cfu/g) of yeasts and molds in fish and fish products

Food	Number of samples	Yeasts (%)[a]		Molds (%)	
		$< 10^4$	$> 10^6$	$< 10^3$	$> 10^4$
Smoked fish	458	79	2	99	0
Frozen fish	105	100	0	100	0
Fish (semi-preserved)	52	100	0	84	8
Mussels (cooked, deshelled)	33	100	0	97	0
Shrimps (cooked, peeled)					
Southeast Asia	73	56	11	95	0
North Sea	73	62	5	95	0

[a]Percentage of number of samples

The number of yeasts and molds in a food may be used as an indicator for poor manufacturing and distributing practices. For that purpose, counts are only meaningful when related to reference values (microbiological criteria). Although the need for microbiological criteria can be reduced by application of the HACCP (Hazard Analysis Critical Control Points) principle (Christian 1982), continuous microbiological monitoring of a production line under conditions of good manufacturing practice will yield reference values

Table 2. Baseline counts (cfu/g) of yeasts and molds in fruit, vegetables and spices

Food	Number of samples	Yeasts (%)[a]		Molds (%)	
		$< 10^4$	$> 10^6$	$< 10^3$	$> 10^4$
Fruit and fruit pulp	116	94	0	93	0
Frozen vegetables					
Blanched	732	99	0	96	1
Cauliflower	22	100	0	77	0
Kale	53	100	0	89	0
Unblanched	59	98	0	98	0
Raw cut vegetables	323	41	7	93	0
Dried vegetables	169	95	0	84	5
Spices	91	99	0	88	3

[a]Percentage of number of samples

Table 3. Baseline counts (cfu/g) of yeasts and molds in various other foods

Food	Number of samples	Yeasts (%)[a]		Molds (%)	
		$< 10^4$	$> 10^6$	$< 10^3$	$> 10^4$
Milk powder	83	100	0	99	0
Pudding powder	87	100	0	100	0
Powdered cheese	160	55	22	71	5
Rye bread	204	97	0	95	0
Pastries					
$a_w \geq 0.90$	786	77	0	96	0
$a_w < 0.90$	78	95	0	87	0

[a]Percentage of number of samples

(guidelines) for raw materials, semi-processed products and end products. Counts exceeding these guidelines may indicate a problem in the production and distribution line such as recontamination or growth. From this point of view, maximum acceptable levels of yeasts and molds are included in the Dutch Food Law for some foods. Examples are presented in Table 4. For dairy products, these levels are rather stringent, while for products such as desserts and prepared salads levels are moderate. For most foods, we do not yet discriminate between yeasts and molds. The numbers given in Table 4 refer to the total numbers of yeasts and molds/g or ml of food. Only in a draft-regulation for pastries is there, for the first time, discrimination between the two types of fungi. Discrimination between yeasts and molds in microbiological criteria is desirable because of the difference in significance. The significance of yeasts is limited to spoilage of food, whereas that of molds may be related to public health, since the presence of a high number of certain molds may indicate the presence of mycotoxins in the food. For this reason, we recommend that yeasts and molds are distinguished when enumerating fungi in foods.

Table 4. Maximum acceptable levels of yeasts and molds (cfu/g or ml) in Dutch Food Law for various foods

Food	Yeasts	Molds
Ice cream with yogurt	10^2	10^2
Yogurt	10^2	10^2
Buttermilk	10^3	10^3
Desserts	10^4	10^4
Prepared salads	10^5	10^4
Pastries[a]	10^4	10^3

[a]Draft-regulation

References

ANON. 1983 International Organization for Standardization. Microbiology General Guidance for the Preparation of Dilutions for Microbiological Examination. ISO-6887.

CHRISTIAN, J. H. B. 1982 Developments in microbiological criteria for foods. Food Technology in Australia 34, 498-499.

HARTOG, B. J. & JANSEN, J. T. 1984 [Examination of the microbiological quality of raw cut vegetables]. De Ware(n) Chemicus. In preparation, in Dutch.

HARTOG, B. J., JANSEN, J. T. & BECKERS, H. J. 1982 [Examination of the microbiological quality of frozen vegetables]. De Ware(n) Chemicus 12, 1-13, in Dutch.

HARTOG, B. J. & KUIK, D. 1982 [Examination of the mycoflora of rye bread]. De Ware(n) Chemicus 12, 217-224, in Dutch.

HARTOG, B. J. & VAN KOOY, J. A. 1982 [Examination of the microbiological quality of smoked fish products]. De Ware(n) Chemicus 12, 97-116, in Dutch.

HARTOG, B. J., WIEDEMEIJER, J., JANSEN, J. T. & NOOITGEDAGT, A. J. 1980 [Examination of the microbiological quality of cream-, custard-, mochacream- and quarg-filled pastry]. De Ware(n) Chemicus 10, 159-172, in Dutch.

MOSSEL, D. A. A., VISSER, M. & MENGERINK, W. H. J. 1962 A comparison of media for the enumeration of moulds and yeasts in foods and beverages. Laboratory Practice 11, 109-112.

NORTHOLT, M. D., EIKELENBOOM, C., HARTOG, B. J., NOOITGEDAGT, A. J. & PATEER, P. M. 1980 [Examination of the microbiological quality and keeping quality of fruitfilled and dry cakes]. De Ware(n) Chemicus 10, 116-124, in Dutch.

NORTHOLT, M. D., & DE BOER, E. 1981 [Mycological examination of cheese powder]. De Ware(n) Chemicus 11, 112-115, in Dutch.

NORTHOLT, M. D., DE BOER, E., EIKELENBOOM, C., HARTOG, B. J. & NOOITGEDAGT, A. J. 1985 Mycological examination of raw materials for industries working up fruit. De Ware(n) Chemicus. 15, 87-89, in Dutch.

WIEDEMEIJER, J. & PATEER, P. M. 1984 [Examination of the microbiological quality of non frozen cooked and peeled shrimp in the retail market]. De Ware(n) Chemicus 14, 1-6, in Dutch.

H. J. BECKERS

Fungal Flora and Counts on Foods

Food products in consumer size packages obtained from a local grocery store were examined for their fungal flora. Almonds and small white beans were in industrial size packages. Samples were prepared by stomaching 50 g plus 450 g of sterile water for 2 min. Samples less than 50 g were analyzed in total (always > 10 g) at a ratio of 1:10 with water. Dilutions were made in 0.1% peptone water, and counts were done by surface-plating onto DRBC medium and incubating upright at 25°C for 5 days. Molds were identified usually to genus level only, directly from the DRBC medium.

Counts are shown in Table 1. Of the 68 samples, 46 had counts < 10^4 cfu/g and 48 < 10^2 cfu/g. These counts were surprisingly low and indicate either good quality raw products or the use of food processing techniques to decrease fungal populations in the raw products. Counts > 10^5 cfu/g were

Table 1. Fungal counts on products purchased from grocery stores (1982–1984)

Product	cfu/g	Identified flora
Dairy products		
Processed cheese	1×10^2	Cladosporium
Shredded Cheddar cheese	2×10^4	Penicillium, yeasts
Fruits		
Frozen blackberries	1.4×10^3	Cladosporium, yeasts
Frozen boysenberries	1.5×10^2	Cladosporium, Penicillium
Thompson seedless grapes	2×10^4	Alternaria, Cladosporium, Penicillium, Rhizopus, yeasts
Thompson seedless grapes	4.5×10^4	Alternaria, Cladosporium
Thompson seedless grapes	8.6×10^3	Penicillium, Rhizopus
Thompson seedless grapes	3×10^2	yeast
Frozen sliced peaches	5×10^1	
Seedless raisins	$< 10^2$	Aspergillus niger
Seedless raisins	5×10^1	Cladosporium, Mucor, Penicillium
Frozen raspberries	2×10^2	Cladosporium
Frozen raspberries	3×10^3	Penicillium, yeasts
Frozen raspberries	5×10^1	Penicillium, yeasts
Fresh strawberries	3×10^2	
Frozen sliced strawberries	5×10^1	Cladosporium, yeasts
Frozen sliced strawberries	2×10^2	Cladosporium, yeasts
Frozen whole strawberries	$< 10^2$	Cladosporium, yeasts
Grains and vegetables		
Pearled barley	1×10^2	Cladosporium, Penicillium, Rhizopus, yeasts
Pearled barley	2.5×10^2	Alternaria
Pearled barley	4.5×10^4	Alternaria
Lima beans	3.4×10^3	Aspergillus terreus, Cladosporium, Penicillium, yeasts
Frozen lima beans	2×10^2	Paecilomyces, yeasts
Pinto beans	$< 10^2$	Aspergillus terreus, Cladosporium, Penicillium
Small white beans, unprocessed	1×10^2	Alternaria, Cladosporium, Penicillium
Frozen cut corn	5×10^1	Cladosporium, yeasts
Yellow cornmeal	7.5×10^2	Aspergillus clavatus, A. flavus, A. glaucus
Yellow cornmeal	3×10^2	Cladosporium, Penicillium
Lentils	1×10^2	Aspergillus flavus, Penicillium
Macaroni	1.5×10^2	Cladosporium
Granola flour mix	2×10^2	Aspergillus, Penicillium, Rhizopus, yeasts
Frozen mixed vegetables	1.5×10^2	Cladosporium, yeasts
Rolled oats	1×10^2	Aspergillus flavus, yeasts
Frozen green peas	1.5×10^2	Cladosporium, Penicillium, yeasts
Green split peas	10^2	Penicillium, yeasts
Yellow split peas, frozen	5×10	
Frozen stuffed potatoes	$< 10^2$	
Long grain brown rice	1×10^2	Alternaria, Aspergillus flavus, Cladosporium, Rhizopus

(continued)

Table 1. Continued

Product	cfu/g	Identified flora
Long grain brown rice	5×10^2	Yeasts
Long grain brown rice	3.3×10^3	
Long grain white rice	1×10^2	
Minute rice	1.5×10^2	Cladosporium, Penicillium
Rice grains	1.4×10^4	Aspergillus flavus, A. niger, Cladosporium
Rice grains	1.1×10^3	Penicillium, Rhizopus, yeasts
Frozen cooked squash	$< 10^2$	Cladosporium
Frozen swiss chard	6×10^2	Aspergillus niger, Cladosporium, yeasts
Frozen turnip greens	5×10^1	Aspergillus fumigatus, Penicillium, yeasts
Wheat flour	5×10^1	Cladosporium, Penicillium
Wheat grains	6.1×10^4	Alternaria, Aspergillus, Cladosporium, Penicillium, yeasts

Meats
| Bologna sausages | 1×10^2 | Cladosporium |
| Frankfurter sausages | 5×10^1 | Cladosporium |

Nuts
Almonds	1.7×10^3	Aspergillus flavus, A. niger, Cladosporium
Almonds	5.6×10^3	Paecilomyces, Penicillium, Rhizopus, Syncephalastrum racemosum, yeasts
Salted peanuts	4×10^1	Cladosporium, Penicillium
Salted peanuts	1×10^2	
Raw Spanish peanuts	6×10^2	Aspergillus flavus, A. niger, Paecilomyces, Penicillium
Pecan pieces	1×10^2	Aspergillus, Pencillium, yeasts
Walnuts	6.5×10^2	Aspergillus flavus, A. niger, Mucor, Penicillium, Rhizopus, yeasts
Walnuts, sliced	3.5×10^2	Alternaria, Aspergillus flavus A. niger, Cladosporium Penicillium, Rhizopus, yeasts

Spices and condiments
Dehydrated chopped onions	2.7×10^3	Aspergillus niger, Penicillium
Dehydrated minced onions	1.5×10^3	A. niger, A. ochraceus, Cladosporium, Penicillium
Dehydrated parsley	5×10^1	Alternaria, Cladosporium
Black pepper	$< 10^2$	Aspergillus, A. niger, Cladosporium
Black pepper	$< 10^2$	Penicillium
Brown sugar	$< 10^2$	
Wheat germ	1.4×10^3	Cladosporium, yeasts

detected in shredded Cheddar cheese, fresh grapes, barley and unprocessed rice and wheat. Counts obtained from unprocessed grains were included to illustrate the influence of processing on fungal counts.

Cladosporium and Penicillium spp. were almost uniformly present in the foods. Other frequently identified organisms were Alternaria, Aspergillus, Rhizopus and yeasts. Sometimes the mycoflora consisted almost entirely of yeasts, e.g., on grapes or rice. At other times the mycoflora was mixed.

A. D. KING, JR.

▶ Baseline Levels of Fungi in Food Ingredients, Beverages, Frozen Foods and Vegetables

Routine screening of foods and food ingredients for microorganisms is an important aspect of any quality control program. Normally samples are screened for a total microbial count, Salmonella, Staphylococcus aureus, Escherichia coli, coliforms and yeast and mold populations. While the significance of many species and genera of bacteria have been established, this is not true of yeast and molds. The acceptable levels of fungi in products have not been correlated with the microbiological quality of the product. Increased concern over acceptable levels of yeasts and molds in foods has led to a concerted effort to determine baseline levels for various products. This paper presents the results of tabulating yeast and mold populations in over 1000 samples of beverages, frozen foods and food ingredients. Data were taken from daily samplings.

Materials and methods

Yeasts and molds were enumerated using pour-plates of acidified PDA. For dilution plating, samples were diluted using sterile 0.1% peptone. Incubation was for 2-5 days at 25°C. When testing for xerophiles, MY40 agar was used in addition to PDA.

Results and discussion

The numbers of yeasts and molds detected in the majority (85%) of 558 samples of noncarbonated, high-acid beverages examined were < 10 cfu/ml (Table 1). The remaining 15% of the products had yeast or mold populations ranging from 1-500 cfu/ml. In many cases, these counts were in excess of

Table 1. Baseline levels of fungi in beverages and juices (ranges, cfu/ml)

Beverage	Number of samples	Yeasts	Molds
Noncarbonated	472	< 10	< 10
High acid	86	1 - 500	1 - 170
Pineapple juice	13	< 10 - 300	< 10 - 70
Prune juice	11	< 10 - 2400	< 10 - 85
Juice concentrate	25	< 20	< 20

Table 2. Baseline levels of fungi in food products and ingredients (ranges, cfu/g)

Product	Number of samples	Yeasts	Molds
Frozen foods	235	< 10 - 1000	< 10 - 1000
Dairy products	24	< 10 - 10	< 10 - 20
Vegetables	25	< 10 - 50	< 10 - 100
Fruits	14	< 10	< 10 - 300
Flour	18	< 10 - 250	< 10 - 650
Starches	20	< 10 - 50	< 10 - 140
Stabilizers	6	< 10	< 10 - 160
Food chemicals	10	< 10 - 40	< 10 - 20
Sweeteners	20	< 10	< 10
Spices	26	< 10 - 10,000	< 10 - 10,000

product specifications, indicating that ingredient quality or plant sanitation was inadequate.

Higher fungal populations were detected in pineapple juice and prune juice than in other high-acid beverages (Table 1). Yeast and mold populations in both of these products were generally > 10 cfu/ml; occasional samples contained undetectable levels of microorganisms. Yeasts and molds, including xerophiles, were undetectable in 25 samples of juice concentrate, perhaps reflecting the greater stability of these products due to reduced a_w/high Brix conditions. Since these juice products or concentrates would have received sufficient heat processing to eliminate most fungal populations, these levels of fungal contamination indicate either inadequate processing or post process contamination.

The results of sampling 398 food products and ingredients are presented in Table 2. Of 235 finished frozen prepared food products, yeast and mold counts ranged from < 10-1000 cfu/g. The largest numbers of yeasts and molds in any of the food ingredients were detected in the spices. Fungal counts ranged from < 10-10,000 cfu/g, depending on the type of spice. Large numbers of fungi in spices may reflect the fact that they were grown and harvested in many parts of the world where poor sanitary conditions prevail (Julseth & Deibel 1974). Fungal counts in flours and starches, sweeteners, dairy products, stabilizers, fruits, vegetables and food chemicals are also presented in Table 2. In most cases, baseline counts for yeasts and molds in these ingredients were < 100 cfu/g. In many instances, these counts were < 10 cfu/g. These data suggest that since fungal populations in processed foods and food ingredients (as well as beverages) are generally low, detection of large numbers of yeasts or molds would be indicative of a potential spoilage situation.

Reference

JULSETH, R. H. & DEIBEL, R. H. 1974 Microbial profile of selected spices and herbs at import. Journal of Milk and Food Technology 37, 414-419.

C. B. ANDERSON
L. J. MOBERG

Baseline Counts for Yeasts in Beverages and Food Commodities

Determination of populations of yeasts and yeast-like organisms offers an attractive method for assessing quality of selected commodities. However, satisfactory interpretation depends on using a standard method and a thorough understanding of the organisms and their effect on the substrate. Hence, even by knowing the numbers and types of yeasts present, one still may not be able to make an accurate assessment or prediction of their significance in a commodity. Baseline counts and their significance vary from commodity to commodity. Tables 1 and 2 show how difficult it can be to recommend baseline counts. It is difficult to suggest such baseline counts for yeasts because safe levels will vary depending on the strain, and strains are affected by the ecological pressures of the local environment. For instance, in a poorly maintained factory, preservative-resistant strains of Zygosaccharomyces bailii may be detected. In addition, the methods used for counts need to be standardized to obtain consistent results.

Table 1. Yeast activity in commodities

| | Range of counts – 10-1000 cfu/g or ml[a] | | | |
Products	Rhodotorula glutinis	Hansenula anomala	Saccharomyces cerevisiae	Zygosaccharomyces bailii
Dairy	S	D	D	D
Chilled meat	D	D	D	D
Frozen foods	D	D	D	D
Brined foods	D	D	D	D
High sugar foods	D	D	S	S
Wines (in bottle)	D	D	S	S
Wines (bag-in-box)	D	S	S	S
Fruit juices				
Metal or glass packs	D	S	S	S
Bag-in-box	D	S	S	S

[a]S = Spoilage; rate depends on many factors; D = dies within a short time or persists for a long period without spoilage symptoms.

Table 2. Tolerance level for Zygosaccharomyces bailii

Stress	Tolerance
High salt	D[a]
High sugar	1 cell per 5 kilograms
Alcoholic beverages[b]/fruit juices, with or without preservatives	1 cell per > 10 liters

[a]See Table 1
[b]Excluding beers

R. R. DAVENPORT

Baseline Counts of Yeasts in Soft Drinks

Consumption of soft drinks has greatly increased in Hungary and is now approaching 100 liters per capita per annum. Some fifty varieties of soft drinks are produced by factories and plants of different capacities and technological standards. Controlling and monitoring product quality has become a major concern of responsible authorities and institutes. After methods were standardized in the mid-1970s, regular monitoring has greatly contributed to the improvement of the microbiological quality of the product.

Besides lactobacilli, yeasts are mainly responsible for spoilage of soft drinks, since under the prevailing ecological conditions only these organisms can grow (Deák et al. 1978; Török & Deák 1974; Deák 1980; for a comprehensive review, see ICMSF 1980). Consequently, their population should be kept as low as possible. However, it is difficult to determine a general baseline count for soft drinks of widely different composition. The Hungarian regulation states that the population of yeasts should not exceed 10/ml (MSZ 21338/1-80).

Materials and methods

Soft drinks were grouped for investigation into four types according to their basic ingredients, i.e., grape, citrus, fruit (mainly berries) and cola. Most drinks were carbonated and produced from juices, nectars and syrups. Determination of yeast populations was made according to the standard method (MSZ 21338/4-80) which employs GYE medium acidified with tartaric acid to pH 4.5 before use. Depending on the expected yeast level, membrane filtration (2 x 10 ml), MPN (3 x 1 ml) or pour-plate (2 x 1 ml) methods were used. Incubation was at 28°C for 3 days.

Results and discussion

Data collected from the laboratories of the Veterinary and Food Control Service are summarized in Tables 1 and 2. The mean yeast count was lowest in cola soft drinks and highest in those produced from fruit juices.

Table 1. Mean values ($\log_{10}$ per ml) and standard deviations of yeast populations in soft drinks of different composition

Statistical parameter	Year	Beverage type[a]			
		Grape	Citrus	Fruit	Cola
Mean	1982	0.36	0.06	0.48	0.06
	1983	0.51	0.49	0.55	0.14
Standard deviation	1982	0.51	0.44	0.58	0.39
	1983	0.38	0.63	0.52	0.45
Number of samples	1982	116	106	229	46
	1983	106	196	195	63

[a]General mean 0.295; 95% confidence limits ± 0.373

Table 2. Distribution of yeast populations in samples of soft drinks

| | Frequency (%) by beverage type | | | | | | | |
| | Grape | | Citrus | | Fruit | | Cola | |
Log_{10} count/ml	1982	1983	1982	1983	1982	1983	1982	1983
< 1 - 0	67	77	79	59	52	56	66	84
0 - 1	24	18	16	18	23	16	34	9
1 - 2	6	0	4	18	8	7	0	7
2 - 3	3	2	0	3	12	6	0	0
3 - 4	0	3	1	1	5	6	0	0

However, all types of soft drinks met requirements of the Hungarian standard. On average, less than 10% of the samples exceeded the limit of 10 yeasts/ml. In addition, the standard deviations were also small, indicating good manufacturing practice.

It is concluded that as a baseline count at the 95% probability level, log_{10} 0.295 ± 0.373 (approximately 0-4 cells of yeasts/ml) can be proposed for wholesome soft drinks.

References

DEÁK, T. 1980 Hygienical and technological consideration of yeasts in the soft drink industry. Acta Alimentaria 9, 90-91.
DEÁK, T., FABRI, I., TABAJDI-PINTÉR, V. & LENDVAI, I. 1978 Unification of methods for the microbiological control of beverages. Acta Alimentaria 7, 418.
ICMSF. 1980 Soft drinks, fruit juices, concentrates and fruit preserves. In Microbial Ecology of Foods, vol. 2 Food Commodities pp. 643-668. International Commission for Microbiological Specifications for Foods, New York: Academic Press.
MSZ 21338/1-80 1980 Non-alcoholic beverages. General technological requirements (Hungarian standard).
MSZ 21338/4-80 1980 Non-alcoholic beverages. Methods of microbiological investigation (Hungarian standard).
TÖRÖK, T. & DEÁK, T. 1974 A comparative study on yeasts from Hungarian soft drinks and on the methods of their enumeration. In Proceedings of the 4th International Symposium on Yeasts, Part 1. pp. 151-152.

T. DEÁK
V. TABAJDI-PINTÉR
I. FABRI

Products discussed in this paper have one basic property in common, i.e., low a_w, which retards growth of microorganisms. Nevertheless, these products may contain microbes that survive processing and persist in this special ecological niche. Of these microbes, it is only molds and yeasts which are potentially harmful from the spoilage point of view in products of low a_w. These microorganisms may start to grow when introduced into food of higher a_w. Hence, the microbiological profiles of sugar, high-fructose syrup and honey, which are frequently used ingredients, are of importance in microbiological food control (ICMSF 1980). While there are abundant data in the literature on sugar, little information is available on the microbiological quality of honey and high-fructose syrup.

Materials and methods

Samples of sugar were collected from ten factories. Granulated, refined, lump (cube) and powder (castor) grades of sugar were investigated. Commercial samples of honey were taken for investigation. In Hungary, there is only one factory which has produced sugar syrup from corn since 1981. Samples of finished product were taken from four storage tanks.

Yeasts and molds were enumerated by the pour-plate or MPN methods. Two milliliters of a 20% (w/v) solution of sugar and syrup were inoculated into each of five GYE agar (pH 5.6) plates and incubated at 28°C for 3-5 days. Data are expressed as count/10 g of sugar or 10 g (dry weight) of syrup, respectively. For honey, 1 ml was inoculated in triplicate into 60% sucrose broth (glucose, 10 g; yeast extract, 10 g; peptone, 10 g; and granulated sugar [sucrose] added to give a final concentration of 60% [w/v] in one liter of medium, ICUMSA 1978).

Results and discussion

In general, low counts of yeasts and molds were detected in these low a_w products. In honey, a mean yeast count of log 1.07/ml was found (Table 1). This was about one log less than the total count of mesophilic aerobic microorganisms. On GYE agar used for the enumeration of the latter, however, yeast colonies developed in about the same magnitude as in 60% sucrose broth, indicating that osmotolerant rather than xerophilic yeasts occurred in honey. The yeast counts did not show a normal, distribution even after logarithmic transformation. From Table 1 it can be seen that the

Table 1. Mean and distribution of counts ($\log_{10}$ cfu/ml) of yeasts in honey[a]

$\log_{10}$ count/ml	Frequency (%)
−1 − 0	21
0 − 0.5	17
0.5 − 1.0	7
1.0 − 1.5	45
1.5 − 2.0	10

[a]Mean 1.07; standard deviation 0.22; number of samples, 48

Table 2. Counts of yeasts and molds in high-fructose syrup

	Yeasts		Molds	
Parameter	1982	1983	1982	1983
Mean	1.34	0.52	0.75	0.31
Standard deviation	–	0.23	–	0.20
Number of samples	18	35	18	35

(header: Log_{10} cfu/10 g (dry weight))

distribution of yeast populations has two modes above and below the mean. This is due to different hygienic levels of the processors.

The production of high-fructose syrup from corn started only a few years ago in Hungary, and manufacturing practices have improved during this period as reflected in the microbiological quality of the product (Table 2). Microbiological limits corn syrup has been standardized to meet the same requirement established for sugar (MSZ 8800-83; Schneider 1979). Accordingly, a limit of 10 cfu/10 g has been set for yeasts and molds. The actual population detected were well below this limit.

In contrast to corn syrup, the production of beet sugar in Hungary has a long tradition. The monitoring of product quality is also a well established practice and is defined in a standard procedure (MSZ 788-84) which conforms to international recommendations (ICUMSA 1978). As Table 3 shows, both the mean and the distribution of counts of yeasts and molds give evidence of good product quality, and this holds true for all four grades of product investigated (Table 4).

In is concluded that the international recommendation of 10 cfu/10 g for yeasts or molds in sugar and similar products can be considered as a well

Table 3. Counts of yeast and molds in granulated beet sugar (log_{10} cfu/10 g)

Counts and statistical parameters	Yeasts		Molds	
	1982	1983	1982	1983
0 – 1	82.2	73.6	76.7	68.2
1 – 2	14.6	15.5	18.7	26.3
2 – 3	3.2	11.0	4.6	5.5
Mean	0.76	1.35	0.47	1.07
Standard deviation	0.29	0.36	0.41	0.38
Number of samples	280	290	305	340

(header: Frequency (%))

Table 4. Counts of yeast and molds in various grades of sugar (log_{10} cfu/10 g)[a]

Grades	Yeasts		Molds		Number of samples
	x	s	x	s	
Granulated	0.77	0.88	0.45	0.66	30
Refined	0.72	0.62	0.62	0.65	26
Lump	0.11	0.34	0.29	0.29	20
Powder	0.34	0.59	0.81	0.61	18

[a]x = mean; s = standard deviation

established baseline count attainable by good manufacturing practice. The same limit, but based on a 1-ml sample, seems appropriate for honey as well.

References

ICMSF 1980 Sugar, Cocoa, Chocolate, and Confectioneries. In Microbial Ecology of Foods, Vol. 2. Food Commodities, ICMSF. pp. 778-818. New York: Academic Press.
ICUMSA 1978 Report of the Proceedings. 17th Session Subject 21. Microbiological tests. pp. 329-340. International Commission for Uniform Methods of Sugar Analysis, Peterborough, England.
MSZ 8800-83 1983 Corn syrup. General technological requirements (Hungarian standard).
MSZ 788-84 1984 Sugar. Methods of microbiological investigation (Hungarian standard).
SCHNEIDER, F. (ed.) 1979 Sugar Analysis. Official and Tentative Methods Recommended by the International Commission for Uniform Methods of Sugar Analysis. Peterborough, England: British Sugar Corp. Ltd.

T. DEÁK
V. TABAJDI-PINTÉR
I. FABRI

 Baseline Counts of Molds in Flour

Wheat flour represents one of the basic ingredients of the traditional diet in Middle Europe. It is mainly used for producing bread and pasta, and is also added to a variety of prepared foods and products. Microorganisms occurring in flour may be of concern both from the public health and spoilage points of view. The presence of molds in flour, though hazardous because of their possible association with mycotoxins and potential for causing spoilage at low a_w, is not as important as staphylococcal enterotoxin or rope-forming bacilli. Mold populations, however, are important indices of manufacturing practice in routine monitoring of microbial quality of flour and cereal products (ICMSF 1980). It is generally accepted that flour may contain molds in the range of 10^2-10^4 cfu/g, though their numbers are

Table 1. Log$_{10}$ mold counts in flour (cfu/g)

Year	BL55	Number of samples	BL80	Number of samples
1979	–	–	2.4	34
1980	2.3	200	2.6	64
1981	2.1	238	2.1	152
1982	2.2	158	2.0	108
1983	2.3	154	2.2	215
Average:	2.21		2.19	

influenced by many factors (Hobbs & Greene 1976).

Materials and methods

Two grades of flour (BL55, a white flour, and BL80, a wholemeal flour) were examined over a 5-year period according to the standard method (MSZ 6369/16-83). Pour-plate counts of molds were determined by plating diluted 10-g samples on CGYE agar incubated at 25°C for 5 days.

Results and discussion

Experience over 5 yr based on several hundred samples shows that the number of mold propagules has a mean value of 160 cfu/g (log 2.2) in flour of both grade BL5 and BL80 (Table 1). The frequency distribution of samples (Table 2) is fairly normal, with about 70% of samples within the range ± one standard deviation from the mean.

Table 2. Distribution of mold counts in samples of flour

Log$_{10}$ cfu/g	Frequency (%)			
	BL55		BL80	
	1982	1983	1982	1983
< 1 – 0	21	–	4	–
0 – 1	4	10	5	9
1 – 2	21	38	29	19
2 – 3	46	43	47	53
3 – 4	8	8	15	18
4 – 5	–	1	–	1
Number of samples	74	158	139	233
Mean	2.19	2.22	2.31	2.27
Standard deviation	0.60	0.84	0.45	0.66

Mold contamination of the two grades of flour tested is low and the mean count varies little from year to year; 200 mold propagules/g appears to be a sound baseline count for the products in question.

References

ICMSF 1980 Cereal and Cereal Products. In Microbial Ecology of Foods Vol. 2. Food Commodities, ICMSF. pp. 669-730. New York: Academic Press.
HOBBS, W. E. & GREEN, V. W. 1976 Cereal and Cereal Products. In Compendium of Methods for the Microbiological Examination of Foods. ed. Speck, M. L. pp. 599-607. Washington, D.C.: American Public Health Association.
MSZ 6369/16-83 Methods for investigation of flour. Microbiological methods (Hungarian standard).

T. DEÁK
V. TABAJDI-PINTÉR
V. NAGEL
I. FABRI

 Baseline Counts for Wheat, Flour and Bran

There have been numerous investigations concerning the microbial flora of wheat, flour and to a lesser extent bran. Most investigations have been undertaken in North America where the wheat growing and harvesting conditions are different from those encountered in Europe and many other parts of the world. The present communication reports results from various surveys which have been undertaken in the United Kingdom which indicate counts of molds and yeasts which can be expected in these commodities in a temperate climate.

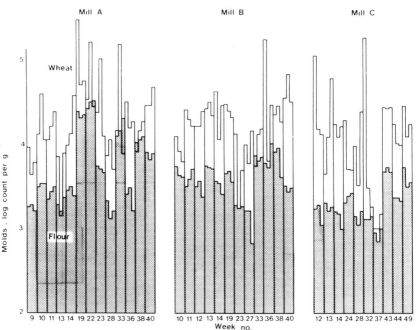

Fig. 1.   Counts of molds in samples of wheat and the derived flour from three mills during February to December, 1980

Materials and methods

In all the tests described, 40 g samples of wheat, flour or bran were mixed in 360 ml of peptone saline diluent using a stomacher to provide the primary dilution. Wheat and bran were soaked in diluent for 30 min and 15 min, respectively, prior to mixing. Counts of molds and yeasts were determined using a pour-plate method with OGYE agar after incubation at 27°C for 5 days.

Results and discussion

Mold counts obtained with samples of breadmaking wheat and its derived white flour from three mills on different occasions during the year are shown in Fig 1. On average, counts in flour are about 17% of those in wheat. It is apparent that the counts are highly variable. Very high mold counts approaching 10^6/g were obtained in some wheat samples. This survey took place in 1980 when the harvesting conditions were reasonably good. In 1977, which was a poor harvest year, many samples of wheat had mold counts in excess of 10^7/g. It should also be mentioned that the wheat examined in this survey was specially selected by the mills for producing breadmaking flour. Even greater variability in the mold count of wheat can be expected in the farm situation.

From 1975-1983, samples of straight run flour (white, with similar ash levels to grade BL55) milled from English wheat shortly after harvest were received from mills and bakeries for microbiological examination. The mean and maximum mold and yeast counts in the flour samples from these surveys are listed in Table 1. The counts varied from year to year according to harvesting conditions. In 1976 and 1981, which were particularly dry harvests, low counts were obtained, whereas in 1977, which was a wet harvest, the counts were high. The number of molds and yeasts in these flours is generally higher than that reported for flours from American or Canadian wheat (Christensen 1946; Hesseltine & Graves 1966; Mislivec et al. 1979; Rogers & Hesseltine 1978).

The results from surveys carried out in the Spring and Autumn of 1981 to determine the counts of molds in 96 samples of bran from five different wheat grists are given in Fig 2. There was little difference between

Table 1. Mold and yeast counts in flours from new harvest UK wheat in the years 1975 to 1983

Year	Number of samples examined	Molds Mean	Molds Maximum	Yeasts Mean	Yeasts Maximum
1975	35	6,500	15,000	–	–
1976	30	1,800	8,000	230	400
1977	19	14,000	57,000	1,300	7,000
1978	34	5,400	14,000	220	2,000
1979	29	4,100	11,000	380	2,100
1980	27	6,000	14,000	290	1,500
1981	32	1,900	6,000	140	500
1983	12	3,200	7,000	450	1,400

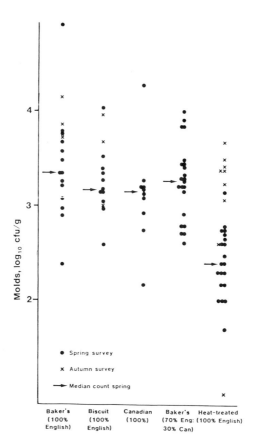

Fig. 2. Distribution of mold counts in brans from different grists

untreated brans from different grists, the mold count generally lying between 10^3 and 10^4/g. In the spring survey, counts in brans from heat-treated wheat were about ten-fold lower than in the untreated brans, but the difference was less in the autumn survey. The few results from the autumn survey suggest that the counts in bran can be expected to be about ten-fold higher than in flour (Table 1). Yeast counts in the brans tended to vary but most untreated samples contained 10^2–10^3/g. The level of count obtained in these surveys is similar to that reported by Spicher & Alonso (1978) in a survey of brans from West Germany.

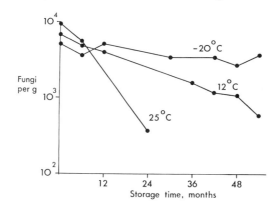

Fig. 3. Counts of fungi in flours stored at various temperatures

The populations of molds and yeasts in wheat, flour and bran with moisture contents below 15% tend to decrease slowly during storage, the rate depending on the storage temperature. This is shown in Fig. 3 which gives the change in average counts of fungi in flours (moisture content 12–13%) during storage at −20, 12 and 25°C. Based on these and other results, the populations of molds and yeasts in dry cereals or cereal products stored at ambient temperature could be expected to be halved after storage for 6 months. This factor may account for some of the differences in counts quoted for these commodities by various workers. However, difference in enumeration techniques between laboratories is likely to be a more important factor.

It is apparent that in any given year the counts of molds and yeasts in wheat, and hence in flour and bran, are likely to vary considerably. Moreover, the general level of counts will fluctuate from year to year according to the harvesting conditions. In this situation it is difficult to set realistic standards for such commodities.

References

CHRISTENSEN, C. M. 1946 Molds and bacteria in flour and their significance to the baking industry. Baker's Digest 21, 21–23.
HESSELTINE, C. W. & GRAVES, R. R. 1966 Microbiology of flours. Economic Botany 20, 156–168.
MISLIVEC, P. B., BRUCE, V. R. & ANDREWS, W. H. 1979 Microbiological survey of selected health foods. Applied and Environmental Microbiology 37, 567–571.
ROGERS, R. F. & HESSELTINE, C. W. 1978 Microflora of wheat and wheat flour from six areas of the United States. Cereal Chemistry 55, 889–898.
SPICHER, G. & ALONSO, M. 1978 Zur frage der mikrobiologischen qualitat von futterkleien und speisekleien. Getreide, Mehl und Brot 23, 178–181.

D. A. L. SEILER

▶ Baseline Counts of Molds in Dried Vegetables

Dried vegetables are increasingly used for soup mixes and as ingredients for preparing foods. They are inherently stable due to their low a_w, but may spoil if their moisture level increases due to improper storage. The normal microflora of dried vegetables is mainly composed of bacteria, among which pathogens may occur. Hence, the microbiological condition of dried vegetables is assessed by examining the bacterial flora. For assessing hygienic conditions at the time of manufacture, analysis for faecal streptococci is suggested (Kampelmacher et al. 1968). Surprisingly little attention has been given to molds, which possess the greatest potential to grow on dried vegetables, and are more suitable indicators than bacteria of the condition of the raw material and sanitation during processing.

Materials and methods

A pour–plate method using CGYE agar was applied for the determination of mold and yeast populations in dried vegetables. Taking into account their water–adsorbing capacity, 5 g samples were added to 95 ml of peptone saline to prepare the initial suspension. After shaking by hand for 10 min, the resulting suspension was allowed to settle for 20 min.

Table 1. Mold counts ($\log_{10}$ cfu/g) of onion slices produced by different factories

Factory	Mean	Standard deviation	Number of samples
1	2.50	0.18	5
2	1.92	0.16	8
3	2.35	0.35	11
4	2.40	0.40	4
5	2.00	0.50	5
6	2.40	0.50	27
7	1.95	0.70	17
8	2.20	1.00	15
9	2.50	1.00	18
10	0.60	1.00	11
11	2.10	1.10	27
Mean:	2.08	0.78	148

Various types of dried vegetables were investigated, including onion, red sweet paprika, horseradish and celery. Most vegetables were produced in two large drying plants belonging to canning factories, but in one study products of smaller manufacturers were also surveyed.

Results and discussion

The major dried vegetable is onion which is produced is the form of slices and powder. Since no blanching is applied during production, microbial quality depends upon the careful sorting of raw material. In the case of onion, this is a difficult task because saprophytic molds often develop under the outer leaves of the bulb. Nevertheless, in plants applying good manufacturing practices, dried sliced onion with fairly low mold counts can be produced (Table 1). Onion powder is produced by grinding uneven slices, off-cut pieces and other remnants; hence, its microbial quality is much inferior to onion slices (Table 2). For some other dried vegetables, no such explanation is available and the rather high mold contamination appears to be an inherent property of these products which cannot be

Table 2. Mold counts ($\log_{10}$ cfu/g) of different dried vegetables[a]

Vegetable	Mean	Standard deviation
Onion slices	2.00	0.56
Onion powder	4.06	0.33
Horseradish flakes	3.35	0.15
Horseradish powder	4.95	0.12
Celery leaves	4.34	0.25

[a]Data are for twelve samples of each vegetable

198

Table 3. Decrease of mold count ($\log_{10}$ cfu/g) with the time of storage

Storage (days)	Sliced onion[a]		Moisture (%)	Storage (days)	Sliced onion[a]		Diced paprika[a]	
	$\bar{x}$	s			$\bar{x}$	s	$\bar{x}$	s
10	2.64	0.30	7.76	0	3.00	0.21	1.89	0.08
39	1.96	0.47	7.34	26	2.63	0.80	1.39	0.12
84	1.89	0.24	8.76	73	1.98	0.33	1.50	0.71
156	1.46	0.27	9.58	116	2.16	0.45	1.39	0.12

[a] $\bar{x}$ = mean; s = standard deviation

substantially reduced by processing. This seems to be the case with horseradish flakes and powder, as well as for celery leaves (Table 2).

It is well known that the number of viable microorganisms in dry products decreases during storage. This occurs particularly in the case of Gram negative bacteria and is attributed to the progressive death of injured cells (Mossel & Corry 1977). Data in Table 3 show that this occurs with molds too, even under conditions where the moisture content of the product increased slightly during storage.

While molds should be of primary concern in dried vegetables, yeasts may also be found in numbers approaching those of molds (Table 4).

In the international trade of dried vegetables, a limit value for molds of 300 cfu/g (log 2.47) is generally established. This value seems appropriate for some products processed under conditions of good manufacturing practice. The limit is also a useful incentive for manufacturers to improve processing hygiene for some other products. However, such a strict limit appears quite unrealistic for yet other dried vegetables.

It is recommended that limiting values should be established only after taking into consideration inherent properties and processing technology used to produce each type of dried vegetable.

Table 4. Mold and yeast counts ($\log_{10}$ cfu/g) of some dried vegetables

Vegetable	Molds[a]		Yeasts[a]	
	$\bar{x}$	s	$\bar{x}$	s
Diced red paprika	2.67	0.91	2.63	0.74
Sliced onion	2.79	0.79	2.30	0.43
Celery flakes	1.79	0.83	1.10	1.02

[a] $\bar{x}$ = mean; s = standard deviation

References

KAMPELMACHER, E. H., INGRAM, M. & MOSSEL, D. A. A. 1968 The Microbiology of Dried Foods. Proceedings 6th International Symposium on Food Microbiology, IAMS. Haarlem: Grafische Industrie.

MOSSEL, D. A. A. & CORRY, J. E. L. 1977 Detection and enumeration of sublethally injured pathogenic and index bacteria in foods and water processed for safety. Alimenta Sonderausgabe 16, 19-34.

T. DEÁK
I. FABRI

▶ Baseline Count of Molds in Paprika

Although spices are minor ingredients in foods, they may be heavily contaminated with microorganisms that may lead to spoilage when introduced into processed foods (Mossel 1955). Molds may represent a major proportion of the microflora and are of interest since they may develop on spices stored at higher humidity and they are indicators of improper sanitary practices which often prevail at the time spices are harvested or processed (ICMSF 1980).

Of the common spices, black and white peppers are notorious for carrying a very high microbial population (Krishnaswamy et al. 1971). Marjoram, ginger, coriander and other spices may also contain high microbial populations. On the other hand, paprika, when processed properly, is only moderately contaminated. In addition, paprika contains substances which inhibit the growth of microorganisms (Gàl 1964). High microbial counts can be effectively reduced by gas treatment or, preferably, irradiation (Farkas 1974; Ingram & Farkas 1977).

In Hungary, special attention is devoted to paprika which is considered a traditional national spice. Factors influencing quality during ripening, harvesting and processing have been studied in detail (Nagy 1977), and the microbial profile of the product is regularly monitored by the stations of the Veterinary and Food Control Service. Data concerning mold contamination of paprika are presented here.

Materials and methods

Mold counts were determined by the pour-plate method in CGYE agar from dilutions of 10-g samples in peptone saline. Plates were incubated at 25°C for 5 days. Various types of paprika produced in two main factories were investigated.

Results and discussion

Mean mold counts and standard deviations of samples investigated in 1974, 1978, 1980 and 1982 are summarized in Table 1. With the exception of 1978, the means were about log 2 cfu/g.

Distribution of counts in paprika is shown in Table 2. On average, the mold count exceeded log 3 cfu/g in less than 20% of samples. The precision of determination of mold counts was assessed in a study involving three

Table 1. Mold counts in paprika ($\log_{10}$ cfu/g)

Year	Mean	Standard deviation	Number of samples
1974	2.87	0.56	40
1978	3.71	0.49	50
1980	2.00	0.82	68
1982	2.27	0.84	66

Table 2. Distribution of mold counts in paprika samples

Count ($\log_{10}$ cfu/g)	Frequency (%)			
	1974	1978	1980	1982
0 – 1	–	–	25.0	9.1
1 – 2	12.5	38.0	26.5	22.7
2 – 3	47.5	42.0	36.7	53.1
3 – 4	27.5	20.0	10.3	15.1
4 – 5	2.5	–	1.5	–

subsamples each of 22 samples. The standard deviation among sample units was only log 0.18/g, while between samples a value of log 1.6/g was calculated. The general mean was log 2.17/g. The mold counts differed in various types of paprika (Table 3). A significantly lower count was obtained in the type produced from raw material without postharvest ripening.

Paprika spice is produced from the red, hollow berries of the paprika plant harvested at the fully ripened stage. Depending on the cultivar as well as on the weather, artificial post-ripening may be necessary to attain a uniform, dark red color. The hotness of the spice depends again on the cultivar and the ratio of seeds added to the fruit wall before grinding.

Table 3. Mold counts ($\log_{10}$ cfu/g) of various paprika grades[a]

Grades	n	$\bar{x}$	s
Delicatessen	8	2.04	0.25
Sweet delicate	12	2.90	0.25
Sweet noble	12	2.95	0.39
Sweet hot	8	2.57	0.11

[a] n = number of samples; $\bar{x}$ = mean count ($\log_{10}$ cfu/g); s = standard deviation

In addition to pathogenic field molds, saprophytic molds may also attack the fruits during ripening, particularly during forced post-ripening. These molds enter via the flower and develop later around the ovary. This explains why molds are an inherent part of the microflora of paprika. Populations can be reduced by processing only to a certain extent. For example, in 1978 conditions favorable for mold development were directly reflected in the quality of the finished product (Table 1). On the other hand, when special cultivars ripened in the field and carefully sorted before processing, top-quality export grades with low mold counts can be obtained (Table 3).

The mold population limit value usually required in international trade, i.e., 300 cfu/g (log 2.47/g), is not easily attainable for most processors. A somewhat less stringent limit, e.g., 500-1000 cfu/g (log 2.7-3.0), would be more realistic without sanctioning low standards in processing technology.

References

FARKAS, J. 1974 Radurization and radicidation of spices. In Aspects of the Introduction of Food Irradiation in Developing Countries. pp. 43-59. Vienna: IAEA Pl-518/6.

GÁL, I. 1964 Capsacidin, eine neue Verbindung mit antibiotischer Wirksamkeit aus Gewürzpaprika. Zeitschrift fur Lebensmittel Untersuchung und Forschung. 124, 333-336.

INGRAM, M. & FARKAS, J. 1977 Microbiology of foods pasteurized by ionising radiation. Acta Alimentaria 6, 123-186.

ICMSF 1980 Spices. In Microbial Ecology of Foods, Vol. 2. Food Commodities pp. 731-751. International Commission for Microbiological Specifications for Foods. New York: Academic Press.

KRISHNASWAMY, M. A., PATEL, J. D. & PARTHASARATHY, N. 1971 Enumeration of microorganisms in spices and spice mixtures. Journal of Food Science and Technology 8, 191-194.

MOSSEL, D. A. A. 1955 The importance of the bacteriological condition of ingredients used as minor components in some canned meat products. Annales de l'Institut Pasteur de l'Lille 7, 171-186.

NAGY, J. 1977 Microbiological and technological consideration about the growth, harvesting and processing of paprika. Post-graduate Thesis, pp. 92. University of Horticulture, Budapest (in Hungarian).

T. DEÁK
I. FABRI

Discussion

FERGUSON

Since nothing was presented on nut specifications, I thought I would present some data. Several years ago, Pillsbury carried out a short survey on the nuts that we purchased as ingredients. The data were combined on pecans, almonds and walnuts. This survey was carried out in 1981 and 1982. The number of samples was 285. We used mycological agar, stomaching and phosphate buffer for diluting. The incubation temperature was 21-25°C for five days. Six percent of the samples had low counts from 0-10, 57% from 20-100, 22% from 10^2-10^3 and 14% 10^3-10^6. These counts reflect both yeasts and molds, but primarily molds.

The specification generally accepted in the U. S. industry is 10^3/g for white flour. I did notice that our data differ quite a bit from those presented by Mr. Seiler and this is probably due to the methodology. We do not presoak and we use a stomacher instead of a high speed blender.

MOSSEL

What exactly is mycological agar?

FERGUSON

It's mycological agar acidified with lactic acid. This is the manufacturer's trade name designation. It is soy protein based agar.

LANE

According to the data presented here, none of the samples follow any pattern.

SEILER

This is true, and I am shattered by the results we are seeing here today. Hesseltine for example here in the U.S. has done a large number of surveys on flours from various mills, graineries, etc. and none of his figures is as low as these that Ferguson presented. Here we are seeing less than ten per gram for flour and I cannot believe that. How can we get as low as ten in flour?

FERGUSON

One thing should be noted. This came from work done years ago and since that time some improvements have been made.

PITT

One of the reasons for these differences may be weather patterns. This might be the reason we see different counts in America versus Australia versus European wheats and grains. Depending on conditions under which the crops grow, there are enormous differences in weather patterns which could affect the flora found on such crops. This could account for a reasonable difference between American grains and English and European grains.

HOWELL

How do you explain differences in counts in flour which have been shipped to different areas?

PITT

There definitely can be an increase in counts during shipment. The reasons are quite straight forward. One thing, you get movement of moisture in cargos. We have seen so much of this.

HOWELL

Then you are saying that the increase in counts that the Europeans get from American grain may be due to fungal growth during shipment?

PITT

Yes. You can definitely get fungal growth and increased counts during shipment.

SEILER

I believe that occurs sometimes, but not always. Sometimes the counts by be reduced with shipment.

HOWELL

I have a hard time believing that.

SEILER

Yes, I would like to see that proven also. But I do believe that counts can go down. Many times, grains from America are held for as much as six months and if held under dry conditions, the counts will steadily decrease with time. As much as a log or two-log reduction in counts may occur. So, therefore, the amount of reduction will depend on shipping and storage time. But I believe the apparent discrepancies result from different methodologies. I believe that one useful outcome of this meeting may be to send samples out on a collaborative basis to see what happens and to see what the difference in results would be.

KING

Instead of microbiological methodology, could there also be processing differences that account for these differences in counts?

SEILER

I would doubt that. I am concerned with lowering these counts, if possible, in flour in order to meet certain specifications. We have tried chlorination and dry heat treatment. The heat treatment does not work for most of the baking flours. However, cleaning methods could make a difference.

BEUCHAT

I am also surprised by some of the numbers reported here. Of the 109 samples we reported yesterday, the average was about 3800-3900 per gram. Some had less than 100, but many were over 10^6. These were foods purchased at the retail level. The counts that we got in most instances were expected.

KING

Yes, the average was 3800 on those samples, but some were as high as 61,000 for wheat flour; 75% of the counts were less than 1000.

SEILER

We've looked at thousands of samples, and we never see counts this low.

PITT

I expect to see great variation in David Seiler's work and European data on counts on commodities, because the conditions in California in which foodstuffs are dried and held are very hot conditions and vary a great deal from the conditions with which he is familiar. This has to be taken into account. I would like the general session now to give an indication to the rest of us as to what we should do with these counts besides publish them. I would like to know if any of the members of the general audience feel that there is anything else we can do other than to make statements like there is high variability depending on where the samples come from and the conditions under which they are treated. Are there more recommendations that we can make from these data or are we going to simply list what we have here, which is of course very valuable? Each of you should address this in your own mind as to what kind of recommendations we can make regarding baseline counts for fungi in foods.

SEILER

We are confronted with some difficulty from all the data that have been presented in drawing conclusions and making recommendations. Don't you also think we should address ourselves to identifying which commodities we must look at more closely and carefully from this point of view? For example, there are commodities which need baseline counts and others which do not? Should we test spices? Some people feel that we should test spices, but we know that we will find a high level in fungi in the spices.

HOWELL

We should think about the end product first, e.g., a_w and other factors involved, and then think about the raw materials used in such products. It is difficult to make recommendations when we present data like this, since the conditions vary considerably and the uses of such commodities are unknown.

SEILER

There are some areas where the recommendations could be clear cut. It seems to me that some sort of control is needed in the beverage field. How we do it, of course, is a different matter. Bob Davenport pointed out that all you need is one cell of a certain species of fungus to produce serious problems.

ANDREWS

With all the differences in methodologies and media, I think we should be careful in recommending baseline counts. I would be very hesitant to make any comment or recommedation until we decide which one method we should use.

SEILER

I agree with you, but possibly at this point we should recommend which commodities we feel would be worth while having baseline counts or some specifications. Even if we cannot recommend a specific level, we could at least define which areas are of concern.

BECKERS

I am also very hesitant to make any recommendations, because I feel that such recommendations could be misused by people outside of science who control the products. Baseline counts could be misused because this is an expert group and we are making recommendations to people who are not experts in mycology. We should be very careful in suggesting any baseline counts. We should just present the data as it is given here and keep in mind the general principles for establishing and and applying microbiological criteria. Is there a need for criteria and how shall we use it as baseline data?

HOCKING

We should address ourselves to the commodities to which baseline counts would be relevant at this stage, without specifying baseline counts because they will vary from country to country.

BECKERS

We have seen the example in the second volume of the ICMFS (Microbial Ecology of Foods) where specifications were given for certain types of foods. These were only suggestions where you have this type of food and you may have these types of counts, but numbers given in that book were misused as specifications.

HOCKING

We should use the numbers to identify those commodities which we should carefully look at and in due course develop baseline counts for specifications.

Recommendations

Clearly, much remains to be done to establish internationally agreed baseline counts of yeasts and molds in foods. First, methods must be standardized as much as possible, for example, by use of one or two media and diluents and one temperature of incubation. The methods of sampling and preparation of the food homogenate should be standardized also, but are likely to vary, depending on the food being examined.

When all this has been done, it may be possible to determine the reasons for the wide variation in fungal counts in various foods, as illustrated for wheat and flour. Differences due to variations in technique used for enumeration can then be distinguished from those due to variations in growing, harvesting, processing or storage. Whether it will then be possible or desirable to establish standard baseline counts for selected commodities and/or for commodities grown and processed in a specific part of the world, remains to be determined.

CHAPTER 6

UNACCEPTABLE LEVELS OF FUNGI

The significance of fungal populations and the setting of unacceptable levels in foods are vexed questions. Unlike bacterial counts, fungal counts may be greatly influenced by contamination with dust or extraneous matter, or degree of sporulation and fragmentation of mycelia, and may not relate directly to actual fungal growth. Nevertheless, food manufacturers are constantly seeking advice on the significance of fungal populations and on the desirability or feasibility of establishing standards for unacceptable counts in foods.

This problem has become increasingly urgent with the recognition that mycotoxins are a significant hazard in some foods. Monitoring fungal contamination of raw materials is a useful but rarely exploited approach to routine quality assurance for mycotoxin contamination and for wholesomeness of finished products. The following contributions offer further insight to criteria which must be considered when setting unacceptable levels of fungi in foods and to the value of such levels to overall quality assurance programs.

▶ Unacceptable Levels of Molds

The question of unacceptable levels of molds in foods will be addressed, not only from a mycological point of view, but from the point of view of an employee of the U.S. Food and Drug Administration. As a mycologist with the FDA, this point has been brought to my attention many times – by other members of the FDA staff and by other federal government agencies, plus by state and local agencies, and universities. Usually questions are phrased, "I have this product and it has a mold count of 'X' colonies per gram. Is it dangerous to use?" The answer necessarily always has been: "That depends precisely upon what mold species are involved." For instance, a commodity could have a high mold count (10^6/g), but because it is primarily due to Aspergillus repens, there should be no great concern. Yet, the count could be less than 10^1/g but happen to be A. flavus or even Coccidioides immitis – a cause for concern.

The FDA is responsible for the regulation of most consumer commodities that are available for interstate use in the U.S. Basically, FDA-regulated consumer commodities fall into four categories: foods, drugs, medical devices and cosmetics. We regularly receive molds and, sometimes, yeasts from all of these commodities for identification – and must make evaluations concerning their unacceptability.

What are the implications of a moldy commodity? Basically there are two:

1. The mold serves as an index of insanitation
2. The mold may serve as a potential, if not actual, human health hazard.

Usually the first category is the case, at least for foods. Sources of insanitation are many but usually involve one or more of the following:

1. Moldy ingredients
2. A dirty factory
3. Improper storage of ingredients or final product
4. Improper transit
5. Defective packaging.

Of interest, knowledge of the precise mold species present in a given commodity can assist one in the determining how the commodity became moldy; e.g., if Paecilomyces variotii was the only mold detected in a "sterile" medical device, indications would be that the sterilization technique was defective.

Regarding hazards, molds can cause infection, elicit allergies and produce toxins. On a commodity-to-commodity basis, however, the hazard potential may differ. For foods, the primary concern is the presence of toxins and allergens. For the other commodities, the primary concern is infection and allergy. When assessing the hazard potential of a given moldy item, several factors must be considered, namely:

1. The specific commodity
2. The manner of use
3. The consumer age group involved
4. Individual health (opportunistic pathogens)
5. The frequency of use
6. The amount used.

These factors, individually and collectively, can alter the hazard potential of any given commodity.

The above should indicate, at least from an FDA point of view, that a total viable count does not necessarily indicate that a given commodity is unacceptable for use. The precise species present need to be determined. The type of commodity involved and the principal users of the given commodity also must be considered.

P. R. MISLIVEC

Unacceptable Levels of Potentially Toxigenic Fungi in Foods

One of the most frequent inquiries faced by food mycologists providing advice to industry is: "We have isolated a potentially toxigenic fungus from this or that commodity – is it important, and if so what should we do about it?" In other words, what constitutes an unacceptable level of a potentially toxigenic fungus in a food?

Clearly, there is no simple answer. We know, on one hand, that an occasional colony on an enumeration medium or an occasional positive on a grain in a direct plating is of no consequence. On the other hand we believe that a heavy infection of a potentially toxigenic fungus is unsatisfactory. Where should the line be drawn? What follows is an attempt to approach this problem.

The growth of molds in foods is not easy to quantify, because molds do not grow as single cells in the manner of bacteria and yeasts, but as filaments. Active biomass in bacteria and yeasts is quite accurately reflected in plate counts. With molds, this is not necessarily so. Fungal hyphae grow deep inside particulate foods, and are not detached by processes such as stomaching. Blending will fragment hyphae, but the counts obtained depend on the time of blending. Counts rise with time as hyphae are broken up, and then decline as further fragmentation produces shorter lengths which leak their contents and lose viability.

While fungal growth consists of hyphae only, counts will be low. But when sporulation ensues, counts will rise rapidly. Pitt (1984) reported that, after an initial lag, viable counts of Aspergillus flavus and Penicillium expansum increased 100 times as rapidly as biomass. His experiments were conducted in laboratory culture, but this relationship between growth and sporulation should also be approximated in foods.

What about toxigenesis? It would solve many of our problems if we could say, for example, that a count of 10^5 from a particular food means a 35% chance of detectable toxin, but of course this is wishful thinking. Dogma says that mycotoxins are only produced by fungi when they are sporulating. In many cases, this is probably so, but in our experience it is entirely possible for toxin production to precede sporulation. In view of these uncertainties, how can we attempt to assess the significance of potentially toxigenic fungi in foods?

This problem can be approached if we make two sweeping generalizations. The first is that, given a food material which has not been subjected to heat or other sterilizing processes, and given it has been in storage for less than a year, then if that commodity is going to contain mycotoxins, it must contain fungi and it must be possible to enumerate them by suitable techniques. And second, some commodities are much more susceptible than others to invasion by potentially toxigenic fungi, and enough information now exists for us to decide which foods and raw materials are at most risk.

Tables 1 and 2 illustrate raw materials and foods considered to be at significant risk from fungal invasion and mycotoxin formation. Most of

Table 1. Foods and raw materials with a high risk of mycotoxin contamination.

Food type	Likely fungi	Likely toxins
Nuts	Aspergillus flavus	Aflatoxins
Corn	A. flavus	Aflatoxins
	Fusarium spp.	Trichothecenes
	F. moniliforme	Zearalenone and unknown toxins
Sorghum	Fusarium spp.	Trichothecenes
	Alternaria spp.	Alternariol, alternariol methyl ether, tenuazonic acid
	Aspergillus flavus	Aflatoxins

Table 2. Foods and raw materials with a moderate risk of mycotoxin contamination.

Food type	Likely fungi	Likely toxins
Wheat (food grade)	Penicillium	Ochratoxin, cyclopiazonic acid, tremorgens, citrinin
Wheat (feed grade)	Fusarium spp. F. graminearum	Trichothecenes Deoxynivalenol
Barley	Penicillium viridicatum, P. verrucosum	Ochratoxin
Flour	Penicillium spp.	Various
Breakfast foods – Corn based Wheat based	Aspergillus flavus Fusarium spp.	Aflatoxins Trichothecenes
Cheese	Penicillium spp.	Various
Processed meats, salamis	Penicillium spp.	Various
Milk and cheese	Residues	Aflatoxin M
Eggs	Residues	Aflatoxins, ochratoxin, trichothecenes

these commodities can be monitored for fungi, but there are some obvious exceptions. Breakfast cereals cannot because they are heat processed, and so must be monitored at the raw material stage or by chemical assay. Cheese and processed meats can be monitored immediately after manufacture, but spoilage of cheese and meat products normally takes place during and after retailing, and so is beyond the manufacturer's control. Spoilage of perishable, refrigerated products is usually visual, and stability really relies on good storage practice. The final two entries in Table 2, eggs and milk including milk products, particularly cheese, may contain toxins present as residues which can only be monitored by chemical methods. Such methods are outside the scope of this paper.

In my opinion, foods and raw materials at high risk (Table 1) should always be monitored mycologically for specific toxigenic fungi, or chemically for particular mycotoxins, or both.

Bearing in mind the exceptions noted above, it should be possible to monitor the foods listed in Tables 1 and 2 by mycological techniques. There are two approaches. First, use a medium selective for particular toxigenic fungi. If interest lies specifically in aflatoxins, use AFPA (Pitt et al. 1983). AFPA can be used by relatively untrained personnel, and results are obtained rapidly. Monitoring can be either by dilution plating or, for particulate foods such as grains, by direct plating. The use of AFPA is still quite new, so that there are no reliable figures yet for the significance of particular population levels of A. flavus and, as pointed out above, there will never be a hard and fast rule. The most practical approach is to monitor sound commodities until an acceptable level is established and then regard higher levels with suspicion.

210

Selective media also exist for _Fusarium_ spp., the most useful being peptone pentachloronitrobenzene (PCNB) agar (Nash & Snyder 1962). Our experience suggests that the published formulation may be far from optimal, but the medium is reported to be effective (Burgess & Liddell 1983). Frisvad (1983) has recently described a selective medium for _Penicillium viridicatum_ which may well find application for some commodities.

The use of selective media is effective if particular toxins are known to be important, but of course has the limitation that there are few effective selective media for fungi. To circumvent this problem, a second approach is needed, viz., to check for potentially toxigenic fungi on routine enumeration media. Carried out systematically, I believe that this procedure can be effective also. The rationale I am proposing here is this: if you use a medium which is selective for toxigenic fungi in general, the presence of one dominant fungus type on dilution plates suggests that it has grown in the commodity and therefore that the commodity may contain toxins.

This rationale requires explanation in more detail. Fungal spores are ubiquitous, but the general run of spores in the air, in dust, in soils and hence in unprocessed foods is a mixed flora of contaminants from these sources. If a variety of colony types is apparent when a commodity is plated out on a fungal enumeration medium, it is probable that these colonies merely represent contaminants, which are of no consequence unless present in extraordinarily high numbers. If dilution plating is carried out routinely on a specific commodity, then a baseline count of the normal mycological flora to be expected in that commodity can be established. The baseline count is an expression of the level of contamination of the commodity with spores from extraneous sources, and not of fungal growth. Hence it is not subject to the vagaries of growth curves and should be reasonably constant for a particular commodity which has been handled and transported under good conditions. However, if a single fungus type predominates on an enumeration medium, the chances are that it has grown in the commodity. At the least, such a result should arouse suspicion.

Note that here we are no longer interested in a viable count as an end in itself, but in the appearance of colonies in plates at a dilution suitable for counting. It would be expected, however, that counts with a single predominant fungus would be higher than normal baseline counts.

As mentioned above, the choice of medium is important. The medium used should produce readily enumerated colonies, free of overgrowth, and should select for genera containing potentially toxigenic fungi, especially _Aspergillus_, _Penicillium_ and _Fusarium_. DRBC is the obvious choice (King et al. 1979).

In summary, it is difficult to rely on viable counts of fungi directly if you wish to monitor for possible mycotoxins in foods. Viable counts can serve a valuable purpose in establishing baseline levels of contamination for specific foods or raw materials, however. Monitoring for potentially toxigenic fungi in foods can be accomplished in two ways, either by using selective media, e.g., AFPA for _Aspergillus flavus_, or by using DRBC and monitoring for the presence of a predominant fungus.

References

BURGESS, L. W. & LIDDELL, C. M. 1983 _Laboratory Manual for Fusarium Research_. Sydney, N. S. W.: University of Sydney.

FRISVAD, J. C. 1983 A selective and indicative medium for groups of *Penicillium viridicatum* producing different mycotoxins in cereals. Journal of Applied Bacteriology 54, 409–416.

KING, A. D., HOCKING, A. D. & PITT, J. I. 1979 Dichloran–rose bengal medium for enumeration and isolation of molds from foods. Applied and Environmental Microbiology 37, 959–964.

NASH, S. M. & SNYDER, W. C. 1962 Quantitative estimations by plate counts of propagules of the bean root rot *Fusarium* in field soils. Phytopathology 52, 567–572.

PITT, J. I. 1984 The significance of potentially toxigenic fungi in foods. Food Technology in Australia 36, 218–219.

PITT, J. I., HOCKING, A. D. & GLENN, D. R. 1983 An improved medium for the detection of *Aspergillus flavus* and *A. parasiticus*. Journal of Applied Bacteriology 54, 109–114.

J. I. PITT

 Unacceptable Levels of Specific Fungi

Heat resistant molds, especially Byssochlamys species

The acceptable level of contamination of a raw material with heat resistant fungal ascospores will depend very much on the type of product into which the material will be incorporated. The heat process to which it can be subjected will also be significant. In our recent experience in Australia, most spoilage problems caused by heat resistant molds have originated from passionfruit juice or pulp, though other fruit such as mango and pineapple are also regularly monitored. There have been no reported instances of this type of spoilage from citrus products. Byssochlamys fulva is the most common heat resistant spoilage mold, although Neosartorya fischeri and Talaromyces flavus also caused occasional problems.

Where passionfruit juice is to be incorporated into frozen desserts such as ice creams and ice confections, or short-life chilled desserts such as fruit salads, pavlovas, cakes and yogurts, there is no need to set a specification for heat resistant mold spores. A requirement for juice with low yeast counts would be more appropriate. Products which are most at risk from spoilage by heat resistant molds are shelf-stable products which receive a relatively light process (such as conventional or UHT pasteurization) and do not contain preservatives, e.g., fruit gel baby foods and aseptically packaged, preservative-free fruit juices.

Practical experience in Australia has shown that, for passionfruit juice, a contamination level of less than two spores per 100 ml gives a negligible spoilage rate in most finished products. Contamination levels of two to five spores per 100 ml are marginal, and more than five spores per 100 ml unacceptable. However, for some products, such as UHT processed fruit juice blends which contain a high proportion of passionfruit juice and do not contain preservative, an even lower level of contamination is required. One manufacturer of this type of product specifies that heat resistant mold spores should be absent from a 100 ml sample of passionfruit juice. Samples which are found to contain heat resistant molds are rejected (Cartwright & Hocking 1984).

Zygosaccharomyces bailii

The preservative resistant yeast, Zygosaccharomyces bailii, first came to the notice of the Australian food industry in the early 1970s, causing

Table 1. Effect of inoculum[a] and preservative level on percentages of cans of carbonated beverage spoiled by S. bailii

| Inoculum numbers | Preservative (mg/kg) | | | | | | |
| | Sorbic acid | | | Benzoic acid | Sorbic + benzoic acid | | |
	400	500	600	400	400	500	600
100/ml	100	90	0	100	56	67	23
10/ml	52	44	0	100	64	81	0
1/ml	68	35	0	100	42	56	0
1/10 ml	47	14	0	73	22	13	0
1/100 ml	0	6	0	55	11	19	0

[a]Inoculum was adapted to 400 mg/kg sorbic acid before inoculation into cans

fermentative and explosive spoilage of a range of acid liquid products such as tomato sauce (ketchup), fruit juices and carbonated beverages containing fruit juice. The yeast was found to be resistant to the maximum permitted levels (400 mg/kg) of sorbate and/or benzoate (Pitt & Richardson 1973).

In liquid products where shelf stability relies either on acidity or on weak acid preservatives or on a combination of these, it is reasonable to establish a zero tolerance for Z. bailii. Table 1 shows that as few as one to five viable yeast cells per container (1/100 ml) can cause spoilage after a comparatively short storage time (2-3 months) at ambient temperature.

In many cases, a spoilage problem may only become apparent after some months of storage, usually as the weather begins to warm up. For example, a manufacturer may spend the winter building up stocks for cordial concentrates to cater for the summer demand. Very low initial levels of Z. bailii may slowly build up causing fermentative spoilage which only becomes obvious when the yeast concentration reaches 10^5-10^6 per ml. If initial yeast levels were low, only one or two containers per carton may be affected, and the manufacturer is then faced with the problem of unpacking all his stock to sort the spoiled product from the unspoiled. He has no guarantee that containers which appear to be unaffected will remain so. He may decide to write off the whole of his stock. Whichever choice he makes, it will mean a substantial financial loss to the company.

Careful monitoring of raw materials for the presence of Z. bailii is necessary in food plants which manufacture products which may be susceptible to spoilage by this yeast. In our experience, the range of products spoiled by Z. bailii includes concentrated orange, apple and pear juices, tomato sauce, mayonnaise, cordial concentrates, water-ice mixtures (containing glucose), fruit toppings, strawberry pulp, cocktail bases, fruit-based carbonated beverages and wines. Scrupulous plant hygiene is necessary to ensure that, should Z. bailii enter a factory in a raw material, it does not become established. Measures such as continuous chlorination of cooling water and steam sanitization of lines, holding tanks and filling machines should be routine. Most Australian manufacturers of products which may be spoiled by Z. bailii are now well aware of the problems this yeast can cause, and take the necessary preventative measures. The only acceptable level for Z. bailii in susceptible products is nil.

References

CARTWRIGHT, P. & HOCKING, A. D. 1984 Byssochlamys in fruit juices. Food Technology in Australia 36, 210-211.
PITT, J. I. & RICHARDSON, K. C. 1973 Spoilage by preservative-resistant yeasts. CSIRO Food Research Quarterly 33, 80-85.

J. I. PITT

▶ Unacceptable Levels for Yeasts

The concept of setting limits for unacceptable levels of specific fungi is the 'ideal' since in practice it is seldom possible or reliable. One can only give guidelines based on study and experience of organisms where they have been found. Habitat features, organism characteristics, significance, identity and counts: these are important in considering unacceptable levels of contamination. It is more pertinent to ask other questions: What is the potential outcome of the determined microflora? Is this microflora indicative of poor factory hygiene or product contamination? How far does the determined microflora reflect the actual microbiological status of the product/product environment? There are technological differences – processing and packaging. Ultimately this can lead to different microbial strains. Does this mean a different level for each different organism(s) and/or circumstance(s)? The examples listed in Table 1 illustrate this dilemma. These observations indicate the difficulty in establishing unacceptable levels. However, some guidelines could be of value:

1. Distinguish between product and product habitat assessments

Table 1. Selected examples of behavior of fungi in apple juice

Fungus	Apple juice (1 liter)		
	Bag-in-box	Tetra pak	Glass/metal container
Penicillium expansum			
1 conidium	No growth	Mycelial mass[a]	No growth
> 10^3 conidia		Mycelial mass	No growth
Debaryomyces hansenii			
> 10^3/ml	No growth	No growth	No growth
Hansenula anomala	Growth	No growth	No growth
Zygosaccharomyces bailii	Growth	Growth	Growth

[a]No conidia formed

214

2. Use significance in screening techniques
3. Use experience with yes/no or low/medium/high levels.

The following suggestions for guidelines/recommendations for commodities are also made:

1. There should be a clear microbiological objective. Priority should be given to either product or product environment or both.
2. Investigations and subsequent assessments should be designed on the objective(s). Mere counting on one medium, followed by conventional taxonomic studies is not usually the answer for technology.
3. Significance/recognition screening techniques are invaluable to give quick reliable 'decision-making' data before any intensive investigations are undertaken. In addition to (1) and (2) above, the techniques used must be related to sampling methodology and commercial constraints.

                                                              R. R. DAVENPORT

▶ Unacceptable Levels of Yeasts in Bottled Wine

Yeasts remaining in bottled wines may start to grow, causing turbidity and sediment but rarely fermentation (Rankine & Pilone 1973; Minarik 1980; Lukacsovics et al. 1981; Kornyei et al. 1983). Many factors have been claimed to cause loss of stability. Data reported in the literature are, however, often contradictory. Hence, a comparative study was conducted to determine the main factors influencing stability of bottled wines.

The objectives of the investigation were to establish the period of microbiological stability of bottled wine, to evaluate some factors influencing stability (e.g., type of wine, content of alcohol, residual sugar, sorbic acid, season of production, number and species of yeast) and to predict the shelf life of wines on the basis of yeast count. Results from analysis of yeast populations are presented here.

Materials and methods

With the cooperation of nine Hungarian wineries, twenty-one different types of wine were investigated during a 3-yr period of bottling and storage. The study was conducted using a mathematical-statistical scheme. Each type of wine was bottled during three or four different seasons and during each bottling occasion several dozens of bottles were put into storage cellars at 8–15°C. Immediately after bottling, five samples were taken for investigation. Sampling was repeated with three bottles after 2 and 4 weeks as well as after 2, 3, 6, 9 and 12 months of storage. When microbiological spoilage occurred, all remaining bottles were examined for stability. A total of 1456 samples were analyzed.

Yeast populations were determined by plating serial dilutions on yeast extract – glucose agar acidified to pH 3.5 with tartaric acid. The pour-plate method or membrane filtration were applied, depending on the level of viable yeast cells. Representative colonies of yeasts from samples of wine that became turbid or sedimented were isolated and identified according to a simplified key (Deák 1981).

Besides microbiological investigations, samples of wine at the beginning and the end of the study were analyzed for pH, alcohol content, residual sugar, titratable acidity and free and bound sulfur dioxide, according to standard procedures.

Results and discussion

Contrary to data in the literature, no significant difference was found in stability of different wine types. White wines and red wines, as well as dry, semi-dry and semi-sweet wines showed no difference in stability, irrespective of their alcohol content, residual sugar or sulfur dioxide content.

Various wines were separated into three groups according to the time of appearance of microbiological spoilage: (1) loss of stability within 3 months; (2) loss of stability within 6 months; and (3) stable for at least 12 months. The frequency of instability of various wine types was evaluated by contingency analysis (Table 1). Using the chi-square test, the calculated value (2.42) was less than the tabulated one (9.49) for four degrees of freedom when $P \leq 0.05$. This indicates that no significant difference can be detected in the stability of various wines after different storage times.

The stability of bottled wines was inversely proportional to the population of viable yeast cells after bottling (Table 2; Figs. 1, 2, 3). When less than 100 cells remained in one liter of wine after bottling, about one-third of the bottles spoiled within 12 months. However, when the initial yeast population exceeded 1000 per liter, nearly all of the bottles spoiled.

Table 1. Frequency of instability according to the types of wine and the storage time

Type of wine	Stability (months, frequency)[a]						
	Less than 3		Less than 6		More than 12		
	k_o	k_e	k_o	k_e	k_o	k_e	Σk_o
Dry	7	5.89	4	2.95	4	6.16	15
Semi-dry	3	4.32	2	2.16	6	4.52	11
Semi-sweet	12	11.79	5	5.89	13	12.32	30
Σk_o	22		11		23		

[a]Frequency expressed as percentage: k_o, observed;

$$k_e, \text{ expected } (k_e = \frac{\Sigma k_o \text{ rows} + \Sigma k_o \text{ columns}}{\Sigma k_o \text{ total}}); \quad \chi^2 = \sum_{1}^{9} \frac{(k_o - k_e)^2}{k_e} = 2.42$$

216

Table 2. Frequency of instability as a function of storage time and
initial cell counts

Initial yeast population (cells/liter)	Frequency (%)		
	3 months	6 months	12 months
< 100	19	26	34
< 1000	38	51	54
> 1000	58	77	98

Closer mathematical analysis showed that the mean count of yeasts in
bottles remaining stable for at least 12 months was significantly less than
that in wines losing stability within 6 months. The mean values and standard
deviations of wines stable for 6 months [$\log_{10} x_1$ = 1.97, s_1 = 1.416,
(n = 33)] and for 12 months [$\log_{10} x_2$ = 0.70, s_2 = 0.866, (n = 23)]
were calculated. The difference between the means is significant with P $\leq$
0.01 probability of error according to the Student's t test,

$$t = \frac{x_1 - x_2}{s_d} = \frac{1.97 - 0.70}{0.305} = 4.15,$$

when t_{tab} = 2.68 (DF = 54 and P $\leq$ 0.01). It can be expected with 95%
confidence limits that those bottles of wine containing 50 $\pm$ 23 yeast cells
per liter would remain stable for 12 months.

The above statistically justified consideration is, however, dependent
upon the physiological properties of the yeast in question. Of the bottles
that became sedimented or turbid, the yeast species most frequently isolated
was Zygosaccharomyces bailii (Table 3) which is well known for its high
alcohol and acid tolerance. Hence, even a few cells of this yeast may start

Table 3. Distribution of yeast strains isolated from spoiled wine

Fungus	Number of isolated strains			
	Dry wines	Semi-dry wines	Semi-sweet wines	Total
Zygosaccharomyces bailii	6	2	17	25
Z. rouxii	–	–	1	1
Debaryomyces hansenii	2	–	–	2
Saccharomyces cerevisiae	2	2	–	4
S. kluyveri	–	–	1	1
Pichia membranaefaciens	2	–	–	2
Candida zeylanoides	2	–	–	2
Total:	14	4	19	37

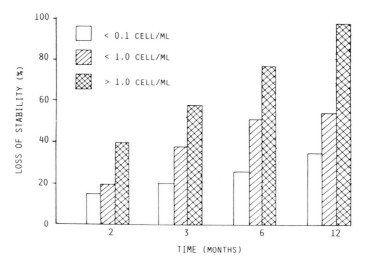

Fig. 1. Frequency of instability of bottled wines. Data are from three
types of wine (55 samples) analyzed in triplicate

growing in bottled wine and cause loss of stability. For this reason, no
definite correlation could be found between yeast population and time of
stability.

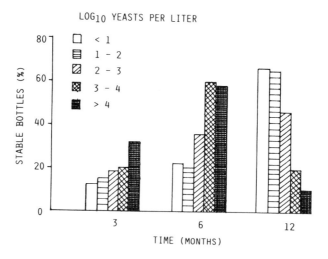

Fig. 2. Stability of bottled wine as a function of initial yeast
population

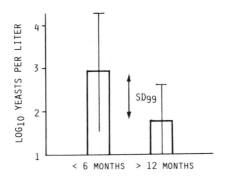

Fig. 3. Predictable stability of wine according to the initial yeast count

References

DEÁK, T. 1981 A simplified scheme for identification of yeasts and its applications in ecological surveys. Trilateral Conference on Yeast (Budapest) 4 pp.

KORNYEI, J., TABAJDI, P. V., DEÁK, T. & REICHART, O. 1983 Investigations of the biological stability of wines. Borgazdasag 31, 103-108. (In Hungarian)

LUKACSOVICS, F., HAJNAL, L. & DEÁK, T. 1981 Microbiological stability of wines of lower alcohol content. Borgazdasag 29, 30-33. (In Hungarian)

MINARIK, E. 1980 Zur Okologie von Hefen und hefeartigen Mikroorganismen abgefullter Weine. Wein-Wiss. 36, 280-285.

RANKINE, B. C. & PILONE, D. A. 1973 Saccharomyces bailii, a resistant yeast causing serious spoilage of bottled table wine. American Journal of Enology Viticulture 24, 55-58.

T. DEÁK
O. REICHART

Discussion

KING

After hearing presentations on baseline counts and unacceptable levels of fungi, it appears that there are three areas that we should consider with regard to mold counts. These are aspects of quality control, toxigenicity and spoilage. When we go into panel discussions we should consider why we are interested in fungi in foods and not be necessarily concerned with the counting aspect. We must take into account that it is not the fungus or its population, but rather what it can do to us or for us that is important.

WILLIAMS

This is very important. We do have baselines to consider, but we should keep in mind that changes are very important and that often we see that change is as important as the number.

KOBURGER

Canadians publish a report on foodborne illness outbreaks every year, and over the last few years there have been about ten reports of illness due to suspect yeasts or molds. Some are classified as being classical gastroenteritis caused by yeasts and molds. I was wondering if anyone else had experienced this.

PITT

In regard to illness being directly caused by fungi in foods, it seems that this is unlikely. I have not come across a problem with yeasts in causing sickness. As for mycotoxins, I have heard of two cases where mold growth in a liquid product caused sickness, one in Australia and the other in the United States. The thought is that we are probably looking at one of the Penicillium neurotoxins. We do not know what the effects of some of these toxins are on humans.

WILLIAMS

In England, I have heard of a couple of such cases over the last year. I have heard the term "psychological emesis" coined, which suggests in many cases that it is the presence of the yeast in the product which causes the illness just because the person visibly sees it or unknowingly swallows it and then becomes ill. It is not the yeast directly causing the disease. It is difficult to say whether this factor accounts for a significant number of these cases.

BECKERS

I am familiar with the data available from the Netherlands. It has been my experience that that type of illness can occur. People have gastrointestinal illness and for some reason relate it to moldy beverage or something else. It is more or less my feeling that it is coincidence and that molds are not the real cause of the illness but something else. In the Netherlands, I believe that we do not have any direct relation between gastrointestinal illness and moldy food or food spoiled by yeasts.

KING

There are two possibilities. The person sees the spoiled food and then becomes ill. On the other hand, do we know enough about mold toxins that may be in a particular food and the acute toxicity they might cause? I am not sure that we have enough data to make these correlations. It may occur in some cases, but probably not in most of them.

MISLIVEC

When someone gets sick in a consumer complaint situation, the agency (FDA) goes to pick up the product and then runs the whole gammet of tests on it. Usually you can't answer the question. They got sick but you really can't give a definite answer to the problem.

SPLITTSTOESSER

I believe that some of the data are suspect. We are talking about the annual reports, and I see that these report illness caused by yeast and mold growth in apple juice. Well, if you drink enough fermented apple juice, you will become ill. I am not blaming those responsible for these reports but I think what happens is that the reports come in from the field and that no central laboratory really takes a close look at them. They are reported in, the data are tabulated and the reports are released. I am not sure there have been any critical reviews of these.

PITT

Just one word of caution. That is, I don't think that we can readily
dismiss this subject. There is one well documented case in which a medical
researcher accidentally drank a beverage contaminated with <u>Penicillium</u> and
he documented the particular symptoms that he got. I do not have any doubt
at all from the sources of this material that this is quite accurate. I
suspect a neurotoxin. The thing that bothers me is that we do not know at
this stage what some of the <u>Penicillium</u> toxins will do. We just don't have
the information. We know that in animals these toxins are very toxic.

MOSSEL

I believe that John Pitt is right when he says there is some information
in the international literature describing diseases of sudden onset related
to the consumption of beverages contaminated with mold. A couple of years
ago in our country a young veterinarian made a survey on what was going on
in the rural areas. The main results of that study were, indeed, as Don
Splittstoesser said, that these cases should be examined extremely carefully
and selectively. I believe that there is acute poisoning of humans and a
distinct possibility that this is due to certain neurotoxins of the
Penicillia, but I have yet to see the first case of a clinical syndrome
elicited by drinking a beverage contaminated by a mold at a level of 10^3
or so colony forming units per milliliter. There have been at least two
cases in which yeast-colonized beverages were revolting and this may have
caused vomiting, which is more a psychological response. So in other words,
there may be a psychosomatic vomiting. I should recommend that in all these
instances, when investigating these outbreaks, we should first distinguish
between yeasts and molds. We should then support Rob Samson and clearly
identify the mold and therefore give the professional mycologist a clue as
to what happened. The yeast pathology is quite doubtful. There is a good
suspicion that molds may do dirty things to us when they have metabolized in
a beverage or food.

SEILER

This is in relation to spoilage of fruit juices. Did you say that
<u>Zygosaccharomyces</u> <u>bailii</u> was the spoilage organism in jams?

HOCKING

When you are talking about jams, you are talking about a completely
different product because of its low water activity, which is the principle
involved in its stability. In that case, you are often looking at
<u>Saccharomyces</u> <u>rouxii</u>.

SEILER

No, we are often looking at <u>Zygosaccharomyces</u> <u>bailii</u>.

HOCKING

In that case, the jam is not at a low enough water activity. Of course
these investigations are somewhat controversial because of the difficulty in
speciation of the yeast.

Recommendations

The following recommendations can be made concerning unacceptable levels
of fungi in foods:

1. It is not possible to generalize for all categories of fungi. Foods which are not pasteurized are likely to contain fungi, but certain fungi are unacceptable in certain foods.

2. In some quality control situations, routine monitoring is adequate. A recognizable change in either count or type of flora or the presence of a single dominant fungus indicates a potential problem.

3. In certain circumstances, the presence of a particular fungus at <u>any</u> level is unacceptable:

   Examples are:

   a. <u>Zygosaccharomyces</u> <u>bailii</u> in liquid, acid, preserved products such as fruit juices, fruit based products, ciders, wines, tomato sauce, mayonnaise and salad dressings.

   b. <u>Xeromyces</u> <u>bisporus</u> in foods which rely on low a_w for stability, except for foods with a_w less than 0.60 and salt-preserved foods.

   c. Heat resistant fungi, particularly <u>Byssochlamys</u>, <u>Talaromyces</u> and <u>Neosartorya</u> in pasteurized fruit juices and fruit-based products which do not contain preservatives.

4. With suspected mycotoxicoses, it is recommended that mycological evaluation be carried out to determine which mycotoxin assays are applicable. However, the mycoflora of products which have received some type of heat treatment may not be indicative of the mycotoxins present.

5. A high level of mold spores in the atmosphere of food processing plants is undesirable.

6. It is recommended that more work be done on foodborne allergenic fungi.

CHAPTER 7

NEW TECHNIQUES FOR ESTIMATING FUNGAL BIOMASS IN FOODS

Motives for investigating alternatives to plate (viable) counts for estimating fungal colonization of foods are as follows:

1. Viable counts estimate the number of colony-forming units (cfu) of fungi, which may be fragments of mycelium of widely varying dimensions, one or several sexual or asexual spores or yeast cells. Factors affecting viable counts include the degree of homogenization to which a sample has been subjected, the technique, medium, diluent and temperature of incubation as well as the history of the food before examination. Mycelial fragments in particular are likely to lose viability rapidly during storage of foods at reduced a_w.

2. Viable counts give no indication of the content of dead fungal biomass, which is useful for retrospective information concerning the quality of raw materials used in processed foods. Even foods that have received minimal treatment may contain significant proportions of non-viable mycelium.

3. When fungal counts are determined to assess the shelf life of products, there is a need to know if the fungi detected are capable of growing in and spoiling the food in question. Hence, tests for specific metabolic activity and/or specific groups of fungi may be more useful than total viable counts.

Investigations of alternative methods have followed several lines. The Howard mold test is a microscopic method that has been used for many years, mainly for examining tomato products, to estimate the level of both viable and dead fungal fragments. Disadvantages are that, like viable counts, it is affected by the degree to which the fungi have fragmented. The technique is also prone to operator error and is very tedious to perform. Suggested alternatives include estimation of chitin and ergosterol, which occur in most, but not all fungi.

Another possibility is to use immunological methods, e.g., antibodies against fungal cell wall components, provided these are present in a wide enough variety of fungi.

There are also possibilities for miniaturizing, replacing or speeding up existing viable count techniques, e.g., by use of impedimetry or ATP determination but, except for using selective culture techniques, little seems to have been done to date to quantify levels of spoilage-associated mycoflora.

There has been considerable interest in recent years in the development of rapid methods to detect fungi in foods and feeds. The presence of fungi in foods and feeds is important because these microorganisms can cause spoilage or toxicity of various commodities and can indicate poor sanitation during processing, storage and distribution. Thus, there is a need for rapid detection because new methods may save time in gathering data, save labor in the laboratory and processing plants and detect the presence of fungi more rapidly than with current methods.

Generally, new methods for detection of fungi have centered around four different areas:

1. Detection of a metabolite product that is produced by fungi but not present in the agricultural commodity being analyzed

2. Microscopic analysis of the commodity for mold filaments that fluoresce or take up selective dyes

3. Chemical analysis of cell wall polymers of fungi that are not present in food or feed, such as chitin

4. Immunological detection of fungi after producing antibodies to fungi in rabbits or other hosts.

Research in my laboratory over the last three years has been directed toward chemical methods for detection of mold in processed foods (Cousin et al. 1984; Lin & Cousin 1985). In this report, I will concentrate on chemical detection of chitin. Since the food industry is constantly using new methods for processing fruits and vegetables, mold may be so finely ground that it is not detectable by Howard mold or rot fragment count procedures. The presence of cell wall material can be detected as chitin regardless of the degree of comminution.

A colorimetric method (Tsuji et al. 1969) as modified by Ride & Drysdale (1971, 1972) and Jarvis (1977) was evaluated for the detection of mold in processed foods. This method, which involves digestion of the sample followed by colorimetric analysis, can be briefly outlined as:

1. Digestion of the sample --
   a. acetone is used to remove soluble glucosamine
   b. alkali is used to deacelate chitin
   c. the resulting chitin is precipitated with ethanol to give chitosan
   d. the chitosan is assayed colorimetrically.

2. Colorimetric assay --
   a. sodium nitrite is used to deaminate the chitosan to yield an aldehyde
   b. potassium bisulfate breaks down the aldehyde
   c. ammonium sulfamate is used to remove excess nitrous acid
   d. 3-methyl-2-benzo-thiazolinone hydrazone (MBTH) produces the color
   e. ferric chloride provides the iron to chelate with MBTH and intensify the color.

Detailed results of our studies are reported elsewhere (Cousin et al. 1984; Lin & Cousin 1985), and only a summary of our observations is presented here. The standard curve for glucosamine is linear over the 0-50 µg range. Above 50 µg, a line of best fit can be developed because the values are repeatable. Six molds (<u>Alternaria alternata</u>, <u>A. solani</u>,

<u>Colletotrichum</u> <u>coccodes</u>, <u>Fusarium</u> <u>oxysporum</u>, <u>Geotrichum</u> <u>candidum</u>, and <u>Rhizopus</u> <u>stolonifer</u>) grown in tomato juice broth at 21°C for 2 weeks served as test material for detectable chitin (expressed as glucosamine). Chitin content in cell walls was age and species dependent. Formation of chitin in the cell was not constant since the curves were nonlinear. The mycelial dry weight did not correspond to glucosamine content. Generally the mycelial dry weights peaked at 7 days and declined thereafter, suggesting autolysis.

Glucosamine was recovered proportionally to the dry weight of molds added to tomato puree. The amount of glucosamine was species and strain dependent. <u>G. candidum</u>, in the machinery mold form, produced more chitin than when grown normally, but coefficients of variability were similar.

Since chitin is also present in the exoskeleton of insects, tomato puree was supplemented with ground <u>Drosophila</u> flies to determine if they interfered with fungal estimation. Less than 2 mg of fruit flies were barely detectable in the tomato puree. Unless insect contamination is severe, fruit fly chitin should not interfere with the assay.

Because the Howard mold count is the current method used to detect mold filaments in food, tomato purees supplemented with mold were analyzed by this method and by the colorimetric procedure. There was a great deal of variation in the Howard mold count at each level of rot and for each trial. When the Howard mold count was compared to the level of glucosamine, similar variation was noted. Hence, a chemical method may need to stand on its own merit and not be compared to a microscopic fragment count.

The digestion of the sample in alkali seemed to be the major obstacle in obtaining repeatable results since digestion of standard samples containing 0.5 mg of <u>A</u>. <u>solani</u> per 2 g of rot-free tomato puree yielded significantly different results for each trial. However, there were no significant differences among the standard glucosamine solutions that were included with the assays.

A high performance liquid chromatographic (HPLC) method was then developed to overcome some of the problems observed with the colorimetric assay. The method involved hydrolyzing the food containing mold with 6N HCl at 121°C for 2 h to release the glucosamine, which was then partially purified by passing through a Dowex-5 cationic resin column. The resulting effluent was derivatized with o-phthalaldehyde and the fluorogenic agent was separated by reversed-phase HPLC and measured with a spectrofluorometer.

Maximum release of glucosamine from tomato puree supplemented with crab shell chitin was observed at 2 h; however, continued hydrolysis resulted in glucosamine destruction. When various amounts of crab shell chitin were added to tomato puree and hydrolyzed, a linear relationship was observed between chitin and glucosamine in the range of 10-100 μg/g of sample. Purified crab shell chitin was used for these experiments since it is commercially available whereas purified mold chitin is not.

Since purified crab shell chitin gave good linear results, fruit samples were hydrolyzed and assayed for glucosamine. The same samples were also assayed for amino acid content. It took 34 min to analyze one sample on an amino acid analyzer compared to 4 min by HPLC.

Different amounts and species of mold were added to tomato puree and analyzed for glucosamine content by HPLC. Linear relationships were observed for samples supplemented with 0.1 to 2.5 mg of mold/g of sample, but different species of molds gave different amounts of glucosamine. Similar linear relationships of added mold to glucosamine were observed when other mold-supplemented processed fruits were assayed.

Research is currently being carried out on the immunological detection of mold in processed foods. Although this work is in the preliminary stages, there appears to be promise for developing methodology in this area.

References

COUSIN, M. A., ZEIDLER, C. S. & NELSON, P. E. 1984 Chemical detection of mold in processed foods. Journal of Food Science 49, 439-445.

JARVIS, B. 1977 A chemical method for the estimation of mold in tomato products. Journal of Food Technology 12, 581-591.

LIN, H. H. & COUSIN, M. A. 1985 Detection of mold in processed foods by high performance liquid chromatography. Journal of Food Protection 48, 671-678.

RIDE, J. P. & DRYSDALE, R. B. 1971 A chemical method for estimating Fusarium oxysporum f. lycopersici in infected tomato plants. Physiological Plant Pathology 1, 409-420.

RIDE, J. P. & DRYSDALE, R. B. 1972 A rapid method for the chemical estimation of filamentous fungi in plant tissues. Physiological Plant Pathology 2, 7-15.

TSUJI, A., KINOSHITA, T. & HOSHINO, M. 1969 Analytical chemical studies on amino sugars. II. Determination of hexosamines using 3-methyl-2-benzo-thiazolene hydrazone hydrochloride. Chemical Pharmaceutical Bulletin 17, 1505-1510.

M. A. COUSIN

▶ New Techniques: Metabolic Measurements

Over the last few years, we have studied a simple and rapid method for the approximate assessment of the microbial colonization of foods that had previously been elaborated (Bomar 1983) and validated (Bartl et al. 1984) elsewhere. It relies on the metabolic activities of microorganisms, the strength of which is related to numbers of cfu present. For this purpose, 10^{-1} suspensions of food samples in diluent are prepared as usual (e.g., by maceration in a shake flask) after which the macerate is centrifuged at about 10^4 x g to concentrate the organisms and at the same time eliminate soluble food constituents which might interfere with the metabolic tests.

Originally nitrate reduction was used as the measuring system and nitrite detection tests were carried out at suitable intervals. Because many types of microorganisms of significance in foods do not reduce nitrate, glucose dissimilation was introduced to complement this test. There are virtually no microorganisms occurring in foods at significant levels that use neither glucose nor nitrate as substrate. The monitoring of glucose dissimilation is as easy as detecting nitrite: both tests can be done with the aid of 'dip sticks' which are commercially available for medical use.

The original methods (Bomar 1983) attempted to assess total colonization only. We have also adapted the method to determinate specific counts using selective media, particularly for the enumeration of Enterobacteriaceae and Staphylococcus aureus. The pellet obtained upon centrifugation of the 10^{-1} food macerate was suspended in lauryl sulfate broth and liquid Baird-Parker medium, respectively, both containing the appropriate concentration of nitrate and glucose. To shorten response times, the incubation temperature was increased to 42°C. Detection times of 4-6 h were required to demonstrate the presence of approximately 10^4 cfu/g, but glucose depletion was generally a slower process than nitrate reduction.

For the approximate assessment of numbers of cfu of yeasts in foods, yeast autolysate with added oxytetracycline and gentamicin as well as nitrate/glucose incubated at 30°C worked satisfactorily. Investigations are in progress to assess numbers of mold propagules in a modified minimal mineral salts medium with polysorbitan 80, glucose, nitrate, oxytetracycline and gentamicin.

This method requires no other instruments than a table top centrifuge and a water bath, does not demand costly reagents, gives reliable results within one working day and can be carried out by staff without advanced training in microbiology.

References

BARTL, B., VANOUSKOVA, S. & ROUBALOVA, V. 1984 Some rapid methods in food microbiology. In Microbiological Associations and Interactions in Food. ed. Kiss, I., Deák, T. & Incze, K. pp. 213-218. Budapest: Akademiai.
BOMAR, M. T. 1983 Rapid determination of critical bacterial counts in foods. Alimenta 22, 189-194.

D. A. A. MOSSEL

▶ Comparison of a Plate-MPN Technique with Surface Plate Count for Mold Enumeration

Koburger & Norden (1975) demonstrated that the MPN technique could be applied to enumeration of fungi using liquid media containing antibiotics. In that study the MPN technique gave consistently higher counts on foods than either the surface- or pour-plate technique. Jarvis (1978) stated that the major objections to the MPN technique were the large quantity of medium required and the low level of statistical significance which can be applied to quantitative results. Tan et al. (1983) reported the use of a plate-MPN technique for bacteria that eliminated the problem of large quantities of media and which was as accurate as the standard technique for bacterial counts. The technique utilized three discrete 0.01-ml samples of an appropriate decimal dilution inoculated onto the surface of each quadrant of a pre-dried medium surface in a standard Petri dish. Discrete spots were observed for growth after incubation and the results were interpreted in a manner analogous to a 3-tube MPN (Speck 1976). This study was done to adapt the plate-MPN technique to mold enumeration and compare the results to surface plate counts.

Materials and methods

Samples. Samples of seven commercial foods (dry gravy mix, chopped walnuts, dry split peas, white flour, yellow corn meal, pecan pieces and "organic" corn meal) were used to compare the techniques. In addition, white flour was inoculated with three levels (ca. 10^2, 10^4 and 10^6/g)

of a mixture of spores from nine different molds: <u>Aspergillus</u> <u>flavus</u>, <u>A.</u>
<u>niger</u>, <u>A.</u> <u>ochraceus</u>, <u>Penicillium</u> <u>martensii</u>, <u>P.</u> <u>roqueforti</u>, <u>P.</u> <u>viridicatum</u>,
<u>Fusarium</u> <u>graminearum</u>, a <u>Cladosporium</u> sp. and an <u>Alternaria</u> sp. Dry spores
of each organism were mass produced on bread cubes in quart Mason jars
according to the method of Sansing & Ciegler (1973). After growth and
sporulation, the bread cube cultures were allowed to dry and the spores were
harvested in 100-200 g of dry flour by dry blending the mixture. The total
number of propagules in the mixture was determined by a surface plate count
technique. Appropriate amounts of the mixture were used to inoculate flour
samples to obtain approximate counts of 10^2, 10^4 and 10^6 propagules/g.

Sample preparation. Samples were prepared by combining 11 g of sample
with 99 ml of sterile phosphate buffer diluent (Speck 1976) and mixing by
either blending or stomaching for 3 min. Subsequent decimal dilutions were
made in phosphate buffer. Test volumes were plated within 1 min of
completing the dilution sequence (Jarvis et al. 1983).

Plate counts. Plate counts were made using a surface plating technique:
0.1-ml volumes of diluted sample were spread over the medium surface using a
sterile bent glass rod. PDA plus 40 µg/ml tetracycline (TPDA; Difco) and
DRBC agar were used.

Plate-MPN technique. Plate-MPN's were also done on TPDA and DRBC
according to the method of Tan et al. (1983). Three discrete inoculum
volumes of 0.01 ml each, using a 20-µl automatic pipettor, were applied to
quadrants of each plate. If counts were expected to be low, the inocula
were increased to 0.1 ml. After incubation, the MPN was determined by
observing spots with growth and consulting a 3-tube MPN table.

Treatment of plates. All plates, surface and MPN, were incubated in an
upright position at 25°C for 5 days. For plate counts only, plates
containing 10-100 colonies were taken as countable plates. If countable
plates could not be obtained using 0.1-ml volumes, counts were repeated
using a 1.0-ml volumes. The 0.01-ml spots on MPN plates were read as
positive or negative just as a tube in the tube MPN technique. All counts,
plate or MPN, were converted to $\log_{10}$ counts for reporting.

Statistical design and analysis. There were three experimental factors
in this study: 1. method of preparation of inoculum (blending vs
stomaching); 2. method of plating (surface vs plate-MPN); and 3. culture
media (PDA + tetracycline vs DRBC). A split plot design was used. The
method for preparation of inoculum was considered as a block. Each food
sample was analyzed in duplicate, and the counts were done three times on
each sample. Organic corn meal was tested six times with duplicate plates
each time. Mean $\log_{10}$ values of counts and differences obtained for each
food sample were subjected to analysis of variance (Steel & Torrie 1980).
The α-value was chosen to be 0.05. The standard deviation of the mean
difference was also determined.

Results and discussion

    Results are summarized in Tables 1 and 2. In most cases the plate-MPN
technique gave counts that were significantly ($P < 0.05$) higher than the
plate count technique. This is an agreement with the observation of
Koburger & Nordan (1975). Overall, the counts were quite comparable
microbiologically between the plate count and plate-MPN procedures since
they were in the same log range. The plate-MPN technique worked equally
well with TPDA or DRBC and with blending or stomaching, and greatly reduced
the amount of media and glassware required. This would facilitate examining

Table 1. Mean $\log_{10}$ yeast and mold counts and differences obtained with surface-plate and a plate-MPN technique on DRBC and PDA + tetracycline using blending for sample preparation

Food	TPDA			DRBC		
	Surface	MPN	Difference	Surface	MPN	Difference
Dry gravy mix	2.52	2.80	−0.28	2.50	2.95	−0.45
Chopped walnuts	3.08	3.67	−0.59	3.31	3.63	−0.32
Dry split peas	1.32	1.81	−0.49	1.34	1.74	−0.40
Bleached flour	2.46	2.63	−0.17	1.84	2.32	−0.48
Yellow corn meal	2.63	2.88	−0.25	2.48	2.65	−0.17
Pecan pieces	2.10	2.44	−0.34	1.90	2.08	−0.18
Organic corn meal	5.42	5.33	0.09	5.41	5.25	0.16
Inoculated flour (10^6/g)	5.19	5.34	−0.15	5.09	5.32	−0.23
Inoculated flour (10^4/g)	3.13	3.56	−0.43	3.16	3.42	−0.26
Inoculated flour (10^2/g)	2.27	2.60	−0.33	2.20	2.71	−0.51
Mean of difference:			−0.294			−0.284
SD of difference:			0.193			0.199

large numbers of samples. On the other hand, the plate-MPN technique does not permit identification or isolation of specific molds.

The yeast and mold counts obtained from the plate-MPN method were significantly higher than counts obtained from surface plate counts in all food samples prepared by either stomaching or blending except the organic corn meal. The increase in sample number from six, as in the other food

Table 2. Mean $\log_{10}$ yeast and mold counts and differences obtained with surface plate and a plate-MPN technique on PDA + tetracycline and DRBC using stomaching for sample preparation

Food	PDA			DRBC		
	Surface	MPN	Difference	Surface	MPN	Difference
Dry gravy mix	2.22	2.63	−0.41	2.23	2.70	−0.47
Chopped walnuts	3.32	3.64	−0.32	3.43	3.?4	−0.11
Dry split peas	1.34	1.82	−0.48	1.23	1.51	−0.28
Bleached flour	2.42	1.81	0.61	2.17	2.52	−0.35
Yellow corn meal	2.45	2.96	−0.51	2.60	2.64	−0.04
Pecan pieces	1.94	1.80	0.14	1.84	2.23	−0.39
Organic corn meal	5.39	5.27	0.12	5.36	5.16	0.20
Inoculated flour (10^6/g)	5.34	5.50	−0.16	5.33	5.58	−0.25
Inoculated flour (10^4/g)	3.34	3.61	−0.27	3.27	3.52	−0.25
Inoculated flour (10^2/g)	2.36	2.85	−0.49	2.62	2.83	−0.21
Mean of difference:			−0.177			−0.215
SD of difference:			0.364			0.193

items, to twelve in the organic corn meal may have contributed to minimizing the higher count effect when the plate-MPN technique was used.

Conclusions

The plate-MPN technique appears to give results comparable with the surface plate count technique, and requires less media, glassware and time to perform. However, no information is obtained concerning the types of molds present using the plate-MPN technique, nor can isolations be made. The plate-MPN technique is useful and valid under certain conditions and in certain situations. More research would be useful to help establish the plate-MPN as a yeast and mold enumeration technique.

References

JARVIS, B. 1978 Methods for detecting fungi in foods and beverages. In Food and Beverage Mycology ed. BEUCHAT, L. R. pp. 471-504. Westport, CT: AVI Publ. Inc.

JARVIS, B., SEILER, D. A. L., OULD, J. L. & WILLIAMS, A. P. 1983 Observations on the enumeration of molds in food and feedstuffs. Journal of Applied Bacteriology 55, 325-336.

KOBURGER, J. A. & NORDEN, A. R. 1975 Fungi in foods VII. A comparison of the surface, pour plate and most probable number methods for enumeration of yeasts and molds. Journal of Milk and Food Technology 38, 745-756.

SANSING, G. A. & CIEGLER, A. 1973 Mass propagation of conidia from several Aspergillus and Penicillium species. Applied Microbiology 26, 830-831.

SPECK, M. L. (ed.) 1976 Compendium of Methods for the Microbiological Examination of Foods. Washington, D.C.: American Public Health Association.

STEEL, R. G. & TORRIE, J. H. 1980 Principles and Procedures of Statistics, a Biometrical Approach. 2nd edn. p. 235. New York: McGraw-Hill Book Co.

TAN, S. T., MAXCY, R. B. & STROUP, W. W. 1983 Colony-forming unit enumeration by a plate-MPN method. Journal of Food Protection 46, 836-841.

J. W. HASTINGS
W. Y. J. TSAI
L. B. BULLERMAN

Published as Paper No. 7608, Journal Series, Agricultural Research Division, Lincoln, NE. Research reported was conducted under Project 16-042.

 Impedimetric Estimation of Molds

The use of impedimetric measurement for estimation of bacteria in foods has been described by Wood et al. (1978). The purpose of the work described here was to investigate the suitability of the Bactometer 32 impedance system for the analysis of molds, both in pure culture and in foods.

Total colony counts of molds in foods, especially moist or wet materials, are subject to many sources of variation. In particular, counts may be affected by the degree of fragmentation of the fungal components (e.g., by comminution during sample preparation), by the extent of sporulation of the mycelia present and by the capacity of fast-growing molds to overgrow slow-growing species.

The advantage of an estimation method based on metabolic activity would be that the result obtained would reflect the overall metabolic activity of mold species present in the chosen substrate, irrespective of biomass or number of fragments present. Such a method should preferably also be economical in materials, convenient to operate, reproducible and more rapid than conventional colony counts which normally take 5 days.

Materials and methods

Equipment. A Bactometer 32 (Bactomatic Inc., Princeton, NJ) was used throughout the present study.

Media. The basal medium was glucose (2%, w/v) with malt extract (Oxoid L39; 2%, w/v), pH 5.5, autoclaved for 10 min at 115°C (GME). During the course of the work the following adaptations were made: (a) addition of agar (1.5%, w/v) prior to autoclaving; (b) adjustment of pH with 0.1M HCl or NaOH; and (c) addition of antibiotics (chloramphenicol and oxytetracycline) at concentrations necessary to give 100 µg/ml of each in the final contents of each test well. GME was used throughout as a diluent.

Spore suspensions. Spore suspensions of thirty-two common molds were washed from 10-day MEA slants using GME broth.

Impedimetry. Wells were inoculated according to methods described by Wood et al. (1978) and detection times were recorded as described by the same authors. When wells contained liquid medium, the total volume in each well was 1 ml. When solidified media were used (0.5 ml/well), the inoculum (normally 0.1 ml/well of the equivalent broth medium) was added after the agar had set. Detection times were compared, where appropriate, with total colony counts of decimal dilutions of the same inocula, using identical nutrient and incubation conditions.

Measurement of impedimetric response by mold species. Each well contained 0.5 ml of GME plus agar. Spore suspension (0.1 ml) was added to the surface of the medium in one well for each of the thirty-two molds tested. Uninoculated medium (0.1 ml) was added to the control wells. Modules were incubated at 26.7°C and impedance responses were recorded automatically by a chart recorder.

Assessment of reproducibility of results by measurements on Aspergillus ochraceus. The method was the same as described above, except that only one mold was used and the incubator temperature was 25°C. Results from experiments carried out on five separate occasions were compared. Similarly, five replicate wells were inoculated with six serial decimal dilutions of spore suspension of A. ochraceus IMI 33910 and incubated at 30°C.

Evaluation of effect of pH value on detection time. GME broth was adjusted with either 0.1M NaOH or HCl to cover a range of pH 3-9 in steps of 0.5 pH units. The broth was filter-sterilized and dispensed (1 ml per well) into modules. An inoculum of 0.05 ml of a spore suspension of A. ochraceus, in the same broth (pH 5.5), was added to each well and 0.05 ml of the uninoculated broth was added to the control wells. Further aliquots of the pH-adjusted, filter-sterilized broths were diluted in the same manner, and the final pH values were recorded. Modules were incubated at 26.7°C.

Use of antibiotics to prevent bacterial growth in mixed cultures and in foods. Filter-sterilized aqueous solutions of chloramphenicol and oxytetracyline were added to autoclaved GME agar at concentrations necessary to give 100 µg/ml of each in the final volume in each well (0.5 ml of medium plus 0.1 ml inoculum). Inocula consisted of serial decimal dilutions of various food substrates or cultures prepared in GME broth. Controls contained no antibiotic.

Bacterial enrichment. To test the system with actively growing bacteria, an enrichment culture was prepared from 10 g of mixed spice incubated for 24 h at 25°C in GME broth (pH 5.5). Colony counts of serial decimal dilutions of the enrichment broth (0.1 ml) were made on tryptic soy agar (TSA), using duplicate spread-plates incubated at 25°C for 4 days. Presence of bacteria was confirmed by Gram staining the enrichment culture and colonies from the dilution plates. Dilutions of the enrichment and of a spore suspension of $\underline{A}$. ochraceus were prepared in sterile broth to give final concentrations of bacteria of 5×10^5 cfu/well and of mold spores of 7×10^4 cfu/well.

Results and discussion

The optimum method of analysis should be characterized by a short detection time, minimum 'baseline' drift and a clear detectable change of signal, followed by an exponential signal as confirmation of detection.

In initial experiments (not reported here), various concentrations and combinations of glucose, sucrose, maltose, malt extract, yeast extract, peptone, yeast nitrogen base, sodium nitrate and ammonium nitrate were evaluated in the culture medium. Media were compared with and without agar and agarose at 1.5% and 0.25% (w/v). $\underline{A}$. ochraceus (IMI 33910) was used throughout these early experiments. The most successful medium was the basal medium (GME). Agar, although improving the quality of the signal, was not essential for achieving a response.

It must be emphasized that the direction in which the impedance signal moves, when growth is detected, is of major importance in the choice of nutrient substrates for culturing molds. The majority of bacterial signals

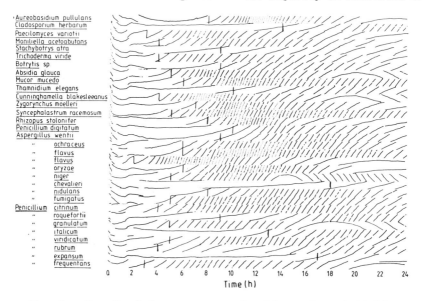

Fig. 1. Impedimetric responses of mold species at 26.7°C

manifest themselves as a decrease in impedance, or 'right-hand' signal. In
the GME medium, however, mold growth is detected by an increase in
impedance, or 'left-hand' signal (Fig. 1). Left-hand signals are engendered
from substrates containing sugars and malt extract when inoculated with
molds. (Malt extract is also a convenient nitrogen source.) Right-hand
mold signals are encountered when yeast extract, peptones and other organic
and inorganic nitrogen sources are used. It is thus necessary to select
basal ingredients of media which do not result in conflicting signals, e.g.,
glucose/malt extract maltose/malt extract or yeast extract/peptone. A
medium such as MYGP (malt extract 'left-hand,' yeast extract 'right-hand,'
glucose 'left-hand' and peptone 'right-hand') would be a very unwise choice
as a substrate for detecting growth by change in impedance.

Impedimetric responses of mold species. The impedance signals given by
the thirty-two strains of mold investigated are shown in Fig. 1. The
objective of this experiment was to ascertain the direction and shape of the
impedance response curves and not the detection times. Viable counts,
therefore, were not determined. All of the molds exhibited clear
'left-hand' signals (increasing impedance), supporting the hypothesis that
mixed populations of molds will not produce mutually antagonistic signals.

The weakest signal was that for Penicillium expansum. It is hoped that
it may be possible to improve the signal by slight modifications to the
growth medium. A promising alternative medium consists of malt extract (2%,
w/v) with maltose (0.5%, w/v), although insufficient trials have so far been
carried out to reach any firm conclusions. P. digitatum, the only common
mold described as unable to utilize maltose (Cochrane 1965), is able to grow
in this medium, presumably by utilizing other sugars present in malt extract.

Reproducibility of results for A. ochraceus. Combined results for five
experiments are given in Fig. 2. An inverse linear relationship is shown
between detection time and the $\log_{10}$ colony count. In addition, the
results of a multiple experiment at 30°C are given (Fig. 3; five replicates
at six dilutions).

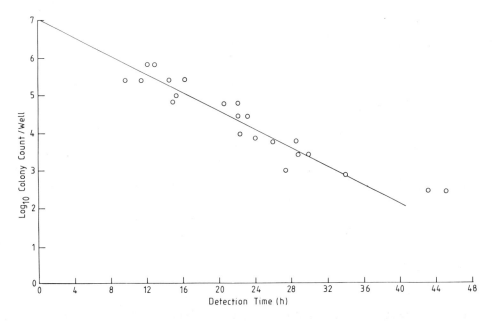

Fig. 2. Relationship between detection time and colony count for Aspergillus
ochraceus at 25°C (Courtesy Journal of Applied Bacteriology)

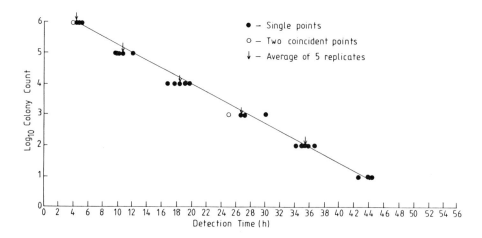

Fig. 3.   Reproducibility of results for **Aspergillus ochraceus** at 30°C

Effect of pH on detection time and growth of A. ochraceus. The results, which were obtained on three successive 2-day periods, are summarized in Fig. 4. Minimum detection times occurred in the region of pH 5.5–7.5 and detection times were approximately doubled at pH 4. Below pH 4, the detection signal changed from an increase to a decrease in impedance, resulting in a loss of detection at pH 3.8. However, bacteria are capable of growth at pH 4 in similar systems and it must be concluded that pH control alone will be unsatisfactory for the selective detection of molds in mixed mold/bacterial populations.

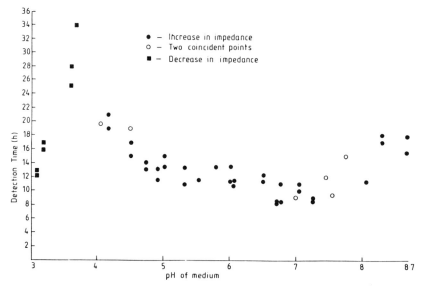

Fig. 4.   Effect of pH of medium on detection time for **Aspergillus ochraceus** grown in glucose/malt extract broth at 26.7°C (inoculum 4 x 10^5 cfu/well)

Table 1. Effect of antibiotics on the growth of <u>Aspergillus</u> <u>ochraceus</u>

Antibiotics included[a]	Colony count (cfu/well)	Detection time (hours) at 25°C
–	7.4×10^4	13.0
–	7.4×10^3	16.6
–	7.4×10^2	22.4
+	7.4×10^4	15.0
+	7.4×10^3	24.0
+	7.4×10^2	34.0
–	Uninoculated	ND[b]
+	Uninoculated	ND

[a]Chloramphenicol (100 µg/ml) plus oxytetracycline (100 µg/ml)
[b]ND = Not detected

<u>Use of antibiotics to prevent bacterial growth</u>. The effects of chloramphenicol (100 µg/ml) and oxytetracycline (100 µg/ml) on growth of <u>A</u>. <u>ochraceus</u> are shown in Table 1 and Fig. 5. Antibiotics caused an increase in detection time and a reduction in the strength of the impedance signal. The relationship between colony count and detection time, however, remained linear.

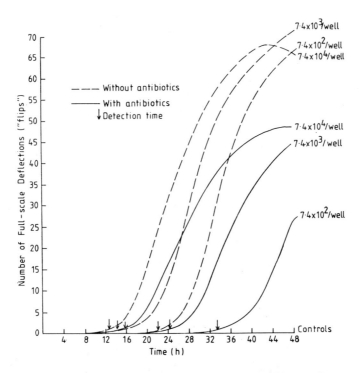

Fig. 5. Effect of antibiotics on growth of <u>Aspergillus</u> <u>ochraceus</u> at 25°C

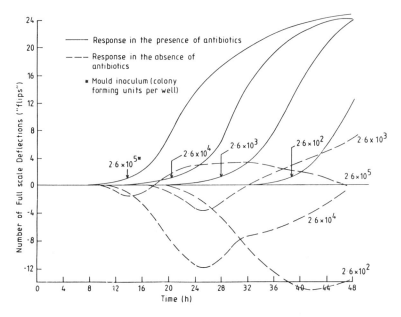

Fig. 6. Detection times for dilutions of moldy cheese, with and without antibiotics (chloramphenicol 100 µg/ml plus oxytetracycline 100 µg/ml) at 25°C

Results for a sample of naturally moldy Cheddar cheese are given in Fig. 6. Detection time in the presence of antibiotics increased linearly with sequential decimal dilutions of the sample (mold counts are included on the graph). In the absence of antibiotics, irregular signals were obtained, possibly as a result of antagonism between the signals from molds and bacteria. The latter, in this system, tend to give signals in the opposite direction.

In the enrichment culture prepared for testing the system with actively growing bacteria, the only organisms which grew were Gram-positive

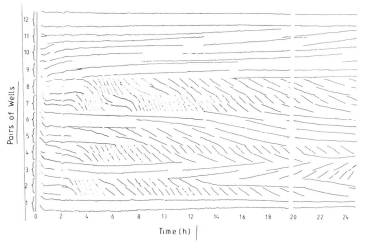

Fig. 7. Chart of experiment described in Table 2

236

Table 2. Effect of antibiotics and pH value on the impedimetric response of
          <u>Aspergillus</u> <u>ochraceus</u>

Pair of wells (see Fig. 7)	Bacterial enrich- ment[a]	<u>A. ochra- ceus</u>[b]	Glucose/ malt extract broth (1.0 ml)	Anti- biotics[c]	pH after inocula- tion	Impedance change
1	−	−	+	−	5.5	None
2	+	−	+	−	5.5	Decrease
3	−	+	+	−	5.5	Increase
4	+	+	+	−	5.5	Decrease
5	+	−	+	−	4.0	Decrease
6	+	+	+	−	4.0	Decrease
7	+	−	+	−	7.0	Decrease
8	+	+	+	−	7.0	Decrease
9	+	−	+	+	7.0	None
10	+	+	+	+	7.0	Increase
11	−	+	+	+	7.0	Increase
12	−	−	+	+	7.0	None

[a] 5×10^5 cfu/well
[b] 7×10^4 cfu/well
[c] Chloramphenicol (100 µg/ml) plus oxytetracycline (100 µg/ml)

spore-forming rods. A copy of the record chart is included as Fig. 7 and
the experimental parameters are summarized in Table 2.

Traces marked 1 and 12 (Fig. 7) were from uninoculated wells and no
impedance change occurred. In trace 2, the bacterial enrichment (pH 5.5) on
its own gave a typical signal of decreased impedance, whereas in trace 3 the
slower and increasing impedance change is typical of a mold. In the mixed
bacterial/mold population (trace 4) the bacterial signal was dominant. At
pH 4.0 (trace 5), bacterial growth still occurred but at a reduced rate, and
the reduction was enchanced in the mixed bacterial/mold population (trace
6). The same pattern was repeated at pH 7.0 (traces 7 and 8) but with a
much stronger bacterial signal.

In the presence of antibiotics, the bacterial response was totally
suppressed (traces 9 and 10) and the mold signal was retained (trace 11),
although delayed (as would be expected from Table 1 and Fig. 5). The mold
response was also present in the mixed bacterial/mold population (trace 10),
although still further delayed (compared with traces 3 and 11).

Control of bacterial growth by antibiotics seems promising. Using a
combination of chloramphenicol (100 µg/ml) and oxytetracycline (100
µg/ml), bacterial signals were totally suppressed, although (as shown in
Fig. 5) at the expense of a delay in mold detection.

Conclusions

Metabolic activity of wide range of molds can be detected
impedimetrically using the Bactometer 32. There is an inverse linear
relationship between the $\log_{10}$ colony count (measured on spore
suspensions) and detection time. The pH of the growth medium affects

237

detection time, roughly doubling the time for a decrease from pH 7 to pH 4 for **A**. ochraceus. Since some bacteria will grow at pH 4, pH control alone would be unsatisfactory for distinguishing between molds and bacteria. Antibiotics successfully suppressed the bacterial signal but detection time for **A**. ochraceus is increased. Further work is needed to assess optimum antibiotic conditions for a wide range of molds. It may be possible to improve detection times by altering antibiotic mixtures as described above.

References

COCHRANE, V. W. 1965 Physiology of Fungi. New York: John Wiley & Sons Inc.
WOOD, J. M., LACH, V. H. & JARVIS, B. 1978 Evaluation of Impedimetric Methods for the Rapid Estimation of Bacterial Populations in Foods. Leatherhead Food R. A. Research Report No. 289.

This paper was first published as Leatherhead Food Research Association, Technical Circular No. 791. (1982).

A. P. WILLIAMS
J. M. WOOD

▶ Modern Detection Methods for Food Spoilage Fungi

Detection of potentially hazardous fungi in foods, beverages and feeding stuffs is a crucial step in food quality control and ultimately in health care (Beuchat 1978). Monitoring fungi in general and mycotoxins and allergens in particular is essential during the many stages of production of raw materials and ultimately of food products since many contaminants are transferred and eventually accumulated.

In general practice, methods based on isolation, culturing and morphological characterization are widely used, sometimes in combination with diagnostic techniques (Beuchat 1978; Samson et al. 1984). However, these approaches are time consuming, since they involve incubation for several days and are mostly applicable only to detecting living fungal propagules. In some cases a direct microscopic count (Howard mold count) can indicate the presence of molds both qualitatively and quantitatively (Gould 1983), but in practice this technique is often not reproducible. During the last decades a number of additional methods have been developed to detect general life processes or fungal metabolites in complex matrices, e.g., in soil (Rosswall 1972), in human and animal tissues or body fluids (Coonrod et al. 1983), in grain and in food products (Beuchat 1978). These approaches fall into four overall categories (Table 1).

Detection of a broad range of fungi (Table 1). The detection of a broad range of fungi depends on the occurrence of certain metabolites. The presence of polyols (e.g., arabinitol) and mannan in human serum is currently being studied as an indicator of invasive mycoses (De Repentigny 1983), especially those caused by yeasts (e.g., candidiasis).

The analysis of fungus-specific volatiles is another promising approach. Zlatkis et al. (1981) pointed out the possibilities in medical diagnosis of using organic volatiles to detect infections.

Table 1. A survey of indirect methods for the detection of fungal contamination

---

Detection of fungi by their life processes:
1. Microcalorimetry
2. Respirometry
3. Detection of ATP
4. Detection of dehydrogenase activity
5. Electrical impedance measurements
6. Fluorescein diacetate (FDA) hydrolysis.

Detection of a broad range of fungi:
1. Detection of polyols
2. Detection of headspace volatiles
3. Pectin esterase detection
4. Detection of chitin, chitosan or glucosamine
5. Detection of ergosterol.

Detection of specific fungi:
1. Detection of specific metabolites, e.g., mycotoxins, pigments
2. Serological techniques.

Overall quality control:
1. Fingerprinting techniques, e.g., pyrolysis.

Spoilage fungi can also be detected by measuring the release of methanol from pectin by the action of the enzyme pectin esterase (Offem & Dart 1983). This method was developed for _Aspergillus_ and _Penicillium_, but it is not reliable for yeast species.

The detection of chitin (Wu & Stahmann 1975) and ergosterol (Seitz et al. 1979) in raw materials (e.g., grain) and food is currently applied to detect general contamination with fungi. Discrimination between viable and dead cells is not possible. Some characteristics of chitin and ergosterol are listed in Table 2.

Chitin is a common component of the fungal cell wall, although its occurrence is limited to less than 1% in the walls of ascomycetous yeasts and the Oomycetes (Bartnicki-Garcia 1968). Contamination with arthropods must be kept in mind since chitin is also the main building block of arthropod cuticles (Rudall & Kenchington 1973). Chitin is usually estimated by determination of the hydrolysis product, glucosamine. Again, non-chitinous glucosamine of bacterial cell wall origin can also contribute in glucosamine assays of hydrolyzed food samples. Therefore, methods in which chitin is converted to chitosan by alkali followed by colorimetric quantification are preferable (Ride & Drysdale 1972). Another way is to estimate both muramic acid and glucosamine (Hicks & Newell 1983). The chemical stability of chitin is an advantage which enables the estimation of fungi in processed and long-term stored foods.

Ergosterol is also an indicator of the presence of fungi with the exception of, again, the Oomycetes (Weete 1980) and some other fungi (Van Eijk & Roeijmans 1982).

Rapid methods are highly desirable both in sample pretreatment and in the actual analysis (Table 3). Extensive fractionation processes should be avoided.

239

Table 2. Chitin and ergosterol as indicators of fungal contamination of raw
         materials and food products

Characteristic	Ergosterol	Chitin
Distribution	Fungal kingdom; some doubtful and scarce reports on plants	Fungal kingdom; arthropods molluscs, annelids, etc.
Exceptions	Oomycetes, _Taphrina_, _Protomyces_	Oomycetes, yeasts
Stability	Thermostable in crystalline state; affected by light and oxygen; not stable in solution.	Stable; autolysis limited
Concentration range	0.1–3% (w/w) of cell	0.5–40% (w/w) of cell
Extraction	Chloroform/methanol (2:1), saponification	Hydrolysis yields glucosamine or chitosan
Detection	TLC, GLC, HPLC, computerized spectrophotometry	Spectrophotometry, HPLC, GLC

Detection of specific fungi (Table 1). An example of target analysis of a highly specific component in a complex matrix (pachybasin in potato tubers) will be discussed here. The same matrix is also analyzed by a non-target approach, viz., pyrolysis mass spectrometry (Py-MS).

Gangrene is an economically important disease of potato tubers caused by the fungus _Phoma_ _exigua_ Desm. var. _foveata_ (Foister) Boerema. A description of the disease is given by Logan (1981). It is presently restricted to

Table 3. Sample pretreatment and instrumental techniques in the
         detection of fungi in food

Sample	Instrumentation[a]
Whole matrices	Pyrolysis (Py-GC-MS and Py-MS), DIP-MS, LAMMA, NMR
Headspace or dissolved volatiles	Purge-and-trap or Tenax-trap followed by GLC
Extracts and hydrolysates	TLC, HPLC, GLC, MS, NMR, UV-VIS hyphenated techniques

[a]Py-GC-MS = pyrolysis gas chromatography-mass spectrometry; Py-MS = pyrolysis mass spectrometry; DIP-MS = direct insertion probe mass spectrometry; LAMMA = laser microprobe mass analysis; NMR = nuclear magnetic resonance spectroscopy; UV-VIS = ultraviolet-visible spectrophotometry

Northern Europe, Australia and the High Andes in South America, and threatens
the export of seed potatoes. The morphologically indistinguishable variety
P. exigua var. exigua is almost cosmopolitan in distribution but less
serious as a pathogen. In the Netherlands the discrimination between the
two varieties is based on TLC detection of the anthraquinone pigment
pachybasin (1-hydroxy-3-methylanthraquinone) in extracts of affected potato
tubers (Mosch & Mooi 1975). The outcome of the diagnostic tests is decisive
for certification and official granting of export licenses.

Materials and methods

All samples and procedures are described in detail by Weijman et al.
(1984b). Isolates of both varieties of P. exigua were obtained from the
Research Institute for Plant Protection (IPO), Wageningen, the Netherlands.
Whole potato tubers were inoculated with a spore suspension, incubated,
extracted with ethanol and concentrated. The concentrate was diluted,
acidified, extracted with ethyl acetate, dried and redissolved in
chloroform/methanol. GC-MS analysis of the solution was performed with a
Hewlett-Packard HP 5993B GC-MS-DS system equipped with a splitless capillary
column injection system and an OV-101 WCOT fused silica capillary column.
Selected ion monitoring (SIM) analyses were performed on mass 238.1 (Weijman
et al. 1984a).

Py-MS samples were prepared by suspending the freeze-dried and powdered
materials in methanol. Suspensions were applied to the ferromagnetic
wires. A description of the instruments and techniques involved in Py-MS
have been given by Meuzelaar et al. (1982). Basically, the Py-MS system
consists of a Curie-point reactor mounted directly in the vacuum of a
quadrupole mass spectrometer. The Curie-point technique involves rapid
inductive heating of a Fe-Ni wire, stabilizing within 0.1 sec at 510°C. The
spectra of the various samples were compared by multivariate analysis using
a set of twenty most characteristic peaks. Two-dimensional non-linear
mapping of the resulting multidimensional relationships was performed by the
procedure of Kruskal (Eshuis et al. 1977).

In Py-MS, the sample is fragmented according to thermochemical
principles and the volatile fragments formed are characterized by mass
spectrometry (14 eV electron impact ionization). The resulting mass
spectrum (pyrogram) can serve as a fingerprint reflecting the overall
chemical composition of the sample. For quality control and diagnostic
purposes, similarities between pyrograms can be calculated using
multivariate and factor analysis techniques (Meuzelaar et al. 1982).

Results and discussion

Typical Py-MS fingerprints of potato tuber tissue infected with P.
exigua var. exigua and P. exigua var. foveata are presented in Fig. 1.
Fungal components contribute insignificantly to the total mass of the
tubers. Consequently, the small quantitative differences between these
spectra can be explained by the variable constitution of potato tubers,
partly caused by reactions following infection (e.g., production of
phytoalexins or enzymes).

Multidimensional relationships of peaks resulting from computerized
multivariate analysis clearly illustrate the ability of Py-MS to
discriminate between different infections. Among the most characteristic
peaks underlying differences are those indicative for phenolic constituents
(Meuzelaar et al. 1982). In this case the phenolic peaks can be used as
single parameters for screening of infections with var. foveata.

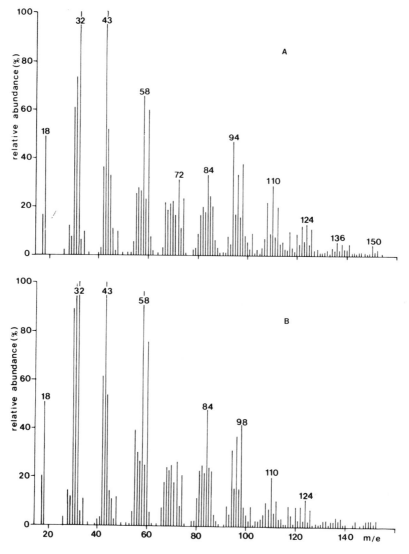

Fig. 1. Pyrolysis mass spectra of _Phoma_ _exigua_ var. _foveata_ isolate 44 (A) and _Phoma_ _exigua_ var. _exigua_ isolate 11 (B)

The GC-MS approach is based on the stability of the target, pachybasin, under 70 ev EI conditions, reflected by the dominance of the molecular ion at m/e 238 in the mass spectrum (Fig. 2). Detection of this ion in the SIM mode of operation leads to a detection limit of about 50 pg. The specificity of GC-MS is illustrated in Fig. 3. Pachybasin could not be demonstrated in control potato tubers or in tubers infected with other potato pathogens.

Fingerprinting technique and quality control. The GC-MS method studied as an alternative approach in the case of gangrene, although more expensive, is faster, more reliable and sensitive compared with methods such as TLC.

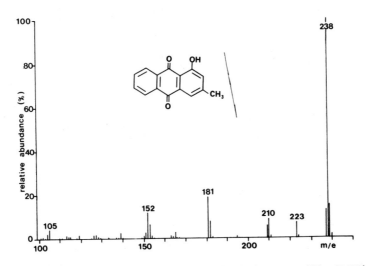

Fig. 2. Mass spectrum of pachybasin reference (70 eV EI)

Gangrene can now be identified in an early stage of development (Weijman et al. 1984b). Py-MS is a non-target approach for rapid discrimination of biological samples. A priori selection of a diagnostic component is not required, since discrimination is based on a set of statistically selected, but unidentified, ions formed after pyrolysis and ionization. Nevertheless,

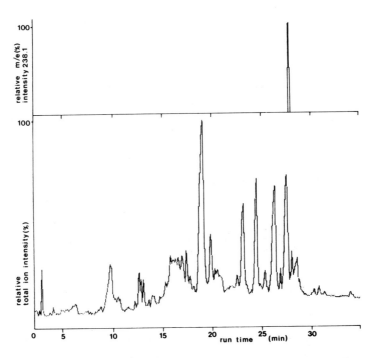

Fig. 3. Total ion gas chromatogram of an extract of potato tissue experimentally infected with Phoma exigua var. foveata isolate 44 and mass chromatogram of m/e 238.1

biochemical interpretation in general terms is possible to some extent by judging sets of covariant mass peak intensities. Thus differences in proteins, phenolics or chitin can be indicated. Modern routine Py–MS instruments can be run unattended with an analysis capacity of 30 samples/h. Commercial instruments have become available recently (Meuzelaar et al. 1982).

To date, GC–MS and LC–MS are promising methods for routine diagnostic application because of general availability, specificity and suitability for automatic analysis and report generation (Waller & Dermer 1980; Gilbert & Self 1981). Recently, a simplified mass spectrometric detector for use with existing gas chromatographs has been introduced for routine diagnostic purposes (Weijmam et al. 1984a). Without special tuning, the mass spectrometer is a factor of 1000 more sensitive as compared with TLC.

Future prospects

Approaches based on the detection of specific antigens and metabolites (e.g., mycotoxins) are highly specific and indicate the presence of fungi at the species level. Other methods are less specific and indicate the presence of microorganisms in general. To restrict these methods to fungi, the detection of at least two components is required, e.g., muramic acid and glucosamine to separate bacterial and fungal elements. Another example is the simultaneous detection of both arabinitol and mannose to restrict for yeast-like fungi. Detection of both ergosterol and glucosamine also limits the effects of variation in single components.

"Hyphenated" techniques (methods using multi-instrumented arrays) (Wilkins 1983) based on capillary GLC and HPLC are most promising for the rapid analysis of the complex matrices frequently encountered in the field of food and beverage technology (D'Amato 1981).

References

BARTNICKI-GARCIA, S. 1968 Cell wall chemistry, morphogenesis and taxonomy of fungi. Annual Review of Microbiology 22, 87–108.
BEUCHAT, L. R. (ed.) 1978 Food and Beverage Mycology. Westport, CT: AVI Publ., Inc.
COONROD, J. D., KUNZ, L. J. & FERRARO, M. J. (eds.) 1983 Direct Detection of Microorganisms in Clinical Samples. Orlando: Academic Press.
D'AMATO, R. F., HOLMES, B. & BOTTONE, E. J. 1981 The systems approach to microbiology. CRC Critical Reviews in Microbiology 9, 1–44.
DE REPENTIGNY, L., KUYKENDALL, R. J. & REISS, E. 1983 Simultaneous determination of arabinitol and mannose by gas liquid chromatography in experimental candidiasis. Journal of Clinical Microbiology 17, 1166–1169.
ESHUIS, W., KISTEMAKER, P. G. & MEUZELAAR, H. L. C. 1977 Some numerical aspects of reproducibility and specificity. In Analytical Pyrolysis. eds. JONES, C. E. R & CRAMERS, C. A. pp. 151–166. Amsterdam: Elsevier Scientific Publ. Co.
GILBERT, J. & SELF, R. 1981 Advances in the analysis of trace organic constituents in the diet with particular reference to mass spectrometry. Chemical Society Reviews 10, 255–269.
GOULD, W. A. 1983 Tomato Production Processing and Quality Evaluation. Westport, CT: AVI Publ., Inc.
HICKS, R. E. & NEWELL, S. Y. 1983 An improved gas chromatographic method for measuring glucosamine and muramic acid concentrations. Analytical Biochemistry 128, 438–445.

LOGAN, C. 1981 Gangrene. In Compendium of Potato Diseases. ed. Hooker, W. J. pp. 57–58. American Phytopathology Society.

MEUZELAAR, H. L. C., HAVERKAMP, J. & HILEMAN, F. D. 1982 Pyrolysis Mass Spectrometry of Recent and Fossil Biomaterials. Compendium and Atlas. Amsterdam: Elsevier Scientific Publ. Co.

MOSCH, W. H. M. & MOOI, J. C. 1975 A chemical method to identify tuber rot in potato caused by Phoma exigua var. foveata. Netherlands Journal of Plant Pathology 81, 86–88.

OFFEM, J. O. & DART, R. K. 1983 Rapid determination of spoilage fungi. Journal of Chromatography 260, 109–113.

RIDE, J. P. & DRYSDALE, R. B. 1972 A rapid method for the chemical estimation of filamentous fungi in plant tissue. Physiological Plant Pathology 26, 7–15.

ROSSWALL, T. (ed.) 1972 Modern Methods in the Study of Microbial Ecology. Bulletin 17 of the Ecological Research Committee of Naturvetenskapliger Forkningsrod (Swedish National Science Research Council), Stockholm.

RUDALL, K. M. & KENCHINGTON, W. 1973 The chitin system. Biological Reviews 48, 597–636.

SAMSON, R. A., HOEKSTRA, E. S. & VAN OORSCHOT, C. A. N. 1984 Introduction to Foodborne Fungi. Baarn, the Netherlands: Centraalbureau voor Schimmelcultures.

SEITZ, L. M., SAUER, D. B., BURROUGHS, R., MOHR, H. E. & HUBBARD, J. P. 1979 Ergosterol as a measure of fungal growth. Phytopathology 69, 1202–1203.

VAN EIJK, G. W. & ROEIJMANS, H. J. 1982 Distribution of carotenoids and sterols in relation to the taxonomy of Taphrina and Protomyces. Antonie van Leeuwenhoek 48, 257–264.

WALLER, G. R. & DERMER, O. C. (eds.) 1980 Biochemical Applications of Mass Spectrometry. First supplementary volume. New York: John Wiley & Sons.

WEETE, J. D. 1980 Lipid Biochemistry of Fungi and Other Organisms. New York: Plenum Press.

WEIJMAN, A. C. M., ROEIJMANS, H. J., VAN EIJK, G. W. & SAKKERS, P. J. D. 1984a Diagnosis of the potato storage disease gangrene by GC-MS. Palo Alto: Hewlett-Packard Application Note (in press).

WEIJMAN, A. C. M., VAN EIJK, G. W., ROEIJMANS, H. J., WINDIG, W., HAVERKAMP, J. & TURKENSTEEN, L. J. 1984b Mass spectrometric techniques as aids in the diagnosis of gangrene in potatoes caused by Phoma exigua var. foveata. Netherlands Journal of Plant Pathology 90, 107–115.

WILKINS, C. L. 1983 Hyphenated techniques for analysis of complex organic mixtures. Science 222, 291–296.

WU, L-C. & STAHMANN, M. A. 1975 Chromatographic estimation of fungal mass in plant material. Phytopathology 65, 1032–1034.

ZLATKIS, A., POOLE, C. F., BRAZELL, R., LEE, K. Y., HSU, F. & SINGHAWANGCHA, S. 1981 Profiles of organic volatiles in biological fluids as an aid to the diagnosis of disease. Analyst 106, 352–360.

A. C. M. WEIJMAN
G. W. VAN EIJK
W. WINDIG
R. A. SAMSON

Discussion

Splittstoesser
What happens when one grows a fungus in a broth with respect to the dye reduction test, that is the old resazurin test? Are there any applications here?

Mossel

I have never tried it for specific mycological purposes, although I can't see why it wouldn't work. The bacteriologists have really never been completely satisfied with dye reduction tests. Don't ask me why because we tried many years ago to rapidly separate good from poor precooked meals distributed throughout the Netherlands. These tests worked remarkably well. We used the old methylene blue test. Does anyone know if the American Public Health Association <u>Compendium</u> recommends any use of dye reduction tests?

Splittstoesser

I know that they do not use it for foods, but it may still have some applications in the dairy industry. It is used mostly for raw milk and not for processed products.

Recommendations

There are a number of alternatives to colony counting for assessing fungal contamination of foods. Most have yet to be widely accepted as practical alternatives to viable count methods, but they have advantages in speed, ability to detect non-viable fungal constituents and possibly also in selectively detecting groups of fungi especially important in the spoilage of the food in question. Further investigations are needed before these alternative methods can be recommended in place of traditional procedures.

CHAPTER 8

TAXONOMIC SCHEMES FOR FOODBORNE FUNGI

The papers assembled in this chapter present an update on the taxonomy of foodborne fungi and give some aids to the classification of these increasingly important microorganisms. Many participants who replied to questionnaires distributed before this workshop responded that they classified fungi only to the genus level. This is no doubt due in part to the difficulty of fungal taxonomy. A large part of traditional mold taxonomy is based upon morphology whereas yeast taxonomy is based more on physiological characteristics. Knowledge of the language used to describe morphological structures is required. Species identification is often necessary, e.g., when an organism is used for a desirable fermentation process. The safety of a product depends upon the mold species present. Certain species produce mycotoxins while closely related species do not. For instance, some isolates of <u>Aspergillus</u> <u>flavus</u> produce aflatoxin while the closely related <u>Aspergillus</u> <u>oryzae</u> does not and has been used for centuries in the Orient for food fermentations. Other reasons for classification of fungi are to predict the spoilage potential of a food, the possible heat resistance of a fungus, the kinds of mycotoxins which might be produced and the level of sanitation during processing. A mixed flora means something different than a single species and can point to a source of contamination.

▶ Compilation of Foodborne Filamentous Fungi

As a result of recent studies on foodborne fungi, it is now possible to list the commonly found taxa. Samson et al. (1984) compiled about 100 of the most common filamentous fungi and yeasts. In this account, we would like to summarize these taxonomic data and also include some of the less common contaminants of food. Some reviews (Beneke & Stevenson 1978), list several fungi, such as several coelomycetous spp., which are weak pathogens of plants and fruit. Most of these taxa are not included in this list.

Isolation and cultivation of molds for identification

Even though about 6000 genera of molds are known (Hawksworth et al. 1983), only a few of these are found as saprophytes in foods. The number of genera isolated from foods also depends on the method of isolation. Visibly moldy foods can be examined directly under a stereoscopic microscope and representative molds can be isolated. In many cases a direct microscopic identification is possible using morphological criteria, if the molds produce sufficient sporulating structures.

If food samples visually appear mold-free, small pieces of food material can be placed on wet filter paper (blotter test), water agar (containing selective agents if desirable) or a selective (maybe diagnostic) medium, or the dilution plating technique can be used. The greatest number of genera and species will probably be obtained using the direct plating technique (Mislivec & Bruce 1977). Some species may grow only on the food product, while other species will grow on the food and the blotter or selective medium. It is very important to observe colonies developing on both the selective medium and the food sample pieces.

The direct identification to genus and in some cases to species level on selective media depends on sporulation taking place and in some cases this may require incubation under 12-h alternating cycles of near ultraviolet light and darkness. For this purpose the use of a growth inhibitor such as rose bengal in the selective medium is not appropriate. The use of macroscopic criteria such as colony color and texture may help in the identification of species and genera.

The incubation temperature is also important. Most species will grow at 25°C (Moreau 1979), but species such as Humicola (Thermomyces) lanuginosa will only grow at incubation temperatures over 30°C while Sporendonema casei will only grow below 18°C (Beuchat 1978). Mycological analyses should therefore be performed according to the probable conditions the original food item would meet and the information wanted (Jarvis et al. 1983).

For correct identification using morphological criteria, appropriate agar media are recommended for optimal cultivation. Table 1 lists the recommended media with reference to the genera. For formulae, consult Stevens (1974) or other books.

Identification of foodborne molds

The importance of identifying foodborne molds to species level cannot be overemphasized. Through species identification much information on resistance of molds towards environmental factors, contamination sources and mycotoxin production becomes available. In most cases the direct plating technique will give a better picture of the qualitative mold flora in foods than the dilution technique (Mislivec & Bruce 1977). Combinations of these techniques with identification of all fungi found may yield useful information on the foods under investigation, help in establishing inoculum potential of the individual species and facilitate comparison of survey results. In some instances, combinations of different sampling procedures with several isolation procedures such as blotter tests, (alternating light), direct plating and dilution plating with and/or without surface disinfection of the food samples as well as different selective and non-selective blotters and media (at different temperatures) would give an optimal estimation of species recovery and relative density. In other cases, a single procedure may suffice. Standardization of these methods for different types of foods under different environmental conditions is overdue. One type of food kept under a given environmental condition (or a succession of environmental conditions) will, in our experience, contain a normal flora or a succession of normal floras which is characteristic. The inoculum and growth potential of the normal flora may cause spoilage or mycotoxin accumulation under prolonged storage. In other cases, deviations from the normal flora indicate a danger of spoilage or mycotoxin production, and proper mycological analyses must be developed according to these conditions. An example may illustrate this: the flora of the barley kept in airtight silos will contain Penicillium roquefortii and Candida spp. The presence of Cladosporium spp. in barley samples kept in airtight siloes for one month or more may indicate that external air has entered the silo.

Table 1. Media[a] recommended for identification schemes

	MEA	Cz(Y)	PDA	OA	CLA	MEA + sucrose
Zygomycotina	+					
Ascomycotina	+		+	+		
Eurotium						+
Deuteromycotina						
Hyphomycetes						
Penicillium and						
Aspergillus	+	+				±
Fusarium			+	+	+	
Other genera	+		+	+		
Coelomycetes	+			+		

[a]Key: MEA = malt extract agar (2%) (for Zygomycotina 4%); Cz = Czapek, CzY = Czapek yeast autolysate; PDA = potato dextrose agar; OA = oatmeal agar; CLA = carnation leaf agar. Most fungi listed in Table 1 will grow excellently on MEA, either containing 2% malt extract (Samson et al. 1984) or 2% malt extract, 2% glucose and 0.1% peptone (Raper & Thom 1949; Raper & Fennell 1965; Pitt 1979). For the final identification to species level, special substrates may be necessary. The Penicillia and Aspergilli are identified on Czapek and malt extract based media, and the Fusaria should be identified using a potato-based medium and a sparse medium like carnation leaf agar (Nelson et al. 1983) or the medium recommended by Nirenberg (1981). Xerophilic fungi may require additional carbohydrate in the medium (20-40% sucrose or glycerol). For other special substrates, see the books by Gams et al. (1980), Pitt & Hocking (1985) and Tuite (1969). Sporulation of many fungi can be stimulated by alternating near UV light (black light) and darkness (e.g., Fusarium, Trichoderma, most dematiaceous fungi and members of the Coelomycetes). Most fungi can be incubated at 25°C and identified after 6-11 days.

Once a significant amount of oxygen has entered the silo, species subgroups such as P. viridicatum II may grow and produce mycotoxins (ochratoxin A).

General literature for identification

In modern taxonomic treatments, several new terms are used. An important term is holomorph which refers to the whole fungus, including the asexual (imperfect = anamorph) state as well as the sexual (perfect = teleomorph) state. For recent glossaries of mycological terms, the reader is referred to Samson et al. (1984) or to the Dictionary of Fungi by Hawksworth et al. (1983).

The classifications of foodborne yeasts will be treated elsewhere in this book. Modern classifications of yeasts are published by Barnett et al. (1983) and Kreger-van Rij (1984).

Identification of genera. For the identification of genera of fungi we recommend two books especially designed for foodborne fungi (Samson et al. 1984; Pitt & Hocking 1985), which describe the most important genera and species found in foods. The books of von Arx (1981) and Barnett & Hunter (1972) should be available for the identification of less frequently encountered genera. Other valuable and informative books concerning fungal genera are The Fungi, volumes IV A and IV B (Ainsworth et al. 1973a,b), Smith's Introduction to Industrial Mycology (Onions et al. 1981), Compendium of Soil Fungi (Domsch et al. 1980) and Genera of Hyphomycetes (Carmichael et al. 1980). Keys or lists of genera containing mycotoxigenic species are given by Wyllie & Morehouse (1977), Moreau (1979) and Samuels (1984).

Identification of species. Many isolates of foodborne molds may be identified according to the books of Samson et al. (1984) and Pitt & Hocking (1985), but manuals, monographs and papers on important genera should be available. Furthermore, there is some disagreement on the taxonomic treatment of the important genus Penicillium. It is strongly recommended that typical living cultures of the important species be obtained for comparison.

As an initial aid we have included here simple diagrammatic keys for foodborne genera of Zygomycotina (mycelium without or with few septa, sexual reproduction resulting in zygospores, asexual spores borne endogenously in sporangia), Ascomycotina (sexual reproduction by ascospores developed endogenously in asci) and Deuteromycotina (sexual reproduction absent, asexual conidia often produced) containing the Coelomycetes (conidia borne in a cavity lined by fungal or fungal/host tissue) and the Deuteromycetes or Hyphomycetes (conidia formed on separate hypha or aggregations, not within discrete conidiomata). In the following list, the common and less common genera are compiled. A short generic description of common genera containing more than one species is included.

Recommended taxonomic schemes for identification of foodborne genera are given in Table 2. The publications listed in this table contain detailed descriptions of important foodborne molds. Many of these works are not available in food microbiology laboratories; therefore, a more restricted list of books containing descrip- tions, keys and additional physiological data is recommended (Raper & Fennell 1965; Ellis 1971, 1976; Pitt 1979; Domsch et al. 1980; Samson et al. 1984).

Table 2. Recommended taxonomic schemes for foodborne genera

Genus	Recommended taxonomic schemes	Additional schemes recommended
Absidia	Zycha et al. (1969)	
Acremonium	Gams (1971)	
Alternaria	Neergaard (1945)	Ellis (1971, 1976)
	Simmons (1967)	
Amylomyces	Ellis et al. (1976)	
Arthrinium	Ellis (1971, 1976)	
Ascochyta	Punithalagam (1979)	
Aspergillus	Raper & Fennell (1965) + Samson (1979)	
Aureobasidium	de Hoog & Hermanides-Nijhof (1977)	
Bipolaris	Ellis (1971, 1976)	Alcorn (1981)
Botrytis	Jarvis (1977)	

(continued)

Table 2.   Continued

Genus	Recommended taxonomic schemes	Additional schemes recommended
Byssochlamys	Stolk & Samson (1971)	Brown & Smith (1957)
Chaetomium	Domsch et al. (1980)	
Chrysonilia	Samson et al. (1984)	
Chrysosporium	van Oorschot (1980)	
Cladosporium	de Vries (1952)	Domsch et al. (1980)
Colletotrichum	Sutton (1980), von Arx (1957)	Kulshrestha et al. (1976)
Cunninghamella	Samson (1969)	
Curvularia	Ellis (1971, 1976)	Benoit & Mathur (1970)
Drechslera	Ellis (1971, 1976)	Chidambaram et al. (1973)
Emericella	Raper & Fennell (1965)	Christensen (1982)
Epicoccum	Schol-Schwartz (1959)	Ellis (1971)
Eupenicillium	Stolk & Samson (1983) or Pitt (1979)	
Eurotium	Blaser (1975)	Raper & Fennell (1965)
Fusarium	Nelson et al. (1983)	Booth (1971), Nath et al. (1970)
Geotrichum	Domsch et al. (1980)	von Arx (1977)
Gliocladium	Morquer et al. (1963)	Domsch et al. (1980)
Humicola	Domsch et al. (1980)	
Lasiodiplodia	Sutton (1980)	
Memnoniella	Jong & Davis (1976)	
Monascus	Hawksworth & Pitt (1983)	
Moniliella	de Hoog (1979)	
Mucor	Schipper (1973, 1975, 1976, 1978)	Zycha et al. (1969)
Myrothecium	Tulloch (1972)	Nguyen et al. (1973)
Neosartorya	Raper & Fennell (1965)	
Neurospora	Frederick et al. (1969)	
Paecilomyces	Samson (1974)	

Penicillium	Two alternatives proposed:
	Samson: Samson et al. (1976)
	          Samson et al. (1977a,b)
	          Stolk & Samson (1983)
	Frisvad: Pitt (1979)
	          Frisvad & Filtenborg (1983)
		Domsch et al. (1980)
		Samson et al. (1984)

Phialophora	Schol-Schwarz (1970)	Domsch et al. (1980)
Phoma	Sutton (1980)	
Pithomyces	Ellis (1971, 1976)	Ellis (1960)
Rhizopus	Schipper (1984)	Zycha et al. (1969)
Scopulariopsis	Morton & Smith (1963)	
Sporendonema	Sigler & Carmichael (1976)	
Stachybotrys	Jong & Davis (1976)	
Stemphylium	Neergaard (1945) Simmons (1967)	Ellis (1971, 1976)
Syncephalastrum	Zycha et al. (1969)	Domsch et al. (1980)
Talaromyces	Stolk & Samson (1971) or Pitt (1979)	
Thamnidium	Zycha et al. (1969)	Domsch et al. (1980)
Trichoderma	Rifai (1969)	Domsch et al. (1980)
Trichothecium	Domsch et al. (1980)	Rifai & Cook (1966)
Ulocladium	Simmons (1967)	Ellis (1971, 1976)
Verticillium	Isaac (1967)	Domsch et al. (1980)
Wallemia	Ellis (1971)	
Xeromyces	Hawksworth & Pitt (1983)	Pitt & Hocking (1982)

Table 3. Diagrammatic key for foodborne genera of Zygomycotina

Characteristic	Genus						
	Mucor	Rhizopus	Absidia	Thamnidium	Syncephalastrum	Cunninghamella	Amylomyces
Merosporangia					+		
Sporangioles				+		+	
Sporangia globose	+	+		+			+
Sporangia pyriform			+				
Rhizoids		+	+				
Sporangiophores especially at rhizoids		+					
Apophysis		+	+				

ZYGOMYCOTINA

For general keys to the Zygomycotina, in particular the Mucorales, see Zycha et al. (1969), Hanlin (1973), Hesseltine & Ellis (1973) and O'Donnell (1979). A diagrammatic key for foodborne genera of Zygomycotina is shown in Table 3, and selected genera are illustrated in Fig. 1.

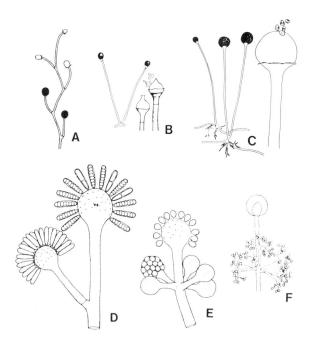

Fig. 1. Foodborne Zygomycota: A. Mucor, B. Rhizopus, C. Absidia, D. D. Cunninghamella, E. Amylomyces, F. Thamnidium

Mucor Mich. Colonies growing rapidly. Hyphae hyaline or colored,
varying from a few millimeters to centimeters in height. Sporangiophores
often branched, always ending in a many-spored sporangium without an
apophysis. Sporangia varying in size; columellae well-developed. Spores
variable in shape, smooth-walled or slightly ornamented. Zygospores without
appendages on the suspensors. Chlamydospores present in some species.

Common species on food are M. plumbeus, M. hiemalis, M. racemosus and M.
circinelloides. For detailed keys, see Schipper (1973, 1976, 1978). The
genus Rhizomucor, including R. pusillus and R. miehii, contains thermophilic
species related to Mucor (Schipper 1978).

Rhizopus Ehrenb. Colonies growing rapidly, with stolons, pigmented
rhizoids and sporangiophores. Sporangia mostly big, whitish when young,
becoming blackish-brown with age. Columellae brown, globose or subglobose,
with an apophysis. Spores short-ellipsoidal, usually irregularly angled,
often striate. Chlamydospores present in some species. Zygospores
Mucor-like. Most species heterothallic.

Common species in food are R. oryzae, R. stolonifer and R. microsporus.
The latter taxon has been separated into four varieties (Schipper & Stalpers
1984) including var. oligosporus and var. rhizopodiformis. For keys, see
Schipper (1984) and Schipper & Stalpers (1984).

Absidia van Tieghem. The most common species is A. corymbifera
(Hesseltine & Ellis 1973).

Syncephalastrum Schroeter. The only common species is S. racemosum Cohn
(Benjamin 1959).

Cunninghamella Thaxter. A foodborne genus, but less common than Mucor
or Rhizopus: C. echinulata, C. elegans (Samson 1969; Lunn & Shipton 1983).

Thamnidium Link. A less common genus in foods: T. elegans (Hesseltine
& Anderson 1956).

Amylomyces Calmette. Also less common genus in foods: A. rouxii (Ellis
et al. 1976).

ASCOMYCOTINA

Most foodborne Ascomycotina belong to the Order Eurotiales (=
Plectomycetes). For a general key to the genera see Benny & Kimbrough (1980)
and von Arx (1981). A diagrammatic key for foodborne genera of Ascomycotina
is shown in Table 4, and selected genera are illustrated in Figs. 2 and 3.

Eurotium Link. Ascomata white to yellow, globose to subglobose,
non-ostiolate, scattered or more or less clustered, commonly accompanied by
the Aspergillus anamorph. Asci globose to subglobose, thin-walled, often
dissolving at an early stage. Ascospores one-celled, small, surface smooth
or roughened, frequently ornamented with equatorial furrows or ridges.
Conidial heads uniseriate, in gray-green shades; conidiophores smooth-walled,
conidia distinctly roughened. Xerophilic, optimally growing on low a_w
media such as MEA + 20% or 40% sucrose.

Anamorph: Aspergillus glaucus group. About 20 species are known. For
a general key see Blaser (1975) and also consult Samson (1979). Common in
food are E. chevalieri, E. amstelodami and E. herbariorum (= E. repens).
Note that the teleomorph is optimally produced on malt extract or Czapek
agar with additional sucrose.

Table 4. Diagrammatic key for foodborne genera of Ascomycotina

Characteristic	Byssochlamys	Talaromyces	Eupenicillium	Eurotium	Neosartorya	Emericella	Monascus	Xeromyces	Chaetomium	Neurospora
Aspergillus anamorph				+	+	+				
Penicillium anamorph		+	+							
Paecilomyces anamorph	+									
Basipetospora anamorph							+			
Chrysonilia anamorph										+
Ascomata on stalk							+	+		
Ascomata without distinct wall	+	+								
Ascomata with hairs and rhizoids									+	
Asci cylindrical or clavate									+	+
Yellowish hülle cells						+				
Ascomata white to cream	+	+	+		+		(+)	+		
Ascomata brown						+	+	+	+	+
Ascomata yellow		+	+	+						
Ascomata dark									+	+
Asci containing two spores								+		

Genus

<u>Eupenicillium</u> Ludwig. Ascomata white or in pale colors, globose to subglobose, pseudoparenchymatous or sclerenchymatous, at first entirely composed of hyaline, polygonal cells with very thick walls; producing ascogonia in the central part, maturing from the center outwards. Peridium persistent, consisting of a few layers of hyaline to slightly colored, polygonal cells. Asci formed either singly, or in chains, or both singly and in chains, globose to ellipsoidal, usually 8-spored, evanescent. Ascospores lenticular or rarely globose, usually with equatorial ridges, yellow, occasionally becoming brown.

Anamorph: <u>Penicillium</u> Link. Several taxa are known (see Pitt 1979; Stolk & Samson 1983). <u>E</u>. <u>ochrosalmoneum</u>, <u>E</u>. <u>euglaucum</u> (= <u>E</u>. <u>hirayamae</u>) and related sclerotial species are found in food.

<u>Talaromyces</u> C. R. Benjamin. Ascomata in white to yellowish colors, (sub) globose, soft, superficial, discrete or confluent, of indeterminate growth. Ascomatal coverings varying from scanty to dense, consisting of a network of hyphae, which may range from very loose-textured to closely knit, usually surrounded by a weft of thin, usually encrusted radiating hyphae, straight or twisted depending on the species. Ascomatal initials of various shape. Asci evanescent, mostly 8-spored, globose to slightly ellipsoidal, borne in chains. Ascospores globose or ellipsoidal, smooth or showing various ornamentations, yellow, rarely becoming reddish.

Anamorph: <u>Penicillium</u> Link; <u>Paecilomyces</u> Bain. About 20 species are known and some taxa are infrequently found in insufficiently heat-treated

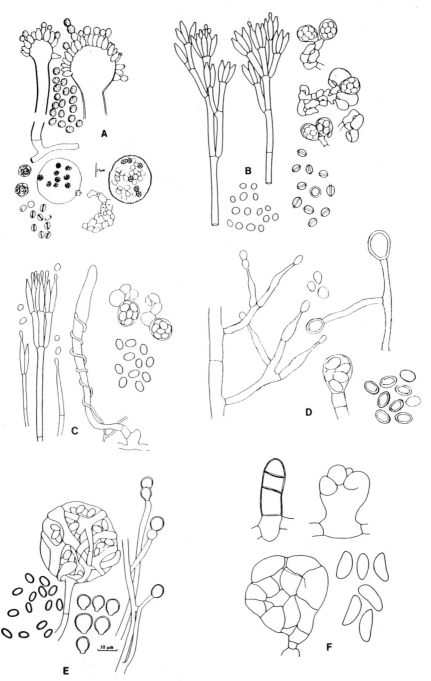

Fig. 2.   Foodborne Ascomycotina: A. _Eurotium_, B. _Eupenicillium_, C.
Talaromyces, D. _Byssochlamys_, E. _Monascus_, F. _Xeromyces_

products (e.g., pasteurized fruit juices). Keys are given by Stolk & Samson (1972) and Pitt (1979).

Byssochlamys Westling. Ascomata discrete and confuent. Wall lacking or very scanty, composed of loose wefts of hyaline, thin, twisted hyphae. Ascomatal initials consisting of ascogonia coiled around swollen antheridia. Asci globose to subglobose, 8-spored, stalked. Ascospores ellipsoidal, smooth, pale yellowish.

Anamorph: Paecilomyces Bain. Four species are known of which B. nivea and B. fulva are commonly encountered. A key is given by Stolk & Samson (1971).

Monascus van Tieghem. Three species are recognized, which Hawksworth & Pitt (1983) have keyed using cultural and microscopic characters.

Xeromyces Fraser. X. bisporus (= Monascus bisporus [Fraser] v. Arx) is obligately xerophilic and is often overlooked because of its failure to grow on commonly used media (Pitt & Hocking 1982).

Chaetomium Kunze. Most species of Chaetomium are found in soil and on plant debris. C. globosum and several other members may cause biodeterioration of food (corn, rice) and feedstuffs (Udagawa et al. 1979, Sekita et al. 1979 and Udagawa 1984). Keys to Chaetomium are found in von Arx et al. (1984) and in Ames (1963).

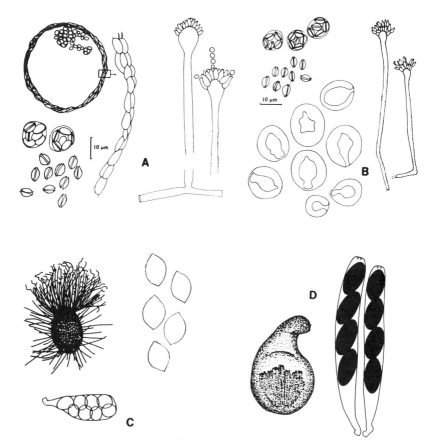

Fig. 3. Foodborne Ascomycotina: A. Neosartorya, B. Emericella, C. Chaetomium, D. Neuorspora

Emericella Berk. & Br. About 20 species are known, of which several are regularly found on food. Synoptic keys are given by Horie (1980) and Christensen & States (1982). The latter authors treated this genus as an anamorphic group, the A. nidulans group, in the sense of Raper & Fennell (1965).

Neosartorya Malloch & Cain. Species of Neosartorya are soilborne and are infrequently encountered in heat-treated products (e.g., pasteurized fruit juices). A key is given by Raper & Fennell (1965) (as the Aspergillus fischeri-group); but also consult Malloch & Cain (1972) for the correct nomenclatural treatment of Neosartorya.

Neurospora Shear & Dodge. About 10 species are known. Neurospora sitophilia, N. crassa and N. intermedia are infrequently found in food (Frederick et al. 1969). Often the Chrysonilia anamorph is abundantly produced.

DEUTEROMYCOTINA

The Deuteromycotina (Fungi Imperfecti) include the most important genera found in foods. Gilman (1957) and Barron (1968) have written books on the subject which have been used in many laboratories. However, these books are outdated and many excellent new books are available on the Deuteromycotina (Kendrick & Carmichael 1971; Subramanian 1971; Ellis 1971, 1976; Domsch et al. 1980; Carmichael et al. 1980). The books by Barnett & Hunter (1972) and von Arx (1981a) provide simple keys and illustrations for many taxa growing in pure culture. The latter two references and the books of Samson et al. (1984) or Pitt & Hocking (1985) will be sufficient in most cases.

The taxonomy of genera in the Deuteromycotina is based on characters of conidiogenesis. For a review of the various patterns of development in conidial fungi, see Cole & Samson (1979). Diagrammatic keys for genera of Deuteromycotina with one-celled and many-celled conidia are shown in Tables 5 and 6, respectively. Selected genera are illustrated in Figs. 4a-4e.

A number of genera of the Hyphomycetes, a major Order of the Deuteromycotina, are of significance to food mycologists:

Acremonium Link. Colonies growing slowly, and are mostly less than 2.5 cm in diameter in 10 days. Hyphae are hyaline. Phialides are mostly awl-shaped, simple, erect, arising from substrate mycelium or from bundled aerial hyphae. Conidia usually one-celled, hyaline or pigmented, mostly in slimy heads.

Teleomorphs: Emericellopsis v. Beyma and others. Most species are saprophytic and isolated from dead plant material and soil, but some species are pathogenic to plants and humans. The genus is world-wide in distribution. For detailed descriptions and keys, see Gams (1971) and Domsch et al. (1980).

Aspergillus Mich. Colonies usually growing rapidly, and are white, yellow, yellow-brown, brown to black or shades of green, mostly consisting of a dense felt of erect conidiophores. Conidiophores consisting of an unbranched stipe, with a swollen apex (= vesicle). Phialides borne directly on the vesicle (uniseriate) or on metulae (biseriate). Conidia in dry chains, forming compact columns (columnar) or diverging (radiate), one-celled, smooth or ornamented, hyaline or pigmented. Hülle cells or sclerotia sometimes present.

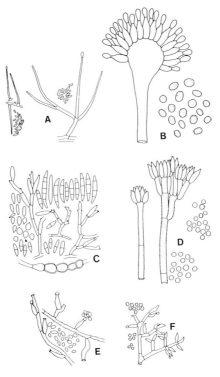

Fig. 4a. Foodborne Hyphomycetes: A. _Acremonium_, B. _Aspergillus_, C. _Fusarium_, D. _Penicillium_, E. _Phialophora_, F. _Trichoderma_

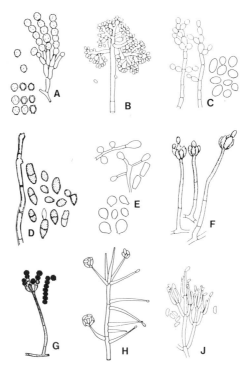

Fig. 4b. Foodborne Hyphomycetes: A. _Scopulariopsis_, B. _Botrytis_, C. _Monilliella_, D. _Cladosporium_, E. _Chrysosporium_, F. _Stachybotrus_, G. _Memnoniella_, H. _Verticillium_, J. _Gliocdium_

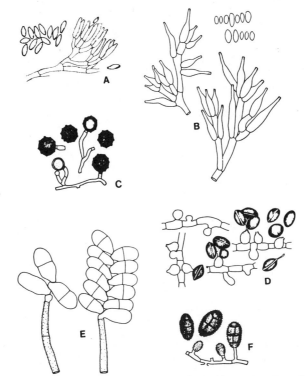

Fig. 4c. Foodborne Hyphomycetes: A. <u>Myrothecium</u>, B. <u>Paecilomyces</u>, C. <u>Humicola</u>, D. <u>Arthrinium</u>, E. <u>Trichothecium</u>, F. <u>Pithomyces</u>

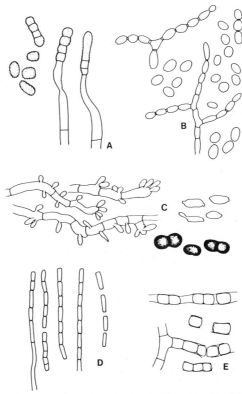

Fig. 4d. Foodborne Hyphomycetes: A. <u>Wallemia</u>, B. <u>Chrysonilia</u>, C. <u>Aureobasidium</u>, D. <u>Geotrichum</u>, E. <u>Sporendonema</u>

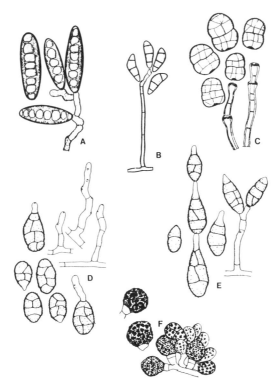

Fig. 4e. Foodborne Hyphomycetes: A. Drechslera, B. Curvularia, C. Stemphylium, D. Ulocladium, E. Alternaria, F. Epicoccum

Teleomorphs: Eurotium, Emericella, Neosartorya and others. For a general key, see Raper & Fennell (1965). For a compilation of Aspergilli described since 1965, see Samson (1979). Several groups have been reinvestigated since then: Aspergillus niger group (Al-Musallam 1980), A. ochraceus group (Christensen 1982); A. nidulans group (Christensen & States 1982), A. flavus group (Christensen 1981). See also Samson & Pitt (1985) for a more recent account on advances in Aspergillus systematics. A key to common foodborne Aspergilli is provided by Samson et al. (1984).

Fusarium Link. Colonies usually grow rapidly and are pale (whitish to cream) or bright-colored in yellow, brownish, pink, reddish, violet or lilac shades. Conidiophores are usually branched, in some species reduced to single phialides, sometimes forming complex pustules (sporodochia) or forming a confluent slimy mass of spores with a fatty or greasy appearance (pionnotes). Phialides often slender and tapering, usually bearing one to many (= polyphialide) fertile opening. Conidia can form false slimy heads, chains or dry masses. Two types of conidia (macro and micro) can be distinguished. In some species both forms can be observed; in other species only the macroconidia. Chlamydospores present or absent, intercalary, solitary, in chains or clusters, formed in hyphae or in conidia. Sclerotia present or absent.

Teleomorphs: Nectria Fr., Plectosphaerella Kleb. and others. Most Fusarium spp. are soil fungi and have a world-wide distribution. Some are plant parasites causing root and stem rot, vascular wilt and fruit rot. Several species are known to be pathogenic to human and animals. Others cause storage rot and are toxin producers. For modern detailed descriptions

Table 5.   Diagrammatic key for genera of Deuteromycotina with one-celled
conidia

Origin	Conidia Type	Light/ Dark[a]	Genus	Description
Phialide	Conidia in heads	D	Stachybotrys	Phialides with a rounded tip
		L/D	Acremonium	Single awl-shaped phialides directly on mycelium
		L/D	Phialophora	Irregularity branched conidiophore, phialides dark with collarette
		L	Verticillium	Phialides in regular verticils
		L	Gliocladium	Phialides as terminal verticils on a branched conidiophore
		L	Trichoderma	Phialides with a short neck, often green conidia, irregularly branched
		L	Myrothecium	Conidiophores in sporodochia, conidia fusiform to ellipsoidal
	Conidia in chains		Aspergillus	Vesiculate stipes[b]
			Penicillium	Green conidia, convergent flask-shaped or acerose phialides
			Paecilomyces	Divergent long-necked phialides
			Memnoniella	Phialides without neck
	Arthro		Wallemia	Conidia formed in quartets
Annelliform	Baso		Scopulariopsis	Conidia with a truncate broad base
Ampulliform	Ceteri		Botrytis	Conidia borne from denticles on terminally swollen conidiogenous cells
Nonsp	Blasto	L	Chrysonilia	Colonies growing very fast, orange
		D/L	Moniella	Both blastic and arthro-conidia
		D	Aureobasidium	Conidia (in slime) formed directly on mycelium
		D	Cladosporium	Conidia and colonies dark brown or green
	Arthro	D/L	Moniliella	Both blastic and arthro conidia
		L	Geotrichum	
		D	Sporendonema	
	Ceteri	L	Chrysosporium	
		D	Humicola	Conidia borne singly, sometimes also phialoconidia present
		D	Arthrinium	Conidia borne singly, light band in conidia

261

[a]D = dark conidia, L = light conidia
[b]Green conidia are produced by members of the Aspergillus glaucus, A. nidulans, A. fumigatus, A. clavatus, A. flavus, A. versicolor and A. restrictus group (Raper & Fennell 1965)

Table 6. Diagrammatic key for genera of Deuteromycotina with many-celled conidia

Conidial origin and type			Genus	Description
Didymo	Misc	Baso	Trichothecium	Often pink reverse
	Nonsp[a]	Blasto	(Cladosporium)	
Phragmo	Phial	Gloio	Fusarium	Often reddish mycelium
	Nonsp	Ceteri	Pithomyces	Solitary conidia
	Radul	Ceteri	Drechslera	Conidia cylindrical, germinating from any cell
			Bipolaris	Conidia fusiform-ellipsoidal, germination bipolar
Dictyo	"Annel"	Ceteri	Stemphylium	Conidiophores percurrent
	Rachi	Ceteri	Ulocladium	Conidium obovoid and non-rostrate
	Nonsp	Blasto	Alternaria	Conidia ovoid with rostrum, often in chains
		Ceteri	Epicoccum	Conidia in sporodochia, visible as black dots
			Pithomyces	Conidia solitary

[a]Species in Cladosporium may produce septate basal conidia (2- or 3-celled)

and keys, the reader is referred to Booth (1971), Gerlach & Nirenberg (1982), Burgess & Liddell (1983) and Nelson et al. (1983). For identification, cultivation both on PDA and especially carnation leaf agar (CLA) is strongly recommended by modern Fusarium workers.

Penicillium Link. Colonies usually growing rapidly, mostly in shades of green, sometimes white. Conidiophores single or bundled, consisting of a single stipe terminating in either a whorl of phialides (simple, monoverticillate) or in a penicillus, containing branches and metulae. Branching pattern either one-stage branched (biverticillate-symmetrical), two-stage branched or three- to more-stage branched (biverticillate-asymmetrical or terverticillate). Conidiophores hyaline, smooth- or rough-walled. Phialides usually flask-shaped consisting of a cylindrical basal part and a distinct neck, or more or less narrow with basal part tapering to a somewhat pointed apex (= lanceolate or acerose). Conidia in long dry chains, divergent or in columns, globose, ellipsoidal, cylindrical or fusiform, hyaline or greenish, smooth- or rough-walled. Sclerotia produced by some species.

Teleomorphs: Eupenicillium, Talaromyces and others. Many species of Penicillium are common contaminants on various substrates and are known as potential mycotoxin producers. Correct identification is therefore important when studying possible Penicillium contamination.

Apart from the out-dated monograph by Raper & Thom (1949), several new classifications (Pitt 1979; Ramirez 1982; Samson et al. 1976, 1977a,b; Stolk & Samson 1983) presently exist, which partly overlap, but which differ in

some aspects. Abe et al. (1982, 1983a,b,c,d) presented revisions of the
sections **Aspergilloides**, **Exilicaulis**, **Divaricatum** and **Furcatum**. Foodborne
taxa have been treated by Samson et al. (1976, 1977a,b) and keys to the most
common species are given by Samson et al. (1984), Onions et al. (1981), Pitt
(1985) and Pitt & Hocking (1985). For a recent account on advances in
Penicillium taxonomy, the reader is referred to Samson & Pitt (1985).

Phialophora Medlar. Colonies usually growing slowly, white, pinkish,
brown or olivaceous-black. Phialides commonly with a distinct collarette,
flask-shaped, cylindrical or reduced (without a basal septum), borne on
branched conidiophores, often in clusters or arising solitarily from
hyphae. Conidia one-celled, globose to ellipsoidal, sometimes slightly
curved, more or less hyaline in slimy heads or chains.

Teleomorphs: **Pyrenopeziza**, **Coniochaeta** and others. **Phialophora** species
have been isolated from decaying wood, foodstuffs (e.g., butter, margarine,
apples), soil, diseased human and animal tissue and also occur as parasites
or saprobes in plant material. For detailed descriptions and keys, see
Schol-Schwarz (1970) and Domsch et al. (1980).

Trichoderma Pers. Colonies usually growing rapidly, initially white,
later usually in green shades due to conidium production. Conidiophores in
tufts, repeatedly branched (irregularly verticillate), bearing clusters of
flask-shaped phialides, with a sterile apical appendage in some species.
Conidia in slimy heads, sometimes hyaline, often green, smooth- or rough-
walled.

Teleomorph: **Hypocrea** (in most species, not produced in culture). A
very common genus, especially in soil and decaying wood. **Gliocladium** (with
strongly convergent phialides) and **Verticillium** (with straight and
moderately divergent phialides) are closely related. For more detailed
descriptions, see Rifai (1969) and Domsch et al. (1980).

Scopulariopsis Bain. Colonies varying from white, creamish, grey or
buff to brown or even blackish, but never green. Conidiogenous cells
cylindrical or with a slightly swollen base, annellate (rings formed after
conidium formation), single or borne on a somewhat penicillate system of
branches. Conidia borne in chains in basipetal succession, smooth- or
rough-walled, one-celled, globose to ovate with a broadly truncate base.

Teleomorph: **Microascus**. For a key and descriptions, see Morton & Smith
(1963) and Domsch et al. (1980).

Botrytis Mich. Colonies broadly spreading, hyaline at first, becoming
light grey to dark brown. Sclerotia frequently formed, both on natural
substrates and in culture. Conidiophores erect, brown, mononematous,
solitary or in groups, branched with branches mostly restricted to the
apical region, with terminal swollen conidiogenous cells, producing numerous
blastoconidia simultaneously on short denticles. Conidia one-celled
(occasionally 2-3 celled), pale brown, globose, ovate or ellipsoidal,
smooth-walled or almost so. Sometimes also a **Myrioconium** phialidic state
with small (sub)globose conidia can be present in culture.

Teleomorph: **Botryotinia** (= **Sclerotinia** Fuckel pro parte). The genus
includes important plant pathogens with a world-wide distribution. For more
detailed descriptions, see Ellis (1971), Jarvis (1977) and Domsch et al.
(1980).

Moniliella Stolk & Dakin. Colonies usually restricted and yeast-like,
smooth, velvety or cerebriform, cream-colored at first, darkening to pale
olivaceous or blackish-brown with age. Budding cells often present,

ellipsoidal to subcylindrical, frequently forming a pseudomycelium (as in yeasts). Hyphae often branched, hyaline or colored, septate, with a strong tendency to fragment. (Blasto)conidia usually one-celled, arising in acropetal chains from undifferentiated hyphae. (Arthro)conidia formed by fragmentation of hyphae, large, often becoming brown and thick-walled after detachment. Chlamydospores sometimes present, one-celled, subglobose, thick-walled, dark brown. For more detailed descriptions and a key, see de Hoog (1979).

Cladosporium Link. Colonies mostly olivaceous-brown to blackish-brown or with a greyish-olive appearance, velvety or floccose becoming powdery due to abundant conidia, rather slow-growing. Conidiophores more or less distinct from vegetative hyphae, erect, straight or flexuous, unbranched or branched only in the apical region, with geniculate sympodial elongation in some species. Conidia in branched acropetal chains, lower conidia often larger and septate (so-called "ramoconidia") upper conidia one-celled, ellipsoidal, fusiform, ovoid, (sub)globose, often with distinct scars, pale to dark olivaceous-brown, smooth-walled, verrucose or echinulate; (blasto)conidia mostly formed on denticles in groups of 1-3 at the apex of the conidiophore, subapically below a septum or on the tip of previously formed conidia.

Teleomorph: *Mycosphaerella* Johanson. A genus with a world-wide distribution. Several species are plant pathogens or saprobic and more or less host-specific on old or dead plant material. For detailed descriptions and keys, see De Vries (1952), Ellis (1971) and Domsch et al. (1980).

Chrysosporium Corda. Most species are isolated from soil or human and animal tissue. However, some species, e.g., C. xerophilum, C. sulfureum and C. farinicola, are found on dried food (Pitt 1966; van Oorschot 1980).

Stachybotrys Corda. About 15 species are known; S. chartarum, in particular, can be common in food (Jong & Davis 1976).

Memnoniella Höhnel. A similar genus to *Stachybotrys* but conidia occur in chains. Three species are known, of which M. echinata sometimes occurs in food (Jong & Davis 1976).

Verticillium Nees. Teleomorph: *Nectria* and related genera. About 35 species are known, mostly found on plant material. Some common species are described by Domsch et al. (1980).

Gliocladium Corda. Teleomorphs: *Nectria*, *Hypocrea*, and others. Common species are G. roseum and G. viride (Domsch et al. 1980).

Myrothecium Tode. Tulloch (1972) accepted and keyed out 8 species.

Paecilomyces Bain. Teleomorphs: *Byssochlamys*, *Thermoascus*, *Talaromyces*. Common foodborne species are P. variotii and the anamorphs of *Byssochlamys* spp. (Samson 1974).

Humicola Traaen. This genus has eight species; some, such as *Humicola* (*Thermomyces*) lanuginosa, may be found in food (Fassatiova 1967).

Arthrinium Kunze. About 20 species are known; A. apiospermum is sometimes found in food (Ellis 1971, 1976).

Trichothecium Link. The only species in this genus, T. roseum, may sometimes be foodborne (Samson et al. 1984; Domsch et al. 1980).

Pithomyces Berk. & Br. About 15 species exist, a well-known species
which produces toxins is P. chartarum (Ellis 1971, 1976).

Drechslera Ito. Teleomorph: Pyrenophora. About 20 species exist,
mostly found on grasses (Ellis 1971, 1976).

Bipolaris Shoemaker. Teleomorph: Cochliobolus. A genus similar to
Drechslera (Ellis 1971).

Curvularia Boedijn. Teleomorph: Pseudocochliobolus. About 30 species
(Ellis 1971).

Stemphylium Wallr. Teleomorph: Pleospora. About six species (Ellis
1971, 1976).

Ulocladium Preuss. This genus includes about 10 species; U. atrum and
U. consortiale are common (Ellis 1971, 1976).

Alternaria Nees. Teleomorph: Pleospora. Many species are known;
common foodborne are A. alternata and A. tenuissima (Ellis 1971, 1976).

Epicoccum Link. Two species are known; E. nigrum (= E. purpurascens) is
very common (Samson et al. 1984; Domsch et al. 1980).

Wallemia Johan-Olsen. W. sebi, the only species, is a very common
xerophile (Domsch et al. 1980; Samson et al. 1984). This fungus is also
known by the names Sporendonema epizoum, Hemispora stellata and Sporendonema
sebi.

Chrysonilia von Arx. Two or three species are known, representing the
anamorphs of Neurospora (von Arx 1981b). Often referred to in the
literature as Monilia.

Aureobasidium Viala & Boyer. Hermanides-Nijhof (1977) keyed and
described 15 species in this genus. A. pullulans is very common.

Geotrichum Link. Teleomorph: Dipodascus. Several species are known,
of which G. candidum is very common (von Arx 1977).

Sporendonema Desmazieres. Two species are accepted by Sigler &
Carmichael (1976). S. casei, found mostly on cheese, is psychrophilic and
grows and sporulates well at 8°C.

As stated above, Coelomycete species are usually not true saprobic
contaminants, but often occur as (weak) plant pathogens. In addition, many
members of the Coelomycetes may not grow on selective laboratory media or
will sporulate poorly. For general references, see von Arx (1981a) for
Coelomycetes sporulating in pure culture. Sutton (1980) provides an
authoritative account of these fungi.

Phoma Sacc. Teleomorphs: Pleospora, Leptosphaeria and others. Common
species like P. exigua and P. herbarum are described by Boerema et al. (1965,
1971), Domsch et al. (1980), Dorenbosch (1970) and Samson et al. (1984).

Lasiodiploida Ellis & Everhart. Teleomorph: Botryosphaeria. L.
theobromae (= Botryodiplodia theobromae) is often found as the anamorph of
Botryosphaeria rhodina from (sub)tropical material (Sutton 1980).

Colletotrichum Corda. C. gloeosporioides (= Glomerella cingulata) is
common. For a key to species, see von Arx (1981a).

Ascochyta Lib. Teleomorph: _Didymella_. About 40 species are known (Punithalingam 1979).

Recommendations

All fungi isolated from foods should be identified to species level if possible. For the identification of most foodborne genera to species level, we recommend the books of Samson et al. (1984), Domsch et al. (1980), Ellis (1971, 1976), Nelson et al. (1983), Pitt (1979), Pitt & Hocking (1985) and Raper & Fennell (1965) or Samson et al. (1976, 1977a,b) and Stolk & Samson (1983). See also Samson & Pitt (1985).

References

ABE, S., IWAI, M. & SUGIHARA, T. 1982 Taxonomic studies of _Penicillium_ I. Species of the section _Aspergilloides_. _Transactions of the Mycological Society of Japan_ 23, 149–163.

ABE, S., IWAI, M. & YOSHIOKA, H. 1983a Taxonomic studies of _Penicillium_ II. Species in the section _Exilicaulis_. _Transactions of the Mycological Society of Japan_ 24, 39–51.

ABE, S., IWAI, M. & TANAKA, H. 1983b Taxonomic studies of _Penicillium_ III. Species in the section _Divaricatum_. _Transactions of the Mycological Society of Japan_ 24, 95–108.

ABE, S., IWAI, M. & AWANO, M. 1983c Taxonomic studies of _Penicillium_ IV. Species in the sections _Divaricatum_ (continued) and Furcatum. _Transactions of the Mycological Society of Japan_ 24, 109–120.

ABE, S., IWAI, M. & ISHIKAWA, T. 1983d Taxonomic studies of _Penicillium_ V. Species in the section _Furcatum_. _Transactions of the Mycological Society of Japan_ 24, 409–418.

AL-MUSALLAM, A. 1980 _Revision of the Black Aspergillus Species._ Dissertation. Utrecht, Netherlands: University of Utrecht.

AINSWORTH, G. C., SPARROW, F. K. & SUSSMAN, A. S. 1973a _The Fungi. An Advanced Treatise_, Vol. 4A. _A Taxonomic Review with Keys_: _Ascomycetes and Fungi Imperfecti_. New York & London: Academic Press.

AINSWORTH, G. C., SPARROW, F. K. & SUSSMAN, A. S. 1973b _The Fungi. An Advanced Treatise_, Vol. 4B. _A Taxonomic Review with Keys_: _Basidiomycetes and Lower Fungi_. New York & London: Academic Press.

ALCORN, J. L. 1981 Generic concepts in _Drechslera bipolaris_ and _Exserophilum_. _Mycotaxon_ 17, 1–86.

AMES, C. M. 1963 _A monograph of the Chaetomiaceae_. U. S. Army Res. Div., Ser. 2. 125 pp.

BARNETT, H. L. & HUNTER, B. B. 1972 _Illustrated Genera of Imperfect Fungi_. Minneapolis: Burgess Publishing Company.

BARNETT, J. A., PANE, R. W. & YARROW, D. 1983 _Yeasts, Characteristics and Identification_. Cambridge Univ. Press. 811 pp.

BARRON, G. L. 1968 _The Genera of Hyphomycetes From Soil_. Baltimore: Williams and Wilkins Co.

BENEKE, E. S. & STEVENSON, K. E. 1978 Classification of food and beverage fungi. In _Food and Beverage Mycology_. ed. Beuchat, L. R. pp. 1–44. Westport: AVI Publ. Co.

BENJAMIN, R. K. 1959 The merosporangiferous Mucorales. _Aliso_ 4, 321–433.

BENNY, G. L. & KIMBROUGH, J. W. 1980 A synopsis of the orders and families of Plectomycetes with keys to genera. _Mycotaxon_ 12, 1–91.

BENOIT, M. A. & MATHUR, S. B. 1970 Identification of species of _Curvularia_ on rice seed. _Proceedings of the International Seed Testing Association_ 35, 99–119.

BEUCHAT, L. R. 1978 _Food and Beverage Mycology_. Westport: Avi Publishing Co.

BLASER, P. 1975 Taxonomische und physiologische Untersuchungen über die Gattung Eurotium Link. ex Fr. Sydowia 28, 1–49.

BOEREMA, G. H., DORENBOSCH, M. M. & KESTEREN, H. A. van 1965 Remarks on species of Phoma referred to Peyronellaea I. Persoonia 4, 47–68.

BOEREMA, G. H., DORENBOSCH, M. M. & KESTEREN, H. A. van 1971 Remarks on species of Phoma referred to Peyronellaea III. Persoonia 6, 171–177.

BOEREMA, G. H. & DORENBOSCH, M. M. J. 1973 The Phoma and Ascocyta species described by Wollenweber and Hochapfel in their study on fruit-rotting. Studies in Mycology, Baarn 3, 1–50.

BOOTH, C. 1971 The Genus Fusarium. Kew, Surrey: Commonwealth Mycological Institute.

BROWN, A. H. S. & SMITH, G. 1957 The genus Paecilomyces Bainier and its perfect stage Byssochlamys Westling. Transactions of the British Mycological Society 40, 17–89.

BURGESS, L. W. & LIDDELL, G. M. 1983 Laboratory Manual for Fusarium Research. Sidney: University of Sydney.

CARMICHAEL, J. W., KENDRICK, W. B., CONNERS, I. L. & SIGLER, L. 1980 Genera of Hyphomycetes. Edmonton: University of Alberta Press.

CHIDAMBARAM, P., MATHUR, S. B. & NEERGAARD, P. 1973 Identification of seed-borne Drechslera species. Friesia 10, 165–207.

CHRISTENSEN, M. 1981 A synoptic key and evaluation of species in the Aspergillus flavus group. Mycologia 73, 1056–1084.

CHRISTENSEN, M. 1982 The Aspergillus ochraceus group: two new species from western soils and a synoptic key. Mycologia 74, 210–225.

CHRISTENSEN, M. & STATES, J. S. 1982 Aspergillus nidulans group: Aspergillus navahoensis, and a revised synoptic key. Mycologia 74, 226–235.

COLE, G. T. & SAMSON, R. A. 1979 Patterns of Development in Conidial Fungi. London: Pitman.

DE HOOG, G. S. 1979 The Black Yeasts, II. Moniliella and allied genera. Studies in Mycology, Baarn 19, 1–34.

DE HOOG, G. S. & HERMANDES-NIJHOF, E. J. 1977 Survey of the black yeasts and allied fungi. Studies in Mycology, Baarn 15, 170–221.

DOMSCH, K. H., GAMS, W. & ANDERSON, T. H. 1980 Compendium of Soil Fungi. Vol. I & II. London: Academic Press.

DORENBOSCH, M. M. J. 1970 Key to nine ubiquitous soil-borne Phoma-like fungi. Persoonia 6, 1–14.

ELLIS, J. J., RHODES, I. H. & HESSELTINE, C. W. 1976 The genus Amylomyces. Mycologia 68, 131–143.

ELLIS, M. B. 1960 Dematiaceous Hyphomycetes. Mycological Papers 76, 1–36.

ELLIS, M. B. 1971 Dematiaceous Hyphomycetes. Kew, Surrey: Commonwealth Mycological Institute.

ELLIS, M. B. 1976 More Dematiaceous Hyphomycetes. Kew, Surrey: Commonwealth Mycological Institute.

FASSATIOVA, O. 1967 Notes on the genus Humicola. 2. Ceska Mykologia 21, 78–89.

FREDERICK, L., UECKER, F. A. & BENJAMIN, C. R. 1970 A new species of Neurospora from soil of West Pakistan. Mycologia 61, 1077–1084.

FRISVAD, J. C. & FILTENBORG, O. 1983 Classification of terverticillate Penicillia based on profiles of mycotoxins and other secondary metabolites. Applied and Environmental Microbiology 46, 1301–1310.

GAMS, W. 1971 Cephalosporium-artige Schimmelpilze (Hyphomycetes). Stuttgart: G. Fischer. 262 pp.

GAMS, W., PLAATS-NITERIND, H. A. van der, SAMSON, R. A. & STALPERS, J. A. 1980 CBS Course in Mycology, 2. ed. Baarn, the Netherlands: Centralbureau voor Schimmelcultures.

GERLACH, W. & NIRENBERG, H. 1982 The Genus Fusarium, A Pictorial Atlas. Mitt. Biol. Bundesanst. Land. u. Forstwissensch., Berlin-Dahlem. 406 pp.

GILMAN, J. C. 1957 A Manual of Soil Fungi. Ames, Iowa: Iowa State University Press.

HANLIN, R. T. 1973 Keys to Families, Genera and Species of the Mucorales. Vaduz: J. Cramer Verlag.

HAWKSWORTH, D. L. & PITT, J. I. 1983 A new taxonomy for Monascus species based on cultural and microscopical characters. Australian Journal of Botany 31, 51-61.

HAWKSWORTH, D. L., SUTTON, B. C. & AINSWORTH, G. C. 1983 Ainsworth & Bisby's Dictionary of the Fungi. Kew, Surrey: Commonwealth Mycological Institute.

HERMANIDES-NIJHOF, E. J. 1977 Aureobasidium and allied genera. Studies in Mycology, Baarn 15, 141-177.

HESSELTINE, C. W. & ANDERSON, P. 1956 The genus Thamnidium and a study of the formation of its zygospores. American Journal of Botany 43, 696-702.

HESSELTINE, C. W. & ELLIS, J. J. 1973 Mucorales. In The Fungi Vol. IVb Ainsworth, G. C., Sparrow, F. K. & Sussman, A. S. pp. 187-217. New York: Academic Press.

HORIE, Y. 1980 Ascospore ornamentation and its application to the taxonomic re-evaluation in Emericella. Transactions of the Mycological Society of Japan 21, 483-493.

ISAAC, I. 1967 Speciation of Verticillium. Annual Reviews of Phytopathology 5, 201-222.

JARVIS, W. R. 1977 Botryotinia and Botrytis species. Taxonomy, physiology and pathogenicity. Canadian Department of Agriculture. Harrow, Monograph 15, 1-195.

JARVIS, B., SEILER, D. A. L., OULD, A. J. L. & WILLIAMS, A. P. 1983 Observations on enumeration of moulds in food and feedingstuffs. Journal of Applied Bacteriology 55, 325-336.

JONG, S. C. & DAVIS, E. E. 1976 Contributions to the knowledge of Stachybotrys and Memnoniella in culture. Mycotaxon 3, 409-485.

KENDRICK, W. B. & CARMICHAEL, J. W. 1973 Hyphomycetes. In The Fungi Vol. IVa. ed. Ainsworth, G.C., Sparrow, F. K. & Sussman, A. S. pp. 323-509. New York: Academic Press.

KREGER-van RIJ, N. J. W. ed. 1984 The Yeasts. A Taxonomic Study. Amsterdam: Elsevier.

KULSHRESTHA, D. D., MATHUR, S. B. & NEERGAARD, P. 1976 Identification of seed-born species of Colletotrichum. Friesia 11, 116-125.

LUNN, J. A. & SHIPTON, W. A. 1983 Re-evaluation of taxonomic criteria in Cunninghamella. Transactions of British Mycological Society 81: 303-312.

MALLOCH, D. & CAIN, R. F. 1972 The Trichocomataceae: Ascomycetes with Aspergillus, Paecilomyces and Penicillium imperfect states. Canadian Journal of Botany 50, 2613-2628.

MISLEVIC, P. B. & BRUCE, V. R. 1977 Direct plating versus dilution plating qualitatively determining the mold flora of dried beans and soybeans. Journal of the Association of Official Analytical Chemists 60, 741-743.

MOREAU, C. 1979 Molds, Toxins and Food. Chichester: John Wiley and Sons.

MORQUER, R., VIALA, G., ROUCH, M. J. FAYRET, J. & BERGE, E. M. G. 1963 Contribution a l'~etude morphogenigue du genre Gliocladium. Bulletin Trimestriel de la Societe Mycologique de France 79, 137-241.

MORTON, F. J. & SMITH, G. 1963 The genera Scopulariopsis Bainier, Microascus Zukal, and Doratomyces Corda. Mycological Papers 86, 1-96.

NATH, R., NEERGAARD, P. & MATHUR, S. B. 1970 Identification of Fusarium species on seeds as they occur in blotter test. Proceedings of the International Seed Testing Association 35, 121-144.

NEERGAARD, P. 1945 Danish Species of Alternaria and Stemphylium. London: Oxford University Press. 560 pp.

NELSON, P. E., TOUSSON, T. A. & MARASAS, W. F. O. 1983 Fusarium Species. An Illustrated Manual for Identification. University Park, Pennsylvania: Pennsylvania State University Press.

NGUYEN, T. H., NEERGAARD, P. & MATHUR, S. B. 1973 Seed-borne species of
Myrothecium and their pathogenic potential. Transactions of the
Brithish Mycological Society 61, 347–354.

NIRENBERG, H. 1981 A simplified method for identifying Fusarium spp.
occuring on wheat. Canadian Journal of Botany 59, 1599–1609.

O'DONNELL, K. L. 1979 Zygomycetes in culture, Palfrey Contributions in
Botany 2. Athens, Georgia: University of Georgia.

ONIONS, A. H. S., ALLSOPP, D. & EGGINS, H. O. W. 1981 Smith's
Introduction to Industrial Mycology. London: Edward Arnold.

PITT, J. I. 1966 Two new species of Chrysosporium. Transactions of the
British Mycological Society 49, 467–470.

PITT, J. I. 1979 The Genus Penicillium and its Teleomorphic States.
Eupenicillium and Talaromyces. London: Academic Press. 634 pp.

PITT, J. I. 1985 Laboratory Guide to Common Penicillium species. North
Ryde, N.S.W.: CSIRO Division of Food Research.

PITT, J. I. & HOCKING, A. D. 1982 Food spoilage fungi. I. Xeromyces
bisporus Fraser. CSIRO Food Research Quarterly 42, 1–6.

PITT, J. I. & HOCKING, A. D. 1985 Fungi and Food Spoilage. Sydney:
Academic Press.

PUNITHALINGAM, E. 1979 Graminicolous Ascochyta species. Mycological
Papers 142, 1–214.

RAMIREZ, C. 1982 Manual and Atlas of the Penicillia. Amsterdam: Elsevier
Biomedical Press. 847 pp.

RAPER, K. B. & THOM, C. 1949 A Manual of the Penicillia. Baltimore:
Williams & Wilkins Co.

RAPER, K. B. & FENNELL, D. I. 1965 The Genus Aspergillus. Baltimore:
Williams & Wilkins Co.

RIFAI, M. A. 1969 A Revision of the Genus Trichoderma. Mycological Papers
116: 1–56.

RIFAI, M. A. & COOK, R. C. 1966 Studies on some didymosporous genera of
nematode trapping Hyphomycetes. Transactions of the British Mycological
Society 49, 147–167.

SAMSON R. A. 1969 Revision of the genus Cunninghamella (Fungi, Mucorales).
Proceedings of the Koninklijke Nederlandse Akademie van Wetenschappen,
ser. C 72, 322–335.

SAMSON, R. A. 1974 Paecilomyces and some allied Hyphomycetes. Studies
in Mycology, Baarn 6, 1–119.

SAMSON, R. A. 1979 A compilation of the Aspergilli described since 1965.
Studies in Mycology, Baarn 18, 1–40.

SAMSON, R. A., STOLK, A. C. & HADLOK, R. 1976 Revision of the subsection
Fasciculata of Penicillium and some allied species. Studies in
Mycology, Baarn 11, 1–47.

SAMSON, R. A., ECKHARDT, C. & ORTH, R. 1977a The taxonomy of Penicillium
species from fermented cheeses. Antonie van Leeuwenhoek 43, 341–350.

SAMSON, R. A., HADLOK, R. & STOLK, A. C. 1977b A taxonomic study of the
Penicillium chrysogenum series. Antonie van Leeuwenhoek 43, 261–274.

SAMSON, R. A., HOEKSTRA, E. S. & van OORSCHOT, C. A. N. 1984 Introduction
to Food-Borne Fungi. 2nd ed. Baarn, the Netherlands: Centraalbureau
voor Schimmelcultures. 1984.

SAMSON, R. A., & PITT, J. I. 1985 Advances in Penicillum and Aspergillus
Systematics. London: Plenum Press.

SAMUELS, G. J. 1984 Toxigenic fungi as Ascomycetes. In Toxigenic fungi.
Their Toxins and Health Hazard ed. Kurata, H. & Ueno, Y. Developments
in Food Science 7. pp. 119–128. Amsterdam: Elsevier.

SCHIPPER, M. A. A. 1973 A study on variability in Mucor hiemalis and
related species. Studies in Mycology, Baarn 4, 1–40.

SCHIPPER, M. A. A. 1975 On Mucor mucedo, Mucor flavus and related
species. Studies in Mycology, Baarn 10, 1–33.

SCHIPPER, M. A. A. 1976 On Mucor circinelloides, Mucor racemosus and
related species. Studies in Mycology, Baarn 12, 1–40.

SCHIPPER, M. A. A. 1978 1. On certain species of Mucor with a key to all accepted species. 2. On the genera Rhizomucor and Parasitella. Studies in Mycology, Baarn 17, 1-70.

SCHIPPER, M. A. A. 1984. A revision of the genus Rhizopus I. The Rh. Stolonifer-group and Rh. oryzae. Studies in Mycology, Baarn 25, 1-19.

SCHIPPER, M. A. A. & STALPERS, J. A. 1984 A revision of the genus Rhizopus II. The Rhizopus microsporus-group. Studies in Mycology, Baarn 25, 19-34.

SCHOL-SCHWARTZ, M. B. 1970 Revision of the genus Phialophora (Moniliades). Persoonia 6, 59-94.

SEKITA, S., YOSHIHIRA, K., NATORI, S., UDAGAWA, S., MUROI, T., SUGIYAMA, Y., KURATA, H. & UMEDA, M. 1979 Mycotoxin production by Chaetomium spp. and related fungi. Canadian Journal of Microbiology 27, 766-772.

SIGLER, L. & CARMICHAEL, J. W. 1976 Taxonomy of Malbranchea and some other Hyphomycetes with arthroconidia. Mycotaxon 4, 349-488.

SIMMONS, E. G. 1967 Typification of Alternaria, Stemphylium and Ulocladium. Mycologia 59, 67-92.

STEVENS, R. B. (ed.) 1974 Mycology Guidebook. Washington: University of Washington Press.

STOLK, A. C. & SAMSON, R. A. 1971 Studies in Talaromyces and related genera. I. Hamigera gen nov. and Byssochlamys. Persoonia 6, 341-357.

STOLK, A. C. & SAMSON, R. A. 1972 The genus Talaromyces. Studies on Talaromyces and related genera. II. Studies in Mycology, Baarn 2, 1-65.

STOLK, A. C. & SAMSON, R. A. 1983 The ascomycetes genus Eupenicillium and related Penicillium anamorphs. Studies in Mycology, Baarn 23, 1-149.

SUBRAMANIAN, C. V. 1971 Hyphomycetes, An Account of Indian Species, Except Cercosporae. New Delhi: Indian Council of Agricultural Research.

SUTTON, B. C. 1980 The Coelomycetes, Fungi Imperfecti with Pycnidia, Acervuli and Stromata. Kew, Surrey: Commonwealth Mycological Institute.

TUITE, J. 1969 Plant Pathological Methods. Minneapolis: Burgess.

TULLOCH, M. 1972 The genus Myrothecium. Mycological Papers 130, 1-42.

UDAGAWA, S., MUROI, T., KURATA, H., SEKITA, S., YOSHIHIRA, K., NATORI, S. & UMEDA, M. 1979 The production of chaetoglobosins, sterigmatocystin, o-methylsterimatocystin and chaetocin by Chaetomium spp. and related fungi. Canadian Journal of Microbiology 25, 170-177.

UDAGAWA, S. 1984 Taxonomy of mycotoxin-producing Chaetomium. In Toxigenic fungi. Their Toxins and Health Hazard ed. Kurata, H. & Ueno, Y. pp. 139-167. Amsterdam: Elsevier.

VAN OORSCHOT, C. A. N. 1980 A revision of Chrysosporium and allied genera. Studies in Mycology, Baarn 20, 1-89.

VON ARX, J. A. 1957 Die Arten der Gattung Colletotrichum Cda. Phytopathologische Zeitschrift 29, 413-468.

VON ARX, J. A. 1977 Notes on Dipodascus, Endomyces and Geotrichum with the description of two new species. Antonie van Leeuwenhoek 43, 333-340.

VON ARX, J. A. 1981a The Genera of Fungi Sporulating in Pure Culture. 3rd edition. Vaduz: J. Cramer Verlag.

VON ARX, J. A. 1981b On Monilia sitophila and some families of Ascomycetes. Sydowia 34, 13-29.

VON ARX, J. A., DREYFUSS, M. & MÜLLER, E. 1984 A revaluation of Chaetomium and the Chaetomiaceae. Persoonia 12, 169-179.

VRIES, de G. A. 1952 Contribution to the Knowledge of the Genus Cladosporium. Diss. Univ. Utrecht, Reprint J. Cramer Lehre (1967).

WYLLIE, T. D. & MOREHOUSE, L. G. 1977 Mycotoxic Fungi, Mycotoxins, Mycotoxicoses. Vol. 1. New York: Marcel Dekker.

ZYCHA, H., SIEPMANN, R. & LINNEMANN, G. 1969 Mucorales. Eine Beschreibung aller Gattungen und Arten dieser Pilzgruppe. Lehre: J. Cramer-Verlag.

R. A. SAMSON
J. C. FRISVAD
O. FILTENBORG

Aids in the Identification of Important Foodborne Species of Filamentous Fungi

Most isolates of foodborne molds can be identified to species level by using solely micromorphological criteria. These anatomical species are often also homogeneous in physiological and chemical characters, even though two different isolates from the same species may differ in characteristics such as preservative resistance or amount of extracellular enzymes produced. Species in four of the most important foodborne fungal genera, Penicillium, Aspergillus, Fusarium and Alternaria, appear to be more heterogeneous, however. Many different taxonomic treatments are available for each of these genera, and species concepts differ considerably among authors as noted in the previous manuscript by Samson et al. These often broad species can be divided into homogeneous subgroups (clusters of isolates based on micromorphology, chemical characters, physiology, etc.), which may or may not gain general taxonomic acceptance depending on future species concepts. The solution to the problem of different species concepts is to identify foodborne fungi according to a specified taxonomic system and provide additional important data such as mycotoxin production, temperature relationships and enzyme production. The recording of colony diameters under specified conditions (Pitt 1984) and determination of parts of secondary metabolite profiles produced on diagnostic substrates (Frisvad & Filtenborg 1983; Frisvad 1984) are easily done in addition to morphological examinations. Guides for the use of criteria other than morphology as aids in the taxonomy of Alternaria, Aspergillus, Fusarium and Penicillium are given below.

Alternaria

Even though there are many Alternaria spp., most of these are substrate specific (Simmons 1981, 1982; Neergaard 1945). The most important species of Alternaria is, however, A. alternata (Fr.) Keissler. It is very important to note that A. alternata is only one member of the unresolved group species A. tenuis Auct. Wiltshire. Members of the latter species group belong to different teleomorphic states: Clathrospora diplospora, C. elynae and Leptosphaeria heterospora, and the taxonomy of the A. tenuis group awaits further clarification. As A. alternata cannot be used for the entire A. tenuis group, heterogeneity must be expected among isolates identified to one of these names (Neergaard 1977; Domsch et al. 1980). Temperature relationships seemed to be alike in most members of the group (Neergaard 1945), however, and the taxonomic significance of mycotoxins and other secondary metabolites in Alternaria taxonomy has not been investigated systematically. The mycotoxins produced by the A. tenuis group and A. alternata are reviewed by Reiss (1983) and Harvan & Pero (1976). These mycotoxins can be detected easily by the methods described by Filtenborg & Frisvad (1980) and Filtenborg et al. (1983). Domsch et al. (1980), Joly (1969) and Neergaard (1945) provide physiological data on Alternaria spp.

Aspergillus

Most isolates of Aspergillus can be readily identified to the groups used by Raper & Fennell (1965) based on conidial color and a few easily recognizable morphological characters, even directly on selective isolation media. Speciation within Aspergillus groups is more difficult but in most cases the species are either very homogeneous in physiological and chemical characters within the group (A. niger group, Eurotium/A. glaucus group, Petromyces/A. ochraceus group, Emericella/A. nidulans group) or the group contains only one or a few more easily recognizable foodborne species (e.g.,

Table 1. Profiles of mycotoxins produced by foodborne _Aspergillus_ spp.

A. clavatus	Ascladiol, clavatol, cytochalasin E & K, kojic acid, kotanin, patulin, tryptiquivalin, tryptiquivalon, xanthocillin-dimethylether
Eurotium spp.	Emodin, physicons, glaucins, gliotoxin?
A. fumigatus	Fumigaclavins, fumigallin, fumigatin, fumitoxins, fumitremorgens[a], gliotoxin, helvolic acid, kojic acid?, spinulosin, tryptiquivalin, tryptiquivalon
N. fischeri var. fischeri	Fumitremorgens[a]
A. ochraceus	Emodin, kojic acid, neoaspergillic acid, ochratoxins, penicillic acid, secalonic acids, viomellein, xanthomegnin
A. niger	Malformins, naptho-pyrones
A. candidus	Candidulin, terphenyllin, xanthoascin
A. flavus	Aflatoxins, aflatrem, aflavinin, aspergillic acids, cyclopiazonic acid, kojic acid, 3-nitropropionic acid, paspalinin
A. parasiticus	Aflatoxins, aspergillic acid, kojic acid, 3-nitro-propionic acid
A. tamarii	Cyclopiazonic acid, kojic acid
A. wentii	Emodin, kojic acid?, 3-nitropropionic acid?, wentilacton, physicons, sulochrins
A. versicolor	Aspercolin, nidulotoxin, sterigmatocystin, versicolorin, versimid, versiol
A. sydowii	Aspermutarubrol, nidulotoxin, sterigmatocystin, sydowic acid, sydowinins
E. nidulans var. nidulans	Asperline, asperthecin, emericellin, emerin, kojic acid, nidulins, nidulol, nidulotoxin, penicillin, shamixanthone, sterigmatocystin
F. flavipes	Asperphenemate
F. nivea	Citrinin
A. terreus	Aspulvinones, asterriquinone, aspterric acid, citreoviridin, citrinin, cytochalasin E, gliotoxin, mevinolin, patulin, terreic acid, terrein, terretonin, territrems

[a]Verruculogen is included in the fumitremorgens

A. flavus group: A. flavus, A. parasiticus, A. tamarii; Neosartorya/A.
fumigatus group: A. fumigatus, N. fischeri; A. clavatus group: A.

clavatus; A. restrictus group: A. restrictus, A. penicilloides; A. versicolor group: A. versicolor, A. sydowii; Fennellia/A. flavipes group: F. flavipes, F. nivea; A. terreus group: A. terreus; A. candidus group: A. candidus; and A. wentii group: A. wentii). Recently some physiological criteria have been introduced in Aspergillus taxonomy, i.e., growth at 37°C and growth rate on Czapek agar (Christensen 1982), and a standardization of such criteria as introduced by Pitt (1973, 1979) may provide some valuable additional auxiliary and confirmatory characteristics in Aspergillus taxonomy. Lists of secondary metabolites produced by the Aspergilli are given by Turner (1971) and Turner & Aldridge (1983). Table 1 summarizes the most important mycotoxins produced by foodborne Aspergillus spp. Physiological data are given by Raper & Fennell (1965), Christensen (1981, 1982), Christensen & States (1982), Al-Musallam (1980), Blaser (1975), Murakami (1971), Murakami et al. (1979) and Domsch et al. (1980).

Fusarium

The use of simple physiological and chemical characters in Fusarium taxonomy has not been tested systematically. Burgess & Lidell (1983) indicate that temperature and water relation characters are possible future auxiliary criteria in Fusarium taxonomy. Colony diameters measured under standardized conditions can be used in Fusarium taxonomy too (Burgess & Lidell 1983; Booth 1971; Gerlach & Nirenberg 1982). Marasas et al. (1984) provide data on mycotoxin production by Fusarium spp. Vesonder & Hesseltine (1981), Turner (1971) and Turner & Aldridge (1983) provide further data on secondary metabolites from the Fusaria. Physiological data are provided by Domsch et al. (1980) and Nelson et al. (1981).

Penicillium

Several authors provide physiological and chemical data for Penicillium spp. (see Niethammer [1949] and Raper & Thom [1949] for early references; Abe 1956; Frank 1966; Engel & Teuber 1978; Pitt 1979; Frisvad 1981; Frisvad & Filtenborg 1983; Frisvad 1984). Three types of simple criteria seem to be of value in Penicillium taxonomy. These are colony diameters measured under standardized conditions of temperature (5, 25 and 37°C) and water activity (CYA, MEA and G25N media), profiles of secondary metabolites (CYA and YES media) and growth on creatine-sucrose agar (CREA) (Frisvad 1981, 1984; Pitt 1979). The criteria of growth at 5°C and 37°C, and of growth rate on G25N seems to be of most value in the subgenera Aspergilloides, Furcatum and Biverticillium, while the use of CREA and YES media is most valuable in subgenus Penicillium.

Most authors agree on the taxonomy of foodborne Penicillium spp. in the subgenera Aspergilloides, Furcatum and Biverticillium. These subgenera contain foodborne species such as P. glabrum (= P. frequentans), P. spinulosum, P. oxalicum, P. citrinum, P. corylophilum, P. funiculosum, P. variabile, P. islandicum, P. rugulosum and P. purpurugenum. The taxonomy of the abundantly occurring subgenus Penicillium spp. (terverticillate or two staged branched Penicillia) is not clarified at present. Table 2 shows that most taxonomic problems are encountered in the P. viridicatum/P. cyclopium series of Raper & Thom (1949) or the anatomical species P. verrucosum of Samson et al. (1976). The species/subgroup system of Frisvad & Filtenborg (1983) based on profiles of secondary metabolites, morphology and few physiological characters can be easily translated to the system of Pitt (1979). The relation of the former system to the anatomical taxonomy of the terverticillate Penicillia of Samson et al. (1976) will be briefly discussed.

Table 2. Foodborne terverticillate (two stage branched) Penicillia according to different authors

Frisvad & Filtenborg (1983)	Pitt (1979)	Samson et al. (1976, 1977a,b)	Raper & Thom (1949)
P. camembertii	P. camembertii	P. camembertii	P. camembertii
			P. caseicolum
P. camembertii II, III	P. camembertii	P. verrucosum	P. biforme
	P. puberulum	v. cyclopium	P. commune
	P. aurantiogriseum		P. cyclopium
	P. viridicatum		P. puberulum
	P. verrucosum		P. palitans
P. echinulatum	P. echinulatum	P. echinulatum	P. cyclopium
			v. echinulatum
P. mali	P. crustosum	P. verrucosum	P. palitans
	P. viridicatum	v. melanochlorum	
P. crustosum	P. crustosum	P. verrucosum	P. crustosum
		v. cyclopium	
P. hirsutum I	P. hirsutum	P. verrucosum	P. corymbiferum
		v. corymbiferum	
P. hirsutum II	P. hirsutum	P. verrucosum	–
		v. cyclopium	
P. hirsutum III	P. hirsutum	P. hordei	–
P. hirsutum IV	P. echinulatum	P. echinulatum	–
	P. hirsutum	P. hordei	
P. aurantiogriseum I	P. aurantiogriseum	P. verrucosum	P. cyclopium
	P. puberulum	v. cyclopium	P. puberulum
			P. martensii
P. aurantiogriseum II	P. aurantiogriseum	P. verrucosum	P. cyclopium
	P. puberulum	v. cyclopium	P. puberulum
			P. martensii
			P. aurantio-
			virens
P. viridicatum I	P. viridicatum	P. verrucosum	P. viridicatum
		v. cyclopium	
P. chrysogenum	P. chrysogenum	P. chrysogenum	P. chrysogenum
			P. notatum
P. roquefortii I, II	P. roquefortii	P. roquefortii	P. roquefortii
P. brevicompactum	P. brevicompactum	P. brevicompactum	P. brevicom-
			pactum
		P. stoloniferum	P. stoloniferum
P. expansum	P. expansum	P. expansum	P. expansum
P. viridicatum II, III	P. viridicatum	P. verrucosum	P. viridicatum
	P. verrucosum	v. verrucosum	P. casei
P. viridicatum IV	P. viridicatum	P. verrucosum	–
	P. crustosum	v. cyclopium	
P. griseofulvum	P. griseofulvum	P. griseofulvum	P. urticae
		v. cyclopium	
P. concentricum I	P. italicum	P. concentricum	–
	P. hirsutum		
P. concentricum II	?	P. cf. concen-	P. martensii
		tricum	
P. italicum	P. italicum	P. italicum	P. italicum
		v. italicum	

P. verrucosum var. verrucosum equals P. viridicatum II and P. viridicatum III, and may be distinguished from the other fungi in the P. viridicatum/P. cyclopium complex by its low growth rate on CYA (< 22 mm after 7 days at 25°C), its moderate production of green conidia (seldom with a bluish tint in the margin of the colonies), growth on nitrite-sucrose agar and cream colored to brown reverse on CYA. P. verrucosum var. verrucosum produces the mycotoxins ochratoxin A, citrinin and oxalic acid. P. verrucosum var. melanochlorum is characterized by a moderate production of very dark green conidia on CYA (the conidia are bluish green in the margin on CYA and bluish green on YES medium), its low growth rate on CYA (18-33 mm after 1 week at 25°C) and the orange yellow reverse on YES medium (see also Söderstöm & Frisvad 1984). P. verrucosum var. melanochlorum (= P. mali Gorlenko et Novobranova) produces cyclopenin and viridicatin. This variety or species is closely related to P. echinulatum and P. camembertii III, and these three species/subgroups are very often found in foods and in the aerial flora of factories. P. verrucosum var. cyclopium can be subdivided into eight subgroups as listed in Table 3. Some characteristics of the subgroups of this variety are listed in Table 4. P. aurantiogriseum II and P. crustosum are characterized by the production of enormous amounts of conidia on CYA (crust formation). P. viridicatum IV always produces strongly plicated colonies on CYA and YES medium. P. hirsutum II produces white coremia on MEA. P. camembertii II always produces a blackish brown pigment in YES medium after 7-21 days of growth.

At present, we recommend that Penicillia be identified according to one taxonomic system (either Pitt 1979, or Samson et al. 1976, 1977a,b with Stolk & Samson 1983, supplemented with Raper & Thom 1949). If a part of the profile of secondary metabolites is included (Frisvad & Filtenborg 1983; Filtenborg et al. 1983; Filtenborg & Frisvad 1980) in each identification, translation from one system to the other is possible. An example may illustrate this: an isolate identified as P. verrucosum var. cyclopium producing griseofulvin equals P. viridicatum IV and will be identified as P. viridicatum in the system of Pitt (1979).

Table 3. Mycotoxins produced by eight subgroups of P. verrucosum var. cyclopium

Subgroup	Mycotoxin
P. aurantiogriseum I	Penicillic acid, S-toxin
P. aurantiogriseum II	Penicillic acid, xanthomegnin, viomellein
P. viridicatum I	Penicillic acid, xanthomegnin, viomellein
P. viridicatum IV	Griseofulvin, viridicatumtoxin
P. hirsutum II	Citrinin, penicillic acid, roquefortine C
P. crustosum	Isofumigaclavin A and B, penitrem A-F, roquefortine C, terrestric acid, viridicatin
P. camembertii II	Cyclopiazonic acid, rugulovasin A and B, cyclopolic acid
P. camembertii III	Cyclopiazonic acid, cyclopenin, palitantin, viridicatin

Table 4. Some characteristics of subgroups of P. verrucosum var. cyclopium

Penicillium spp.	CREA[a]	Diam[b]	Reverse[c]	Conidiogenesis[c]
P. aurantiogriseum I	w/+	(28–)32–38	Yellow-brown	Strong, bluish green
P. aurantiogriseum II	–/w/+	22–27(–33)	Yellow	Weak, bluish green
P. viridicatum I	w/+	24–35	Yellow	Weak, green
P. viridicatum IV	w/+	25–34	Yellow	Strong, green
P. hirsutum II	w/+	32–48	Yellow	Weak, bluish green
P. crustosum	++	34–48	Yellow	Strong, grayish green
P. camembertii II	++	24–35	Cream-brown	Weak to moderate
P. camembertii III	++	22–33(–37)	Yellow	Strong, bluish green

[a]Growth on CREA (see Frisvad 1981, 1983)
[b]Diameter (mm) on CYA after 7 days of growth at 25°C
[c]Reverse color or degree of conidiogenesis and color on YES medium

In conclusion, we recommend that one taxonomic system is used for each genus. For the possible comparison of qualitative results of mycological analysis of foods, we recommend that taxonomic authorities are always stated and additional characteristics are given for the important fungi isolated (i.e., range of colony diameters, mycotoxins produced, etc.).

References

ABE, S. 1956 Studies on the classification of the Penicillia . Journal of General and Applied Microbiology 2, 1–344.
AL-MUSALLAM, A. 1980 Revision of the black Aspergillus species. Utrecht: University of Utrecht. Dissertation. 92 pp.
BLASER, P. 1975 Taxonomische und physiologishce Untersuchungen über die Gattung Eurotium Link ex Fr. Sydowia 28, 1–49.
BOOTH, C. 1971 The Genus Fusarium. Kew, Surrey: Commonwealth Mycological Institute. 237 pp.
BURGESS, L. W. & LIDDELL, C. M. 1983 Laboratory Manual for Fusarium Research. Sydney: University of Sydney. 162 pp.
CHRISTENSEN, M. 1981 A synoptic key and evaluation of the Aspergillus flavus group. Mycologia 75, 1056–1084.
CHRISTENSEN, M. 1982 The Aspergillus ochraceus group: two new species from western soils and a synoptic key. Mycologia 744, 210–255.
CHRISTENSEN, M. & STATES, J. S. 1982 Aspergillus nidulans group: Aspergillus navahoensis, and a revised synoptic key. Mycologia 74, 226–235.
DOMSCH, K. H., GAMS, W. & ANDERSON, T. H. 1980 Compendium of Soil Fungi. Vol. I & II. London: Academic Press. 859 + 405 pp.

ENGEL, G. & TEUBER, M.   1978   Simple aid in the identification of
   Penicillium roqueforti Thom.   Growth in acetic acid.   European Journal
   of Applied Microbiology and Biotechnology 6, 107–111.
FILTENBORG, O. & FRISVAD, J. C.   1980   A simple screening-method for
   toxigenic molds in pure cultures.   Lebensmittel Wissenschaft und
   Technologie 13, 128–130.
FILTENBORG, O., FRISVAD, J. C. & SVENDSEN, J. A.   1983   Simple screening
   method for moulds producing intracellular mycotoxins in pure cultures.
   Applied and Environmental Microbiology 45, 581–585.
FRANK, H. K.   1966   Ein Beitrag zur Taxonomie der Gattung Penicillium Link.
   München: Technische Hochschule MÜnchen.   Dissertation.   109 + 59 pp.
FRISVAD, J. C.   1981   Physiological criteria and mycotoxin production as
   aids in the identification of common asymmetric penicillia.   Applied and
   Environmental Microbiology 41, 568–579.
FRISVAD, J. C.   1983   A selective and indicative medium for groups of
   Penicillium viridicatum producing differenct mycotoxins in cereals.
   Journal of Applied Bacteriology 54, 409–416.
FRISVAD, J. C.   1984   Expressions of secondary metabolism as fundamental
   characters in *Penicillium* taxonomy.   In Toxigenic Fungi.   Their Toxins
   and Health Hazard.   ed. Kurata, H. & Ueno, Y.   pp. 98–106.   Amsterdam:
   Elsevier.
FRISVAD, J. C. & FILTENBORG, O.   1983   Classification of terverticillate
   Penicillia based on profiles of mycotoxins and other secondary
   metabolites.   Applied and Environmental Microbiology 46, 1301–1310.
GERLACH, W. & NIRENBERG, H.   1982   The Genus Fusarium, A Pictorial Atlas.
   Metteilungen aus der Biologischen Bundesanstalt für Land–und
   Forstwirtschaft Berlin–Dalem 209, 1–406.
HARVAN, D. J. & PERO, R. W.   1976   The structure and toxicity of the
   Alternaria metabolites.   In Mycotoxins and Other Fungal Related Food
   Problems.   ed. Rodricks, J. V.   pp. 344–355.   Advances in Chemistry
   Series 149.   Washington D. C.:   American Chemical Society.

JOLY, P.   1969   Essais d'application de méthodes de traitement numériques des
   information systématiques.   I.   Etude du groupe des *Alternaria* sensu
   lato.   Bulletin Trimestriel de la Societe Mycologie Francais 85,
   213–233.
MARASAS, W. F. O., NELSON, P. E. & TOUSSOUN, T. A.   1984   Toxigenic Fusarium
   Species: Identity and Mycotoxicology.   University Park, Pennsylvania:
   The Pennsylvania State University Press.
MURAKAMI, H., YOSHIDA, K. & NORO, F.   1979   Tables of mycological characters
   of the representative strains of the black *Aspergilli*.   Journal of the
   Brewing Society, Japan 74, 466–470.
NEERGAARD, P.   1945   Danish species of Alternaria and Stemphylium.   London:
   Oxford University Press.
NEERGAARD, P.   1977   Seed Pathology.   Vol. I & II.   London: The MacMillan
   Press.
NELSON, P. A., TOUSSOUN, T. A. & COOK, R. J.   1981   Fusarium: Diseases,
   Biology, and Taxonomy.   University Park, Pennsylvania: The
   Pennsylvannia State University Press.
NIETHAMMER, A.   1949   Die Gattung Penicillium Link.   Stuttgart: Verlag Eugen
   Ulmer.
PITT, J. I.   1973   An appraisal of identification methods for *Penicillium*
   species: novel taxonomic criteria based on temperature and water
   relations.   Mycologia 65, 1135–1157.
PITT, J. I.   1979   The Genus Penicillium and its Teleomorphic States
   Eupenicillium and Talaromyces.   London, New York, Toronto, Sydney, San
   Francisco: Academic Press.
PITT, J. I.   1984   The value of physiological characters in the taxonomy of
   Penicillium.   In Toxigenic Fungi.   Their Toxins and Health Hazard.   ed.
   Kurata, H. & Ueno, Y.   pp. 107–118.   Amsterdam: Elsevier.
RAPER, K. B. & FENNELL, D. I.   1965   The Genus Aspergillus.   Baltimore:
   Williams and Wilkins Co.

RAPER, K. B. & THOM, C.   1949   <u>A Manual of the Penicillia</u>.   Baltimore:
    Williams and Wilkins Co.
REISS, J.   1983   Alternaria mycotoxins in grains and bread.   <u>Zeitschrift für</u>
    <u>Lebensmittel Untersuchung und Forschung</u>   176, 36–39.
SAMSON, R. A., ECKARDT, C. & ORTH, R.   1977a   The taxonomy of <u>Penicillium</u>
    species from fermented cheeses.   <u>Antonie van Leeuwenhoek</u>   43, 341–350.
SAMSON, R. A., HADLOK, R. & STOLK, A. C.   1977b   A taxonomic study of the
    <u>Penicillium</u> <u>chrysogenum</u> series.   <u>Antonie van Leeuwenhoek</u>   43, 261–274.
SAMSON, R. A., STOLK, A. C. & HADLOCK, R.   1976   Revision of the subsection
    Fasciculata of <u>Penicillium</u> and some allied species.   <u>Studies in</u>
    <u>Mycology</u>, Baarn   11, 1–47.
SIMMONS, E. G.   1981   <u>Alternaria</u> themes and variations.   <u>Mycotaxon</u>   13,
    16–34.
SIMMONS, E. G.   1982   <u>Alternaria</u> themes and variations.   <u>Mycotaxon</u>   14,
    17–43 & 44–57.
STOLK, A. C. & SAMSON, R. A.   1983   The ascomycete genus <u>Eupenicillium</u> and
    related <u>Penicillium</u> anamorphs.   <u>Studies in Mycology</u>, Baarn   23, 1–149.
SÖDERSTRÖM, B. & FRISVAD, J. C.   1984   Separation of closely related
    asymmetric Penicillia by pyrolysis gas chromatography and mycotoxin
    production.   <u>Mycologia</u>   76, 408–419.
TURNER, W. B.   1971   <u>Fungal Metabolites</u>.   London: Academic Press.   446 pp.
TURNER, W. B. & ALDRIDGE, D. C.   1983   <u>Fungal Metabolites</u> II.   London & New
    York: Academic Press.   631 pp.
VESONDER, R. F. & HESSELTINE, C. W.   1981   Metabolites of <u>Fusarium</u>.   In
    <u>Fusarium: Diseases, Biology, and Taxonomy</u>.   ed. Nelson, P. E.,
    Toussoun, T. A. & Cook, R. J.   pp. 350–364.   University Park,
    Pennsylvania: The Pennsylvania State University Press.

J. C. FRISVAD

▶ **A Simplified Scheme for the Identification of Yeasts**

In ecological studies of foods it is often necessary to identify a great
number of yeasts. For the identification of any one yeast strain, some 90
morphological and physiological tests are used (Barnett et al. 1983). This
is a hardly feasible task when hundreds of strains are concerned. Hence, it
is desirable to devise simplified identification schemes enabling rapid and
easy handling of numerous isolates. There have been several attempts at and
proposals for such methods (Barnett & Pankhurst 1974; Barnett et al. 1979;
Davenport 1979; Fuchs & Dolan 1982). The computer-generated keys still
require numerous tests while other methods make use of ready-made
identification kits. Our attempt was to devise a key as short as possible
by applying tests performable in factory laboratories.

The first version of our simplified scheme was formulated in 1981. It
was based on the description given in the 2nd edition of Lodder's manual
(Lodder 1970). To these data we added characters of some 130 yeast species
validly described by the end of 1980. In all, we considered data of 460
species.

Recently a new comprehensive treatise on yeasts has appeared (Barnett et
al. 1983) supplying full descriptions of 469 species whose characteristics
have been determined in an identical manner. The 3rd revised edition of <u>The</u>
<u>Yeasts</u> (Kreger van Rij 1984) has also been published. It describes some 500
yeast species. Based on these recent data, we have also revised our
simplified identification scheme.

To devise a simplified identification key, one should select the most efficient diagnostic tests. The responses of yeasts to physiological and biochemical tests applied in identification are usually described in a plus-or-minus fashion. However, some responses are variable and assigned v, d or ±. Obviously, the most efficient test would be one that would divide a given group into two groups (positive and negative) with no variable responses. In practice, unfortunately, no such test can be found. The efficiency (e) of any one test can be calculated, according to Gower & Barnett (1971) from the expression:

$$e = (p - 0.5)^2 + (q - 0.5)^2 + r^2$$

where p is the ratio of species with positive responses, q is the ratio of species with negative responses and r denotes the ratio of variable responses. Out of all possible tests, the most efficient is the one for which the value of e is smallest. This formula was used to select tests for developing an identification key.

Materials and methods

Identification of yeasts is based on results of selected tests. Out of the numerous standard tests, the following are used in one or another key: assimilation of (i.e., aerobic growth on) maltose, raffinose, galactose, cellobiose, trehalose, xylose, erythritol, mannitol and nitrate (designated by M, R, G, Ce, Te, X, Er, Mt and NO_3, respectively), fermentation of glucose (d), growth in the presence of 0.01% cycloheximide (cyc) and production of urease (ure). These tests are supplemented by the microscopic investigation of cell morphology. This is done from an isolated colony when the purity of the strain is checked after repeated streaking on agar plates. Using this procedure, the formation of sexual spores can frequently be observed. However, a direct search for spore formation, which is time consuming and uncertain is not necessary when using the simplified keys. Often it is also useful to examine the capability of a yeast to produce true hyphae or pseudohyphae or both, as well as to form arthroconidia. This can be done on a slide culture.

For the assimilation of carbon sources, the auxanographic method is used. This enables testing four substrates in one Petri dish. The assimilation of nitrate and the fermentation of glucose can be made in liquid media. Other tests can be performed by any established procedure. Details of procedures and composition of media can be found in several books (Barnett et al. 1979, 1983; Lodder 1970; Kreger van Rij 1984) and are not described here.

Results and discussion

On the basis of data collected for the 460 accepted species, the efficiency of various fermentation and assimilation tests was first calculated (Table 1). For example, the fermentation of glucose was reported positive for 242 species, negative for 161 and variable for the other 57. Thus, the efficiency of glucose fermentation was calculated as 0.039. Overall, the most efficient test proved to be maltose assimilation. Although there were only three species with variable nitrate assimilation, the efficiency of this test is relatively low because it divides the yeasts into two rather unequal groups.

The construction of the key started with maltose assimilation. This approach divided the 460 species into two groups consisting of 268 positive and 217 negative yeasts, as the 25 variable species were added to each

Table 1. Responses of yeasts to and efficiency of the identification test

	Responses			
Test[a]	+	–	Variable	Efficiency
d	242	164	57	.039
M	243	192	25	.011
S	275	162	23	.034
G	243	160	57	.040
R	155	279	29	.049
Er	118	322	20	.099
NO$_3$	114	343	3	.121

[a]d = fermentation of glucose; assimilation = M, maltose; S, sucrose;
G, galactose; R, raffinose; Er, erythritol; NO$_3$, nitrate

group. The next most efficient test was then selected for the division of each branch. Although sucrose assimilation appeared efficient for

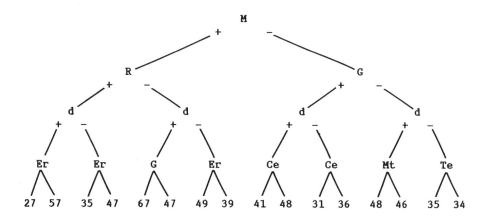

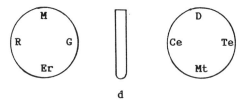

Master key of identification

d

Fig. 1. Master key of identification. Figures denote the number of species falling into each of the 16 groups. The first stage of identification can be performed by using two Petri dishes and one test tube. Assimilation tests: M = maltose; R = raffinose; G = galactose; Er = erythritol; Ce = cellobiose; Mt = mannitol; Te = trehalose; D = glucose (control); d = fermentation of glucose

Table 2. Distribution of isolates from a wine bottling plant

Number of group	Number of strains	Identified species
1	5	2
2	20	4
4	1	1
5	16	5
6	3	2
10	6	4
12	3	2
13	19	3
14	38	6
15	1	1
16	6	3
Total:	118	33

separating the whole group of yeasts studied, its application in the second level after maltose proved to be redundant, since nearly all maltose-positive yeasts assimilate sucrose too, and maltose-negative species responded similarly to sucrose. We found the raffinose test most efficient for maltose-positive species and galactose for maltose-negative ones.

The selection of tests was continued successively till sixteen main groups were formed. The building of the key is shown in Fig. 1. In this form, our scheme differs from similar studies in at least two respects. First, it makes use of an extremely small number of tests. When selection of the most efficient test is rigorously applied, the number of tests involved increases greatly. Some keys generated by computer programs make use of 40-50 or more tests (Barnett et al. 1979, 1983). In contrast, we attempted to minimize the number of tests by choosing, whenever it was practicable, a test that occurred elsewhere on a different branch of the key, even if it was slightly less effective. Finally, we used eight tests only.

Four tests are enough for separating each of the 16 branches; the remaining four tests can be used to complete the identification of each strain. This is the second unique part of our scheme, i.e., it is partitioned in two stages. In the first stage, by applying the standard eight tests, we arrive at 16 groups, each containing on the average 40 species. In the second stage two to seven additional tests are used, selected respectively for each group. All in all, 10-15 tests are used in the two stages, thus eliminating at least 80% of the otherwise necessary diagnostic tests. This two-step identification scheme is economic, saving both labor and materials. Even time can be saved if it is considered that only a single test tube and two Petri dishes are needed for one strain. Thus, a great number of isolates can be examined at one time.

An example is given to illustrate the scheme. During an investigation of a wine bottling plant, some 118 strains were collected from samples taken from the bottling lines. Having subjected them to the simplified scheme, the strains were separated into eleven groups. All but two contained more than one strain which could be separated into more than one species (Table 2).

Based on the eight tests applied in the first stage, together with a microscopic investigation of morphology, we arrived at species identity in

Table 3. Details of the key

Group 14:   M- G- d+ Mt-
(other tests: R, Ce, Te, Er)[a]   Identity   Additional tests

1. Septate hyphae +	2	
Septate hyphae –	3	
2. Arthroconidia +	*Geotrichum fici*	
Arthroconidia –	*Endomycopsella vini*	spores +
3. Cellobiose +	4	
Cellobiose –	10	
4. Raffinose +	*Saccharomycodes ludwigii*	
Raffinose –	5	
5. Bipolar budding	*Hanseniaspora*	
	guilliermondii	growth 37°C +
	H. occidentalis	growth 30°C + sucr +
	H. uvarum	growth 30°C + sucr –
	H. valbyensis	growth 25°C +
Multilat. budding	6	
6. Trehalose +	*Brettanomyces naardensis*	
Trehalose –	7	
7. Nitrate +	8	
Nitrate –	9	
8. Spores +	*Hansenula mrakii*	xylose +
Spores –	*Candida berthetii*	xylose –
9. Pseudohyphae +	*Pichia norvegensis*	rhamnose –
	P. sargatensis	rhamnose +
Pseudohyphae –	*Candida dendrica*	
10. Raffinose +	11	
Raffinose –	15	
11. Trehalose +	12	
Trehalose –	13	
12. Conjugation +	*Zygosaccharomyces*	
	florentinus	
Conjugation –	*Saccharomyces*	
	cerevisiae	
13. Splitting +	*Schizosaccharomyces*	
	malidevorans	
Splitting –	14	
14. Spores +	*Saccharomyces*	
	cerevisiae	
Spores –	*Candida lactis-condensi*	nitrate +
	C. stellata	nitrate –
15. Trehalose +	16	
Trehalose –	19	
16. Pseudohyphae +	*Brettanomyces*	
	custerianus	
Pseudohyphae –	17	
17. Conjugation	*Zygosaccharomyces bailii*	
Spores free	*Kluyveromyces delphensis*	
Not as above	18	
18. Growth 37°C +	*C. glabrata*	cells large (> 5 μm)
	C. castellii	cells small (< 5 μm)
Growth 37°C –	*C. fructus*	xylose +

[a]See Table 4 for explanation of abbreviations

Table 4. Final identification of wine strains, group 1: M+R+d+Er+[a]

Strain number	Further tests							Confirmatory tests					Identity
	G	Ce	Te	Mt	psh	Spore test	60%	Rh	Me	L	MG	vit	
27	+	+	+	+	+	round	–	+	+	+	+	–	Debaryomyces hansenii
31	+	+	+	+	–	round	–	+	+	+	+	–	D. hansenii
38	+	+	+	+	+	round	–	–	–	–	+	–	D. hansenii
61	+	+	+	+	–	hat	+	–	–	–	+	+	Hansenula anomala
63	+	+	+	+	–	hat	+	–	–	–	+	+	H. anomala

[a]Assimilation tests: M = maltose; R = raffinose; Er = erythritol; Mt = mannitol; L = lactose; G = galactose; Ce = cellobiose; Te = trehalose; Rh = rhamnose; Me = melibiose; MG = alpha-methyl-glucoside; d = fermentation of glucose; psh = formation of pseudohyphae; 60% = growth in the presence of 60% (w/v) glucose; vit = growth without vitamins

Table 5. Final identification of wine strains Group 10: M–G+d+Ce–

Strain number	Further tests							Confirmatory tests						Identity
	R	Er	Te	Mt	psh	Spore test	60%	NO$_3$	g	s	m	r	me	
57	+	–	–	–	–	round	–	–	+	+	+	+	–	S. cerevisiae
76	–	+	–	+	+	round	–	–	+	–	–	–	–	P. farinosa
80	–	–	–	+	–	round	+	–	–	+	–	–	–	Z. bailii
81	–	–	+	+	–	round	+	–	–	+	–	–	–	Z. bailii
87	–	–	–	+	–	round	+	–	–	+	–	–	–	Z. bailii
97	+	–	+	+	–	round	+	–	+	+	–	+	–	T. delbrueckii

[a]Abbreviations as in Table 4; fermentation tests: g = galactose; s = sucrose; m = maltose; r = raffinose; me = melibiose; NO$_3$ = assimilation of nitrate.

the majority of cases by using a single additional test. A part of the key is shown as an example in Table 3. When several more tests were carried out for confirmation, the results corresponded to previous identification (Tables 4 and 5). All of the 118 strains were split into 23 species of which seven were represented by a single strain only (Table 6).

Since the elaboration of the first version of our scheme, substantial changes have been carried out in yeast taxonomy (Kreger-van Rij 1984). A few more tests have also been applied in identifications (Barnett et al. 1983). Some of them proved to be very efficient when we reevaluated their usefulness (Table 7). Maltose assimilation is still the most efficient,

Table 6.　Yeast species identified from a wine bottling plant

Species	No. of strains
Saccharomyces cerevisiae	16
Zygosaccharomyces bailii	19
Z. bisporus	10
Z. rouxii	2
Debaryomyces hansenii	10
Torulaspora delbrueckii	4
Pichia fermentans	6
P. membranaefaciens	12
P. etchellsii	3
P. humboldtii	2
P. farinosa	1
Hansenula anomala	2
Candida brumptii	6
C. valida	7
C. lambica	3
C. tropicalis	2
C. utilis	1
C. vini	1
C. zeylanoides	1
Endomyces fibuliger	2
Endomycopsella vini	1
Geotrichum capitatum	1
Rhodotorula rubra	1

Table 7.　Efficiency of identification tests reevaluated on the basis of recent data (Barnett et al. 1983)

Test[a]	Responses			Efficiency
	+	–	v	
M	226	202	41	.013
S	263	176	30	.023
cyc	176	230	63	.034
Ce	261	142	66	.062
37°	143	258	68	.062
G	214	163	92	.064
cit	225	151	93	.071
LA	143	239	87	.073
R	125	288	56	.082
d	195	164	110	.085
NO_3	119	343	7	.115
ure	113	354	2	.132

[a]Assimilations:　M = maltose; S = sucrose; Ce = cellobiose; G = galactose; cit = citrate; LA = L-arabinose; R = raffinose; NO_3 = nitrate; d = fermentation of glucose; cyc = growth in the presence of 0.01% cycloheximide; 37° = growth at 37°C; ure = splitting of urea

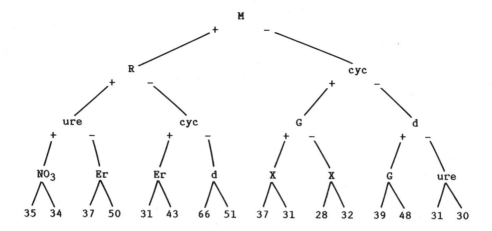

Master key of identification II

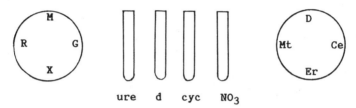

ure   d   cyc NO_3

Fig. 2.   Revised version of master key. Two Petri dishes and four test tubes are needed to initiate identification for one strain. Abbreviations as in Fig. 1: X = assimilation of xylose; NO_3 = assimilation of nitrate; cyc = growth in the presence of 0.01% cycloheximide; and ure = splitting of urea

while the utility of glucose fermentation is the worst because a higher ratio of species proved variable in this test. Of the new tests, cycloheximide resistance has proven to be very efficient. Besides nitrate assimilation, splitting of urea is also a character of outstanding utility for giving just a few variable responses; moreover, a simple and quick method is available for its determination.

With these new data at hand we revised our simplified scheme (Fig. 2). The number of tests to be performed at the first stage has somewhat enlarged, and 2-10 tests are required at the second stage. The revised key has been checked using known strains from our collection and it performed reliably. It is now being tried in practice. In addition, a computer program is being developed to facilitate the use of the key.

The simplified keys so far described can be applied for identifying the whole group of yeasts consisting by now of some 500 species. An even more simplified scheme can be worked out if it is to be applied to a particular group of yeasts occurring in a restricted habitat. Keys of this kind have been described, e.g., for yeasts occurring in wine, on strawberries and in clinical materials (Barnett et al. 1979, 1983; Buhagiar & Barnett 1971). In contrast to these, the advantage of the proposed key for foodborne yeasts is that it involves very few tests.

Table 8.   List of 100 yeasts most frequently occurring in foods

Yeast[a]	Food types[b]								
	1	2	3	4	5	6	7	8	9
Brettanomyces anomalus					+	+			
B. bruxellensis (t = Dekkera)					+	+			
B. claussenii				+	+	+			
B. intermedia (t = Dekkera)	+				+	+	+		
B. naardensis					+				
Bullera alba	+						+	+	
Candida apicola		+	+	+	+				
C. boidinii		+		+	+				
C. cantarellii	+				+				
C. catenulata	+				+		+	+	
C. diddensiae			+	+			+		
C. etchellsii	+	+	+						
C. friedrichii			+						+
C. glabrata	+	+	+	+	+		+		+
C. inconspicua	+				+		+		
C. intermedia			+	+	+	+	+		
C. javanica (t = Filobasidium capsulig.)	+				+	+			+
C. lactis-condensi	+	+	+				+		
C. lipolytica (t = Yarrowia)	+			+	+		+		+
C. magnoliae			+	+					
C. mesenterica						+	+	+	
C. norvegica				+	+	+			
C. parapsilosis	+	+			+	+	+		
C. pulcherrima (t = Metschnikowia)	+				+	+		+	
C. reukaufii (t = Metschnikowia)	+		+	+	+		+		
C. rugosa	+	+			+		+	+	
C. sake	+			+	+	+	+		+
C. scottii (t = Leucosporidium)	+				+		+		
C. solani					+	+			
C. sorboxylose	+				+				
C. steatolytica				+	+				
C. stellata	+			+	+	+			
C. tenuis	+				+	+			
C. tropicalis	+	+	+	+	+	+	+	+	+
C. utilis (t = Pichia jadinii)		+			+				+
C. vartiovaarai				+	+	+			
C. versatilis	+	+	+		+	+			+
C. vini	+				+	+	+	+	
C. zeylanoides	+				+		+		
Citeromyces matritensis (a = Candida globosa)	+	+	+		+				
Clavispora lusitaniae (a = Candida lusitaniae)	+			+					
Cryptococcus albidus	+			+	+		+	+	+
C. infirmo-miniatus (t = Rhodosporidium)	+								+
C. laurentii	+			+	+		+		+

(continued)

Table 8.   Continued

Yeast[a]	Food types[b]								
	1	2	3	4	5	6	7	8	9
Debaryomyces hansenii (a = Candida famata)	+	+	+	+	+	+	+	+	+
D. marama					+	+			
Endomyces fibuliger					+				+
Geotrichum candidum	+						+	+	
G. capitatum						+			+
Hanseniaspora guilliermondii (a = Kloekera apis)	+				+				
H. uvarum (a = Kloekera apiculata)	+	+		+	+	+			
H. valbyensis (a = Kloekera japonica)	+	+		+	+	+			
H. vineae (a = Kloekera africana)	+				+	+			
Issatchenkia orientalis (a = Candida krusei)	+	+	+	+	+	+	+	+	+
I. terricola	+			+	+				
Hyphopichia burtonii (a = Candida variabilis)	+				+				+
Kluyveromyces marxianus (a = Candida kefyr)	+	+			+	+		+	+
K. thermotolerans	+	+			+				+
Lodderomyces elongisporus			+	+					
Pichia angusta	+		+	+				+	+
P. anomala (a = Candida pelliculosa)	+	+	+	+	+	+	+		
P. carsonii					+		+	+	
P. etchellsii	+	+			+			+	
P. farinosa	+				+	+			+
P. fermentans (a = Candida lambica)	+	+	+	+	+	+		+	+
P. guilliermondii (a = Candida guilliermondii)	+	+	+	+	+	+	+	+	+
P. humboldtii (a = Candida ingens)	+				+	+			
P. kluyveri	+	+							
P. membranaefaciens (a = Candida valida)	+	+	+	+	+	+	+	+	+
P. nakasei (a = Candida citrea)	+			+	+				
P. ohmeri (a = Candida membranaefaciens)	+	+	+			+			+
P. onychis	+					+			+
P. pijperi		+					+		
P. subpelliculosa	+	+	+		+	+			
Rhodotorula aurantiaca	+				+	+			
R. glutinis (t = Rhodosporidium)	+	+		+	+		+	+	+
R. minuta (t = Rhodosporidium)	+	+			+		+	+	+
R. mucilaginosa	+		+	+	+	+	+	+	+
Saccharomyces cerevisiae	+	+	+	+	+	+	+		+
S. dairensis		+					+	+	

(continued)

Yeast[a]	Food types[b]								
	1	2	3	4	5	6	7	8	9
S. exiguus (a = Candida holmii)	+	+		+	+	+	+	+	+
S. kluyveri				+	+				
S. unisporus					+	+			
Saccharomycodes ludwigii	+				+				
Schizosaccharomyces japonicus				+	+				
S. malidevorans	+				+				
S. octosporus	+				+				
S. pombe	+		+	+	+	+			+
S. shibatanus (t = Sporidiobolus)	+				+		+		
S. roseus	+		+		+	+		+	
Sporobolomyces salmonicolor (t = Sporidiobolus)	+				+				
Torulaspora delbrueckii	+	+	+	+	+	+	+	+	+
Trichosporon cutaneum	+				+				+
T. pullulans					+	+	+	+	+
Wickerhamiella domercqiae	+		+		+				
Williopsis californica	+			+	+		+		+
Zygosaccharomyces bailii	+		+	+	+	+			+
Z. bisporus	+	+	+	+	+				
Z. microellipsoides	+			+	+				
Z. rouxii	+	+	+	+	+	+		+	+

[a]When a species is known both in teleomorphic (t) and anamorphic (a)
 states, the name of its more common state is used

[b]1 = Fruits, vegetables and other raw plant materials; 2 = brined, pickled
 and fermented foods; 3 = concentrates, syrups, jam, honey; 4 = soft drinks,
 fruit juices, non-alcoholic beverages; 5 = must, wine, cider; 6 = beer; 7 =
 meat, meat products, poultry; 8 = milk, cheese, other milk products; 9 =
 miscellaneous other foods

Based on data from the literature, a list has been compiled for yeasts
most frequently occurring in foods (Table 8). Identification of these
selected yeasts can be made by using the key in Table 9. As the number of
species considered is relatively low (about 100, or 20% of the total number
of yeast species), the selection of tests and the build-up of the key is
somewhat different from those covering the whole group of yeasts. In all,
ten tests are used to form eight groups consisting on the average of fifteen
yeast species. Within these groups, final identification can be made using
those tests not considered for a given group, supplemented with morphological
characters. Additional tests are necessary in a few cases only to
distinguish between two species. It should be noted, however, that this
scheme is based on selected yeast species and results differing from those
in the key are likely to occur. An extended list of yeasts that have been
isolated from foods or are likely to occur in them contains some 220
species. This number is high enough to be treated by the key described for
the whole group of yeasts.

In summary, by carefully selected diagnostic tests, it is possible to
devise simplified yeast identification schemes which make it possible to
handle a great number of strains simultaneously, while saving labor and

Table 9.   A simplified key for yeasts most frequently occurring in foods

Key to Groups

1.	Erythritol +	Group I	
	Erythritol –	2	
2.	Urease +	Group II	
	Urease –	3	
3.	Nitrate +	Group III	
	Nitrate –	4	
4.	Raffinose +	5	
	Raffinose –	6	
5.	Cellobiose +	Group IV	
	Cellobiose –	Group V	
6.	Cellobiose +	Group VI	
	Cellobiose –	7	
7.	Mannitol +	Group VII	
	Mannitol –	Group VIII	

Character	Positive	Negative

Group I (Erythritol +, 19 species)

Character	Positive	Negative
1. urease	2	6
2. arthroconidia	3	4
3. nitrate	Trichosporon pullulans	Trichosporon cutaneum
4. raffinose	5	Candida lipolytica
5. ballistoconidia	Bullera alba	Cryptococcus laurentii
6. nitrate	7	10
7. maltose	8	Candida boidinii
8. galactose	Pichia anomala	9
9. raffinose	Pichia sub- pelliculosa	Pichia angusta
10. raffinose	11	15
11. galactose	12	Endomyces fibuliger
12. glucose fermented	13	14
13. hyphae+arthrocon	Hyphopichia burtonii	Candida friedrichii
14. spores 1 to 2	Debaryomyces hansenii	Debaryomyces marama
15. maltose	16	18
16. galactose	17	Candida mesenterica
17. growth at 37°C	Candida diddensiae	Candida tenuis
18. galactose	Pichia farinosa	Candida cantarellii

Group II (urease +, 18 species)

Character	Positive	Negative
1. pigmented	2	10
2. nitrate	3	7

(continued)

Table 9. Continued

Character	Positive	Negative
3. maltose	4	Sporobolomyces salmonicolor
4. raffinose	5	Rhodotorula aurantiaca
5. galactose	6	Sporobolomyces roseus
6. starch production	Cryptococcus infirmo-miniatus	Rhodotorula glutinis
7. raffinose	8	Rhodotorula minuta
8. xylose	9	Sporobolomyces pararoseus
9. starch production	Cryptococcus laurentiae	Rhodotorula mucilaginosa
10. splitting cells	11	15
11. glucose fermented	12	Trichosporon cutaneum
12. raffinose	13	Schizosaccharomyces octosporus
13. true hyphae	Schizosaccharo-myces japonicus	14
14. maltose	Schizosaccharo-myces pombe	Schizosaccharomyces
15. nitrate	16	17
16. true hyphae	Candida scottii	Cryptococcus albidus
17. raffinose	18	Candida japonica
18. ballistoconidia	Bullera alba	Cryptococcus laurentii

Group III (nitrate +, 14 species)

Character	Positive	Negative
1. acetate produced	2	5
2. spores	3	4
3. cellobiose	Dekkera intermedia	Dekkera bruxellensis
4. hyphae	Brettanomyces anomalus	Brettanomyces claussenii
5. raffinose	6	10
6. maltose	7	9
7. cellobiose	8	Citeromyces matritensis
8. galactose	Candida versatilis	Candida utilis
9. mannitol	Candida magnoliae	Candida lactis-condensis
10. galactose	11	13
11. maltose	Candida etchellsii	12
12. glucose fermented	Candida magnoliae	Wickerhamiella domercqiae
13. spores	Williopsis californica	14
14. maltose	Candida vartio-vaarai	Candida norvegica

Group IV (raffinose +, cellobiose -, 11 species)

Character	Positive	Negative
1. glucose fermented	2	11
2. xylose	3	8
3. galactose	4	− Pichia onychis

(continued)

Table 9. Continued

Character	Positive	Negative
4. spores	5	6
5. spores round	Kluyveromyces marxianus	Pichia guilliermondii
6. growth at 37°C	7	Candida intermedia
7. true hyphae	Candida steato-lytica	C. guilliermondii
8. maltose	9	Saccharomycodes ludwigii
9. mannitol	Pichia ohmeri	10
10. spores	Dekkera inter-media	Brettanomyces claussenii
11. pseudohyphae	Pichia carsonii	Debaryomyces hansenii

Group V (raffinose +, cellobiose -, 11 species)

Character	Positive	Negative
1. conjugation	2	4
2. protuberances	Torulaspora delbrueckii	3
3. cells large	Zygosaccharomyces bailii	Zygosaccharomyces microellipsoides
4. spores free	5	7
5. glucose fermented	6	Pichia carsonii
6. maltose	Kluyveromyces thermotolerans	Kluyveromyces marxianus
7. spores	8	9
8. maltose	Saccharomyces kluyveri (lysine+) Saccharomyces cerevisiae (lysine-)	Saccharomyces exiguus (actidione+) Saccharomyces cerevisiae (actidione-)
9. mannitol	Candida apicola	Candida stellata

Group VI (raffinose -, cellobiose +, 17 species)

Character	Positive	Negative
1. mannitol	2	14
2. galactose	3	12
3. glucose fermented	4	10
4. spores	5	7
5. spores free	6	Clavispora lusitaniae
6. spores round	Kluyveromyces marxianus	Pichia etchellsii
7. growth at 37°C	8	9
8. true hyphae	Candida tropicalis	Metschnikowia reukaufii
9. pseudohyphae	Candida sake	Metschnikowia pulcherrima
10. spores	11	Brettanomyces naardensis
11. growth at 37°C	Pichia etchellsii	Pichia carsonii
12. maltose	13	Pichia pijperi
13. growth at 37°C	Clavispora lusitaniae	Candida solani
14. acetate produced	15	16

(continued)

291

Table 9. Continued

Character	Positive	Negative
15. spores	Dekkera intermedia	Brettanomyces claussenii
16. maltose	Hanseniaspora vineae	17
17. growth at 37°C	Hanseniaspora guilliermondii	Hanseniaspora valbyensis (25°C) Hanseniaspora uvarum (30°C)

Group VII (raffinose −, cellobiose −, mannitol +, 14 species)

Character	Positive	Negative
1. arthroconidia	Geotrichum candidum	2
2. conjugation	3	6
3. protuberances	Torulaspora delbrueckii	4
4. growth in 1% acetate	5	Zygosaccharomyces rouxii
5. cells large	Zygosaccharomyces bailii	Z. bisporus
6. spores	7	9
7. glucose fermented	8	Pichia carsonii
8. xylose	Lodderomyces elongisporus	Saccharomyces cerevisiae
9. maltose	10	12
10. glucose strongly ferm	11	Candida catenulata
11. growth at 37°C	Candida parapsilosis	C. sake
12. galactose	Candida rugosa	13
13. trehalose	Candida zeylanoides	C. vini

Group VIII (raffinose −, cellobiose −, mannitol −, 18 species)

Character	Positive	Negative
1. xylose	2	3
2. glucose strongly ferm	Pichia fermentans	Candida sorboxylosa
3. glucose fermented	4	14
4. conjugation	5	7
5. protuberances	Torulaspora delbrueckii	6
6. cells large	Zygosaccharomyces bailii	Zygosaccharomyces bisporus
7. galactose	8	9
8. spore 1	Saccharomyces unisporus	Saccharomyces cerevisiae (cells large) Saccharomyces dairensis (cells small)
9. maltose	Dekkera bruxellensis	10
10. pseudohyphae strong	11	13
11. spores free	Pichia kluyveri	12
12. growth without vita	Issatchenkia orientals	Issatchenkia terricola

(continued)

Table 9. Continued

Character	Positive	Negative
13. growth at 37°C	Candida glabrata	Pichia nakasei
14. galactose	15	16
15. arthroconidia	Geotrichum capitatum	Pichia humboldtii
16. pseudohyphae	Pichia membranae-faciens	Candida inconspicua

material. The results of the identification by these simplified schemes are reasonably accurate. Their use is suggested for ecological studies. For the purpose of taxonomic work, one should still rely upon all the tests required by accepted practices.

References

BARNETT, J. A. & PANKHURST, R. J.   1974   A New Key to The Yeasts
    Amsterdam: North-Holland Publ. Co.
BARNETT, J. A., PAYNE, R. W. & YARROW, D.   1979 A Guide to Identifying and
    Classifying Yeasts. Cambridge: Cambridge University Press.
BARNETT, J. A., PAYNE, R. W. & YARROW, D.   1983 Yeasts: Characteristics and
    Identification. Cambridge: Cambridge University Press.
BUHAGIAR, R. W. M. & BARNETT, J. A.   1971 The yeasts of strawberries.
    Journal of Applied Bacteriology 34, 729-739.
DAVENPORT, R. R.   1979 Experimental ecology and identification of
    microorganisms. In Identification Methods for Microbiologists, Society
    of Applied Bacteriologists Technical Series No. 14, 2nd ed.
    pp. 258-277. New York: Academic Press.
FUCHS, P. C. & DOLAN, T.   1982   Performance of yeast identification
    systems. American Journal of Clinical Pathology 78, 664-667.
GOWER, J. C. & BARNETT, J. A.   1971 Selecting tests in diagnostic keys
    with unknown responses. Nature, London 232, 491-493.
KREGER VAN RIJ, N. J. W., ed.   1984   The Yeasts. A Taxonomic Study
    3rd ed. Amsterdam: Elsevier Sci. Publ.
LODDER, J., ed.   1970 The Yeasts. A Taxonomic Study   2nd ed.
    Amsterdam: North-Holland Publ. Co.

T. DEÁK

Discussion

Pitt
    I am impressed with your system Dr. Deák. It seems quite useful. What does the large R stand for?

Deák
    It stands for raffinose.

King

This seems to simplify the scheme for identifying yeasts and minimize the materials needed.

Deák

One further advantage of this system is that one can see differences in foodborne yeasts.

Hocking

Let me ask you about the plate test. Is that a well test?

Deák

A suspension of the yeast is fixed into the agar medium and then the carbon source is added in powder form.

Hocking

A similar type of test can be run by cutting a well into the medium and then adding a solution of your carbon source.

Andrews

You did not test for sorbitol? I was under the impression that in Barnett's key for identification of yeasts, in the simplified key, key 15, he uses sorbitol as his primary test for the first major split. Were you able to make the split without using it?

Deák

Yes, we used mannitol assimilation, since if one is positive for one sugar, then usually it will be positive for the other.

Andrews

Therefore, you used mannitol instead of sorbitol.

Deák

Yes, that is correct. We also tested sorbitol, but we found that mannitol was more efficient.

Andrews

Is this key just for food yeasts?

Deák

This key seems to work well for all yeasts regardless of source. This key is based on the 460 species described in Barnett's book.

Samson

It is too bad that this key could not be compiled just to identify only the foodborne yeasts, since there is only a limited number of yeasts which are truly foodborne. That would be a very practical key for food microbiologists.

Andrews

That would be the advantage applying the key only to the food yeasts. You could probably get down to specific identifications using less tests than are called for in Barnett's key, which I believe has 38 tests. If you could get it down to half of that, it would make life very much more bearable.

Deák

A key restricted to foodborne yeasts has been elaborated and will be included in the text of the Proceedings.

Formulae for Mycological Media

The following is a list of formulae for media commonly used to enumerate yeasts and molds in foods. Unless otherwise stated, all ingredients should be combined, heated to dissolve agar and sterilized by autoclaving at 121°C for 15 min. Distilled water should be used for all media. Antibiotic solutions should be filter sterilized if added to sterile base media. pH tolerance is ±0.2.

Many of these media are commercially available. Descriptions of historical development, use and limitations are provided by Difco (1984) and Oxoid (1982). Formulae of commercial products may differ in some instances from those listed in reports describing their development. Such changes have been made to improve the utility and performance of media.

1. **AFPA:** Aspergillus flavus and parasiticus Agar (Pitt et al. 1983)

Yeast extract	20	g
Peptone, bacteriological	10	g
Ferric ammonium citrate	0.5	g
Dichloran (0.2% in ethanol, w/v)	1	ml
Chloramphenicol	0.1	g
Agar	15	g
Water	to 1000	ml

Final pH 6.0–6.5

This medium is suitable for the enumeration of Aspergillus flavus/parasiticus. Dichloran (2,6-dichloro-4-nitroanaline) and chloramphenicol can be added before sterilization.

2. **ATGY:** Acidified Tryptone Glucose Yeast Extract Agar

Glucose	20 g
Yeast extract	5 g
Agar	15 g
Water	1000 ml

Final pH 4.5

To obtain a reaction of pH 4.5, acidify the sterile medium at a temperature not greater than 50°C with sterile 10% tartaric acid.

3. **CREA:** Creatine Sucrose Agar (Frisvad 1983)

Creatine (1 H_2O)	3	g
Sucrose	30	g
KCl	0.5	g
$MgSO_4 \cdot 7H_2O$	0.5	g
$FeSO_4 \cdot 7H_2O$	0.01	g
K_2HPO_4	1.3	g
Bromocresol purple	0.05	g
Agar	15	g
Water	1000	ml

Final pH 8.0   (adjusted after medium is autoclaved)

4. **CYA:** Czapek Yeast Autolysate Agar (Pitt 1979)

Sucrose	30	g
Yeast extract	5	g
$NaNO_3$	3	g
KCl	0.5	g
$MgSO_4 \cdot 7H_2O$	0.5	g
$FeSO_4 \cdot 7H_2O$	0.01	g
K_2HPO_4	1	g
Agar	20	g
Water	1000	ml

Final pH 6.0–6.5

If distilled water is used, it is recommended to add 0.01 g of $ZnSO_4 \cdot 7H_2O$ and 0.005 g of $CuSO_4 \cdot 5H_2O$ per liter medium. Copper ions are essential for the production of melanoproteins, which color the conidia of most Penicillia green.

5. **DG18:** Dichloran 18% Glycerol Agar (Hocking and Pitt 1980)

Glucose	10	g
Peptone	5	g
KH_2PO_4	1	g
$MgSO_4 \cdot 7H_2O$	0.5	g
Glycerol, A. R.	220	g
Agar	15	g
Dichloran (0.2% in ethanol, w/v)	1.0	ml
Chloramphenicol	0.1	g
Water	1000	ml

Final pH 5.6

Add minor ingredients and agar to 800 ml of water. Steam to dissolve agar, then make to 100 ml with water. Add glycerol (18%, final concentration) and sterilize by autoclaving. The final a_w is 0.955.

6. DRBC: <u>Dichloran Rose Bengal Chlortetracycline (Chloramphenicol) Agar</u>
(King et al. 1979)

Glucose	10	g
Peptone, bacteriological	5	g
KH_2PO_4	1	g
$MgSO_4 \cdot 7H_2O$	0.5	g
Rose bengal (5% aqueous soln., w/v)	0.5	ml
Dichloran (0.2% in ethanol, w/v)	1	ml
Chlortetracycline (0.01% soln., w/v)	1	ml
Agar	15	g
Water	1000	ml

Final pH 5.6

Dichloran (2,6-dichloro-4-nitroaniline)(2 µg/ml, final
concentration) and chlortetracycline (10 µg/ml) are added to the
sterile medium just before pouring. A later publication (Hocking
1981) reports replacement of chlortetracycline with chloramphenicol
(100 µg/ml). In this case, autoclaving (15 min, 121°C) all
ingredients together does not compromise the performance of the
medium.

7. DRYES: <u>Dichloran Rose Bengal Yeast Extract Sucrose Agar</u> (Frisvad
1983)

Yeast extract	20	g
Sucrose	150	g
Dichloran	0.002	g
Rose bengal	0.025	g
Chloramphenicol	0.1	g
Agar	20	g
Water to	1000	ml

Final pH 5.6 (adjusted after medium is autoclaved)

Prepare as described for DRBC above.

8. GTC: <u>Glucose Tryptone Chloramphenicol Agar</u>

Glucose	40	g
Tryptone	10	g
Chloramphenicol (0.1% soln., w/v)	1	ml
Agar	15	g
Water	1000	ml

Final pH 5.6

Chloramphenicol can be added before sterilizing the medium.

9. GYE: <u>Glucose Yeast Extract Agar</u>

Glucose	20	g
Yeast extract	5	g
Agar	20	g
Water	1000	ml

Final pH 5.6

To obtain a reaction of pH 3.5, acidify the sterile medium at a temperature not greater than 50°C with sterile 10% tartaric acid. The medium should not be heated after addition of acid. Chloramphenicol (100 μg/ml, final concentration) may be added to medium (final pH 5.6) before sterilization to control growth of bacteria.

10. **GYES**: Glucose Yeast Extract Sucrose Agar

Sucrose	450 g
Glucose	20 g
Yeast extract	5 g
Agar	20 g
Water	505 ml

Final pH   7.0

The reduced a_w (0.91) of this medium selects for xerophilic fungi.

11. **G25N**: 25% Glycerol Nitrate Agar   (Pitt 1979)

Glycerol (analytical grade)	250	g
Yeast extract	3.7	g
$NaNO_3$	2.25	g
KCl	0.375	g
$MgSO_4 \cdot 7H_2O$	0.375	g
$FeSO_4 \cdot 7H_2O$	0.0075	g
K_2HPO_4	0.975	g
Agar	12	g
Water	750	ml

Final pH   6.0-6.5

12. **MEA**: Malt Extract Agar

Formula 1 (Oxoid)

Malt extract	30 g
Peptone	5 g
Agar	15 g
Water	1000 ml

Final pH   5.4

Formula 2 (Raper and Thom formula used by Pitt 1979)

Malt extract	20 g
Peptone	1 g
Glucose	20 g
Agar	20 g
Water	1000 ml

Final pH   4.6-5.0

13. **MSA:** Malt Salt Agar

Malt extract	30 g
Sodium chloride	75 g
Agar	15 g
Water	1000 ml

Final pH   7.0

Malt salt agar may be formulated to contain sodium chloride at a concentration less than or exceeding 7.5%, depending upon the food being analyzed and the major genera of molds suspected to be present. Chloramphenicol (100 µg/ml) may be added before sterilization to control the growth of halotolerant bacteria.

14. **NO_2S:** Nitrite Sucrose Agar

$NaNO_2$	5	g
Sucrose	30	g
KCl	0.5	g
$MgSO_4 \cdot 7H_2O$	0.5	g
$FeSO_4 \cdot 7H_2O$	0.01	g
K_2HPO_4	1	g
Agar	15	g
Water	1000	ml

Final pH   6.0–6.5

15. **OGY:** Oxytetracycline Glucose Yeast Extract Agar (Mossel et al. 1970)

Glucose	10	g
Yeast extract	5	g
Oxytetracycline (0.1% soln., w/v)	100	ml
Agar	15	g
Water	1000	ml

Final pH   7.0

Oxytetracycline solution should be added to sterile medium. Gentamicin (50 µg/ml, final concentration) can replace or be used in combination with oxytetracycline to control the growth of Enterobacteriaceae. Commercial oxytetracycline glucose yeast extract agar has been modified to contain 20 g of glucose and 12 g of agar per liter of water.

16. **PCA:** Plate Count Agar

Tryptone	5	g
Yeast extract	2.5	g
Glucose	1	g
Agar	15	g
Water	1000	ml

Final pH   7.0

Addition of chloramphenicol and/or chlortetracycline at concentrations not exceeding 100 µg/ml has been used successfully to select for yeasts and molds in foods.

17. **PCNB:** <u>Pentachloronitrobenzene Agar</u> (Nash & Snyder 1962)

Peptone	15	g
KH_2PO_4	1	g
$MgSO_4 \cdot 7H_2O$	0.5	g
Streptomycin	0.3	g
Pentachloronitrobenzene	1	g

Final pH 5.2

This medium was developed for estimating <u>Fusarium</u> propagules in
soils but may have also utility for food. Pentachloronitrobenzene
is added as a 75% wettable powder, and the authors state that best
results are obtained by allowing poured plates to dry in a dark,
cool, dry place for 3-5 days before application of the sample.

18. **PDA:** <u>Potato Dextrose Agar</u> (Booth 1971)

Potatoes, infusion from	200	g
Dextrose	15	g
Agar	20	g
Water	1000	ml

Final pH 5.6

To obtain a reaction of pH 3.5, acidify the sterile medium at a
temperature not greater than 50°C with sterile 10% tartaric acid.
The amount of acid added to the medium is stated on the package
label of commercial products. Medium should not be heated after
addition of acid.

19. **PRYES:** <u>Pentachloronitrobenzene Rose Bengal Yeast Extract Sucrose</u>
<u>Agar</u> (Frisvad 1983)

Yeast extract	20	g
Sucrose	150	g
Pentachloronitrobenzene	0.1	g
Chloramphenicol	0.1	g
Rose bengal	0.025	g
Agar	20	g
Water to	1000	ml

Final pH 5.6

PCNB (pentachloronitrobenzene) is sparingly soluble in water. A
fine powder should be used and the medium should be agitated very
thoroughly during preparation. Commercial 50% active preparations
(Brassicol and Quintozen) can be added instead of PCNB.

A combination of chloramphenicol (50 µg/ml) and chlortetracycline
(50 µg/ml) is more effective in inhibiting bacteria than 100
µg/ml of chloramphenicol alone. Chlortetracycline should be added
to molten heat-sterilized base medium adjusted to 45-50°C.

The pH of the medium is adjusted to 5.6 after autoclaving.

20. RBC: Rose Bengal Chlortetracycline Agar (Jarvis 1973)

Peptone	5	g
Glucose	10	g
KH_2PO_4	1	g
$MgSO_4 \cdot 7H_2O$	0.5	g
Rose bengal (0.5% aqueous soln., w/v)	10	ml
Chlortetracycline (0.1% soln., w/v)	1	ml
Agar	20	g
Water	1000	ml

Final pH 7.2

Rose bengal and chlortetracycline solutions should be added to sterile medium. Modifications of this medium include replacement of chlortetracycline with chloramphenicol. This medium can be autoclaved after addition of all ingredients.

21. SDA: Sabouraud Dextrose Agar

Peptone	10	g
Dextrose	40	g
Agar	15	g
Water	1000	ml

Final pH 5.6

22. YES: Yeast Extract Sucrose Agar

Yeast extract	20 g
Sucrose	150 g
Agar	20 g
Water	1000 ml

Final pH 6.0–6.5

Chloramphenicol and chlortetracycline may be added to YES agar to give final concentrations of 50 µg/ml.

References

BOOTH, C. 1971 Fungal culture media. In Methods in Microbiology ed. Booth, C. pp. 49–94. London: Academic Press.

DIFCO MANUAL 1984 Dehydrated Culture Media and Reagents for Microbiology, 10th ed. 1155 pp. Difco Laboratories, Inc., Detroit, Michigan 48232, USA.

FRISVAD, J. C. 1983 A selective and indicative medium for groups of Penicillium viridicatum producing different mycotoxins in cereals. Journal of Applied Bacteriology 54, 409–416.

HOCKING, A. D. 1981 Improved media for enumeration of fungi from foods. CSIRO Food Research Quarterly 41, 7–11.

HOCKING, A. D. & PITT, J. I. 1980 Dichloran-glycerol medium for enumeration of xerophilic fungi from low-moisture foods. Applied and Environmental Microbiology 39, 488-492.

JARVIS, B. 1973 Comparison of an improved rose bengal-chlortetracycline agar with other media for the selective isolation and enumeration of moulds and yeasts in foods. Journal of Applied Bacteriology 36, 723-727.

KING, A. D., JR, HOCKING, A. D. & PITT, J. I. 1979 Dichloran-rose bengal medium for enumeration and isolation of molds from foods. Applied and Environmental Microbiology 37, 959-964.

MOSSEL, D. A. A., KLEYNEN-SEMMELING, A. M. C., VINCENTIE, H. M., BEERENS, H. & CATSARAS, M. 1970. Oxytetracycline-glucose-yeast extract agar for selective enumeration of moulds and yeasts in foods and clinical material. Journal of Applied Bacteriology 33, 454-457.

NASH, S. M. & SNYDER, W. C. 1962 Quantitative estimations by plate counts of propsagules of the bean root rot Fusarium in field soils. Phytopathology 52, 507-572.

OXOID MANUAL 1982 Culture Media, Ingredients, and Other Laboratory Services, 5th ed. 352 pp. Oxoid Ltd., Basingstoke, Hampshire RG24 OPW, England.

PITT, J. I. 1979 The Genus Penicillium and Its Teleomorphic States Eupenicillium and Talaromyces. London: Academic Press.

PITT, J. I., HOCKING, A. D. & GLENN, D. R. 1983 An improved medium for the detection of Aspergillus flavus and A. parasiticus. Journal of Applied Bacteriology 54, 109-114.

PARTICIPANTS AND ATTENDEES

ANDERSON, C. B., Del Monte Corporation, Walnut Creek, California, U. S. A.

ANDREWS, S., South Australian Institute of Technology, South Australia, Australia

BECKERS, H. J., National Institute of Public Health, Bilthoven, The Netherlands

BEUCHAT, L. R., University of Georgia, Experiment, Georgia, U. S. A.

BULLERMAN, L. B., University of Nebraska, Lincoln, Nebraska, U. S. A.

CONNER, D. E., University of Georgia, Experiment, Georgia, U. S. A.

COUSIN, M. A., Purdue University, West Lafayette, Indiana, U. S. A.

DEÁK, T., University of Horticulture, Budapest, Hungary

DIJKMANN, K. E., University of Utrecht, Utrecht, The Netherlands

EKE, D., Marmara Scientific and Industrial Research Institute, Gebze, Turkey

FERGUSON, R., Pillsbury Co., Minneapolis, Minnesota, U. S. A.

FRISVAD, J., Institute of Biotechnology, Food Technology, Lyngby, Denmark

HOCKING, A. D., Commonwealth Scientific and Industrial Organization, North Ryde, New South Wales, Australia

HOWELL, M., Ministry of Agriculture, Fisheries and Food, London, U. K.

KING, A. D., JR., United States Department of Agriculture, Albany, California, U. S. A.

KOBURGER, J. A., University of Florida, Gainesville, Florida, U. S. A.

LANE, A. L., Difco Laboratories, Detroit, Michigan, U. S. A.

LENOVICH, L. M., Hershey Foods Corporation, Hershey, Pennsylvania, U. S. A.

MISLIVEC, P. B., United States Food and Drug Administration, Washington, D. C.

MOSSEL, D. A. A., University of Utrecht, Utrecht, The Netherlands

OLSON, K. E., General Mills, Inc., Minneapolis, Minnesota, U. S. A.

PITT, J. I., Commonwealth Scientific and Industrial Research Organization, North Ryde, New South Wales, Australia

SAMSON, R. A., Centraalbureau voor Schimmelcultures, Baarn, The Netherlands

SEILER, D. A. L., Flour Milling and Baking Research Association, Chorleywood, Herts, U. K.

SPLITTSTOESSER, D. F., Cornell University, Geneva, New York, U. S. A.

WILLIAMS, A. P., The British Food Manufacturing Industries Research Association, Leatherhead, Surrey, U. K.

AUTHOR INDEX

This index does not include any reference to the names of authors of publications listed in bibliographies. It contains the names of all authors and co-authors of papers contained in this Proceedings.

Aspermutarubrol, 272
Aspulrinones, 272
ATGY (acidified tryptone glucose
       yeast extract agar), 119
  formula, 295
Aureobasidium, 17, 110, 250, 259,
       261, 265
  pullulans, 232, 265
AYEGA, 167

Barley, 20, 27, 33–40, 44–46, 111,
       133–134, 137–139, 183, 210
Baseline counts, 177–178, 180, 187,
       190, 192, 194
  for beverages, 185, 187
  for flour, 192, 194
  for frozen foods, 185
  for honey, 190
  for soft drinks, 188
  for sugar, 190
  for vegetables, 185, 197
  for wheat, 194
Basipetospora, 254
Beans, 20, 111, 182
Beef, 120–121
Beer, 67, 288
Benzoic acid, 213
Beverages, 22, 164–165, 185–187,
       189, 220–221, 288
Bipolaris, 250, 262, 265
Biverticillium, 273
Black pepper, 50, 53, 86, 88, 90,
       93, 138–139, 184
Blackberries, 183
Blending
  evaluation of, 4–9, 10–11
Botryodiplodia theobromae, 265
Botryosphaeria, 265
  rhodina, 265
Botrytis, 133, 232, 250, 258, 261,
      263
Boysenberries, 183
Bran, 103–105, 109–110, 194–197
Bread, 144, 179, 181
Brettanomyces anomalus, 286, 290
  bruxellensis, 286
  claussenii, 286, 290–292
  custerianus, 282
  intermedia, 286
  naardensis, 282, 286, 291
Buckwheat, 133
Bullera alba, 286, 289–290
Buttermilk, 181
Byssochlamys, 150, 154, 177, 222,
      251, 254–256, 264
  fulva, 151, 153–154, 212, 256
  nivea, 151

Camomile, 138–139
Candida, 38–39, 81, 108, 137, 248
  apicola, 286, 291

berthetii, 282
boidinii, 286, 289
brumptii, 284
cantarellii, 286, 289
castellii, 282
catenulata, 286, 292
citrea, 287
dendrica, 282
diddensiae, 286, 289
etchellsii, 286, 290
famata, 71
friedrichii, 286, 289
fructus, 282
glabrata, 282, 286, 293
glaebosa, 71
globosa, 286
guilliermondii, 71, 73, 287, 291
holmii, 288
inconspicua, 286, 293
ingeniosa, 71
intermedia, 286, 291
japonica, 290
javanica (t = Filobasidium
      capsulig.), 286
kefyr, 71, 73, 287
lactis-condensi, 282, 286, 290
lambica, 284, 287
lipolytica, 289
lipolytica (t = Yarrowia), 286
lusitaniae, 71, 286
magnoliae, 286, 290
membranaefaciens, 287
mesenterica, 286, 289
norvegica, 286, 290
parapsilosis, 286, 292
pelliculosa, 287
pseudotropicalis, 86
pulcherrima (t = Metschnikowia),
    286
reukaufii (t = Metschnikowia), 286
rugosa, 286, 292
sake, 71, 73, 286, 291–292
scottii, 290
scottii (t = Leucosporidium), 286
solani, 286, 291
sorboxylosa, 286, 292
steatolytica, 286, 291
stellata, 282, 286, 291
tenuis, 286, 289
tropicalis, 71, 73, 284, 286, 291
utilis, 284, 290
utilis (t = Pichia jadinii), 286
valida, 284, 287
vartiovaarai, 286, 290
versatilis, 286, 290
vini, 284, 286, 292
zeylanoides, 71–73, 217, 284, 286,
    292
Candidulin, 272
Cauliflower, 179–180
Cayenne, 50–51, 53–55, 160

Hansenula anomala, 71, 73, 187, 214, 283–284
  mrakii, 282
  subpelliculosa, 71
Hay, 27
Heat resistant molds, 150–155, 212
  incidence, properties, 153–155
  techniques for enumeration, 150–155
Hemispora stellata, 265
Honey, 190, 288
Horseradish, 198
Howard mold test, 223–225, 238
Humicola, 251, 259, 261, 264
Humicola (Thermomyces) lanuginosa, 248, 264
Hyphomycetes, 250, 258–260
Hyphopichia burtonii, 287, 289
Hypocrea, 263–264

Ice cream, 181
Impedimetric estimation, 230
Impedimetric responses, 232–233
Impedimetry, 223, 231
Injured fungi, 168
Injured cells, 159, 169–170, 175
Injury, 159–176
Isofumigaclavin, 275
Issatchenkia orientalis, 287, 292
  terricola, 287, 292

Jam, 162–163, 288
Jams, 157, 221
Juice, 124, 150, 214, 220

Kale, 180
Kloeckera, 81
  apiculata, 119
Kluyveromyces delphensis, 282
  lactis, 86
  marxianus, 287, 291
  thermotolerans, 287, 291
Kojic acid, 272

Lasiodiplodia, 251, 265
  theobromae, 265
Legume flours, 20, 111
Legumes, 78–79
Lemonade, 67
Lentils, 20, 111, 183
Leptosphaeria, 265
  heterospora, 271
Leucosporidium, 71, 286
Linseed, 139
Lodderomyces elongisporus, 287, 292

MA (malt agar), 152
Macaroni, 106–108, 183
Maize, 18
MALT, 109–111
Malt extract agar, 298

Malt salt agar, 143, 299
Mango, 212
MEA (malt extract agar), 85–88, 91, 133, 162, 175
  formula, 298
Meat, 64, 70, 123, 288
Meats, 11, 120, 122, 184, 187, 210
Memnoniella, 251, 258, 261, 264
  echinata, 264
Mesophilic mycoflora, 138–142
Metschnikowia, 286
  pulcherrima, 86, 291
  reukaufii, 291
Microascus, 263
Milk, 179, 210, 246, 288
Milk powder, 15–16, 101–102, 181
Mixed herbs, 138–139
Monascus, 251, 254–256
  bisporus, 256
Moniella, 261
Monilia, 265
Moniliella, 251, 258, 261, 263
  acetoabutans, 232
Most probable number (MPN), 166–167, 176, 188, 190, 227–230
MPN technique, 227–230
MSA (malt salt agar), 109–110, 133–134
  formula, 299
Mucor, 37, 43, 52, 71–72, 76, 81, 94, 96, 99, 101, 108, 110, 126, 133, 159–161, 183, 251–253
  circinelloides, 72–73, 139, 253
  hiemalis, 253
  mucedo, 232
  plumbeus, 86,89, 92, 253
  racemosum, 72, 253
  rouxii, 119
Mucorales, 37, 80
Mussels, 180
MY40, 185
MY40G, 143–145
MY60G, 143–145
Mycological agar, 143, 202–203
Mycosphaerella, 264
Myrioconium, 263
Myrothecium, 251, 259, 261, 264

Nectria, 260, 264
Neosartorya, 222, 251, 254, 256–257, 260, 272
  fischeri, 151, 153, 212
  fischeri var. fischeri, 272
Nephrotoxin, 133
Neurospora, 81, 126, 251, 254, 256–257, 265
  crassa, 257
  intermedia, 257
  sitophilia, 257

311

NO$_2$S (nitrite sucrose agar), 299
Noodles, 107–108
Nuts, 11, 20, 78–79, 111, 128, 161, 177, 184, 202, 209
OAES (Ohio Agriculture Experiment Station agar), 26, 84–85
Oatmeal, 103–105
Oats, 20, 111, 139–140, 183
Ochratoxin, 33, 39, 45–46, 132–133, 135, 210, 272
OGGY, 67, 69, 70–74, 120, 122–123
OGY (oxytetracycline glucose yeast extract agar), 20–22, 26, 51–53, 55, 63, 66–67, 69, 71–74, 84–85, 89–91, 103–106, 109–115, 117–120, 122–123, 126, 148–149, 159–161, 164–165
  formula, 299
OGYE, 195
Oilseeds, 128
Onions, 139–140, 198–199
Onion powder, 97–99
Orange juice, 90, 124
Oxytetracycline glucose yeast extract agar, 299

Pachybasin, 240–243
Paecilomyces, 80–81, 156, 183–184, 251, 254, 256, 259, 261
  variotii, 38, 137, 140, 208, 232, 264
Paprika, 50–51, 53–54, 160, 199–202
Parsley, 139–140, 184
Passionfruit, 212
Pasta, 78–79, 103–107
Pastries, 67, 179, 181
Patulin, 272
PCA (plate count agar), 20–21, 111–112, 120, 122, 124–125, 172, 175
  formula, 299
PCNB (pentachloronitrobenzene agar), 156
  formula, 300
PDA (potato dextrose agar), 4–5, 8, 20–22, 47–48, 63, 77, 94, 96–97, 103, 124, 126, 152, 154, 164–165, 175, 185, 228–229, 249
  formula, 300
Peaches, 183
Peanuts, 128–131, 184
Peas, 7–8, 20, 94–95, 111, 183, 227, 229
Pecans, 7, 49–51, 53–54, 95, 160, 184, 202, 227, 229
Penicillia, 6–7, 17, 58, 82, 133–134, 137, 141, 249, 273
Penicillic acid, 272, 275

Penicillium, 4–5, 13, 17, 24, 42–43, 71–72, 81, 98–99, 107–108, 127–128, 133–134, 136, 140, 152, 183–185, 210–211, 220–221, 239, 249–251, 254, 258, 261–263, 271, 273
  aurantiogriseum, 7, 37–39, 47, 134–135, 141, 274–276
  biforme, 274
  brevicompactum, 72–73, 141, 274
  camembertii, 141, 274–276
  casei, 274
  caseicolum, 274
  chrysogenum, 29–31, 72, 141, 274
  citrinum, 38, 72–73, 137, 141, 232, 273
  commune, 274
  concentricum, 274
  corylophilum, 72–73, 273
  corymbiferum, 274
  crustosum, 141, 274–276
  cyclopium, 274–275
  digitatum, 232–233
  echinulatum, 274–275
  expansum, 86, 92, 134, 209, 214, 232–233, 274
  frequentans, 119, 232, 273
  funiculosum, 273
  glabrum, 141, 273
  granulatum, 232
  griseofulvum, 141, 274
  hirsutum, 134, 141, 274–276
  hordei, 274
  implicatum, 18–19
  islandicum, 273
  italicum, 232, 274
  mali, 141, 274
  martensii, 94, 228, 274
  melinii, 141
  nigricans, 72–73
  notatum, 23, 274
  oxalicum, 273
  palitans, 274
  puberulum, 274
  purpurescens, 45
  purpurogenum, 273
  roquefortii, 7–8, 38, 47, 72, 94, 137, 141, 228, 232, 248, 274
  rubrum, 232
  rugulosum, 273
  spinulosum, 273
  steckii, 72–73
  stoloniferum, 274
  urticae, 274
  variabile, 273
  verrucosum, 42, 132, 210, 274–276
  verrucosum var. cyclopium, 72–73
  viridicatum, 8, 33–37, 39, 45–47, 94, 132–135, 141, 210–211, 228, 232, 274–276
  selective medium for, 132

    bisporus, 145, 157, 222, 256
Xerophiles, 127, 185–186
Xerophilic, 162
Xerophilic fungi, 106, 109, 249
Xerophilic molds, 143
Xerophilic yeasts, 146–147, 157, 177
Xerotroph, 157

Yarrowia, 286
    lipolytica, 71, 73
Yeast extract sucrose agar, 301
YES (yeast extract sucrose agar),
        45, 132
    formula, 301
Yogurt, 90, 124, 181, 212

Zearalenone, 209
Zygomycotina, 249–250, 252
Zygorynchus moelleri, 232
Zygosaccharomyces bailii, 177, 187,
        212–214, 217, 221–222, 282–
        284, 288, 291–292
    bisporus, 284, 288, 292
    florentinus, 282
    microellipsoides, 288, 291
    rouxii, 71–73, 119, 148–149, 217,
        284, 288, 292